Saving the World Before Breakfast
A Better Green New Deal

Roger Carlson

December 15, 2020

Except as otherwise marked, all material in this book is copyright protected. None of the copyrighted material in this book may be reproduced and sold or otherwise used without the express permission of the author.

Copyright © 2020 Roger Carlson
Saving the World before Breakfast Press

All rights reserved by author

ISBN 978-1-7362623-0-6

Contents

Preface—How You Might Fit vii

How I Fit viii

1 Introduction to the Green New Deal 1
   1.1   The New Deal Background 1
   1.2   The Green New Deal Political Background 3

2 What Are the Global Warming Threats? 7
   2.1   How Global Warming (and Cooling) Happens 7
   2.2   The Pluses and Minuses of Increased Heat in General 11
   2.3   The Bottom Line: Food Supply Is the Only Crucial Danger 15

3 A Quick Summary of the Green New Deal? 17

4 The Good of the Green New Deal 19
   4.1   Take a First Cut at an Overall Plan and Work from There 19
   4.2   Regenerative Agriculture Is Huge 19
   4.3   Zero-Sum Accounting Works—What Is the Net Warming? 19
   4.4   Considering Resiliency as Part of the Picture 19
   4.5   Yes, Commercial Aviation Contributes to Global Warming 20
   4.6   The Power of Positive Insanity 20

5 The Bad of the Green New Deal 25
   5.1   A Perceived Socialist Bait and Switch Leads to Loss of Credibility ... 25
   5.2   The Short Deadlines Are Unworkable and Undermine Credibility 26
   5.3   Governments and Big Business Do Stunts and Disasters; Entrepreneurs and Outsiders Are More Likely to Make Technology Revolutions 29
   5.4   The Long-Term Perils of National Mobilizations 34
   5.5   Only Nuclear Fission Can Provide Enough Noncarbon Power .. 37
   5.6   Change the World or Forget It 40
   5.7   People Will Only Support a Green New Deal If There Is a Visible Threat ... 41
   5.8   The Perils of Trend Extrapolation Amid Human Responses 47
   5.9   Patience Should Be a Green New Deal Virtue 49

6 Ground Rules for a Better Green New Deal 53
6.1 Invent, Develop, and Only Then Mass Produce 53
6.2 Always Build—*Semper Facilitat*— Rather Than Tear Down 54
6.3 Green New Deal Innovation Do's and Don'ts 56

7 Food, Fear, and Climate Fixes from the Land 59
7.1 Beyond the Plow: Regenerative Agriculture 60
7.2 Try a Little Variety: Food, Fiber, and Materials 64
7.3 Plant Breeding Miracles the Neo-Prairie, and the Ten-Foot Moon Shot 72
7.4 Eat Your Veggies 82
7.5 Rise of the Robo Farm Machines 85
7.6 Controlled-Atmosphere Agriculture: Oases, Hanging Gardens, Greenhouses, Vertical Farms, and the Shoebox Paradox 89
7.7 Eat Your Little Veggies—Sprouts 94
7.8 Magic Shrooms—Bacteria and Fungi Alone Could Feed the World! 95
7.9 Minilivestock: Chickens Down to Bugs, McBugs, McCrawleys, & McWigglies—Grazers of Small Prairies and Small Ponds 102
7.10 Biochar, Remaining Materials, Extracted Juices, and Leaf-Protein Concentrate in Biorefineries 105
7.11 The Unnoticed Frontier: Lawns and Other Turf-Grass Areas 110
7.12 Waste Not, Want Not 115
7.13 The Sweet and Woody Solution: Artificial Meat, Fruits, and Vegetables 119
7.14 Carbonated Fields 124
7.15 Conclusion: People Can Grow Enough Food and Help Cool the Climate 125

8 Fixes from Energy and Material 127
8.1 Energy Power, Limits, Threats, and Possibilities 127
8.2 Problems, Limitations, and Fixes from Nuclear Fission to Compete 130
8.2.1 Problems with The Best Energy Hope: Nuclear Fission Reactors 130
8.2.2 The Basic Lesson on Nuclear Fission Reactors 134
8.3 The Return to Atom Splitting Power with Better Next-Generation Nuclear Fission Reactors 140
8.3.1 The Background Primer on Next-Generation Nuclear Reactors 141
8.3.2 Ways to Develop NexGen Reactors 142
8.4 Nuclear Fission Energy—Electricity, Heat, and Synthetic Fuels 165

- 8.4.1 Electricity, Space Heating, and Carbon Dioxide for Synthetic Fuel 165
- 8.4.2 Process Heat Uses 166
- 8.4.3 Using Mass-Produced Hydrogen and Heat 169

8.5 Overchoice in Energy Economies—The Hydrogen, Methanol, Methane, Electricity, … Ammonia Economy? 173

8.6 Controlling Fire Emissions with Fountains into the Sky 187

8.7 Key Toolmaking Technologies 198
- 8.7.1 Computer Power for Calculation, Simulation, Analysis, and More 199
- 8.7.2 The Range from Servo Controls Leading to Robotics 200
- 8.7.3 Advanced Materials Processing 205
- 8.7.4 Prefabricated and Modular Construction 211
- 8.7.5 Supercritical Carbon Dioxide Turbines—Another Supporting Energy Revolution 212
- 8.7.6 Electronic Remote Connection—The Internet, Webinars, and Offices Anywhere 213
- 8.7.7 Green-Energy Storage 214
- 8.7.8 Materials for Absorbing. Reflecting, Generating, and Using Energy 224
- 8.7.9 Putting It All Together: Hybrid Systems of Combustion, Ultracapacitors, … and Batteries 238

8.8 Electric, Autonomous, Transportation Utilities—Elon Musk Strikes Again 240

8.9 Electric Airplanes … Really! 241

9 Fears and Fixes from Sea and Sky 245

9.1 Ocean Worries—The Seven Deadly Fins 245
- 9.1.1 Over Fishing and Bad Fishing 246
- 9.1.2 Global Warming—Including Sea-Level Rise 248
- 9.1.3 Reduced Ocean Productivity Due to Increased Stratification 248
- 9.1.4 Decreased Carbon Drawdown from the Atmosphere 249
- 9.1.5 Pollution and Trash 251
- 9.1.6 People Crowding Infrastructure into Beaches, Wetlands, and reefs 252
- 9.1.7 Earthquakes, Volcanoes, Giant Waves, Large Objects from the Sky, and Nuclear-Bombs 257

9.2 The Greenwater Revolution on the Continental Shelf 267
- 9.2.1 The Potential of the Continental Shelf and Barriers to Its Use 268
- 9.2.2 Starting the Greenwater Revolution 271
- 9.2.3 Organizing and Making the Greenwater Revolution 275

- 9.3 Marine Farming and Infrastructure for the New Atlantis 278
  - 9.3.1 It's Not Landfill Material, It's Floats for Ocean Farms 282
  - 9.3.2 The First Stone of New Atlantis--Floating Concrete 283
  - 9.3.3 The Second Stone of New Atlantis—Regenerating Corals and Other Limestone-Accreting Species 286
  - 9.3.4 The Seagoing Line Shack/Winnebago—We All Live in a Yellow Submarine 289
  - 9.3.5 The Third Stone of New Atlantis—Carbonated Igneous Rock for Power, Food, and Tsunami Defense 292
  - 9.3.6 What to Do 305
- 9.4 A Step Even Further: The Blue-to-Green Revolution 306
  - 9.4.1 Fertilizing the Bluewater Deserts and the Ammonia Economy at Sea 308
  - 9.4.2 Motor Ships and Factory Ships for Power and Infrastructure at Sea 315
  - 9.4.3 Additional Technologies Needed for Archipelagoes 319
  - 9.4.4 Putting It All Together–A Complete Archipelago 328
  - 9.4.5 The Bottom Line for Green Oceans as Well as Agriculture and Energy 348

10 Conclusions 351
- 10.1 A Start on the Better Green New Deal 351
- 10.2 Beyond the Better Green New Deal? 360

11 Original Green New Deal Resolution & FAQ 363
- 11.1 Original Green New Deal Resolution 363
- 11.2 Original Green New Deal Fact Frequently Asked Questions (FAQs) 371

12 Annotated Bibliography 379

Acknowledgements 401

Preface—How You Might Fit

Do you wake up in the night worrying that the world may end soon because of global warming and corollary catastrophes such as rising sea level and more intense storms? Conversely, do you get berated by environmentalists for even the slightest doubts about climate doom? In either case, this book is for you.

The Green New Deal (GND) is a wildly socialist set of policies popularized in the United States starting in 2019. The four key good and bad aspects of the GND are:

1. Human industry is impacting the world climate at an accelerating rate, and that impact must be dealt with inside of twelve years lest there be painful consequences, perhaps to a civilization-destroying level.

2. The GND proposal set loses traction because of many socialist policies sprinkled throughout that are totally unrelated to world climate. For instance, universal health care, higher education for all, and construction only by union labor. These policies may or may not be good, but they would not impact world climate.

3. Some of the worries are overwrought or are totally bogus. For instance, there were initial comments that humanity would be doomed if the GND were not just started but actually completed within 12 years.

4. The (admittedly preliminary) GND did not address many additional technological and-social initiatives that might be even better ways to deal with the possibility of too much global warming.

Still, the GND did put all the industry-energy-climate issues in one overall discussion. It provides a good starting point.

This book reviews the GND strengths and weaknesses. It can provide worriers some comfort from the fears, and provide skeptics some rhetorical support. At the same time, it supports serious concerns about global warming as well as global chilling that could come as a surprise. Finally, it describes three major technology initiatives to help ease or even reverse global warming. These three initiatives are described in the major technology chapters of

(7) Food, Fear, and Climate Fixes from the Land: Regenerative and advanced agriculture (briefly discussed in the GND)

(8) Fixes from Energy and Material: Inherently safe next-generation nuclear fission reactors (not covered in the GND) and many of the component technologies that also apply in the other two major technological initiatives

(9) Fears and Fixes from Sea and Sky: Ocean development (not discussed in the GND)

These three alone would be sufficient to ease or even stop global warming.

Although they are largely different from each other, the three have important overlaps. For instance, advancing robotics and 3D printing of materials are keys to progress in land, next-generation nuclear, and ocean development; so, they are all in this book.

You may be interested in all three technology initiatives, or you might be interested in just one or two. You can choose which sections to read.

Likewise, some of you may wish to follow references provided to get more detailed information, much of it readily available on the Internet, to check or possibly build on the ideas presented. Others of you may read through just for general interest; you can skip on by the references, enjoy, and learn. I appreciate you both.

Lastly, as an optimistic caveat, the major themes of agriculture, nuclear-related energy, and ocean expansion are only three of the many directions the future may hold. Focusing on these three allowed me to contain them in one book, but there are many other possible directions. Some of you readers may be the ones to identify and develop some of those other directions ... perhaps even better ones.

How I Fit

Hard times are often worse the second time around. This was for me. My family lived through the Great Depression, the Dust Bowl and World War II; any and all of which seemed to those people living them that they might be the end of the world. Via the good-and-bad old-time stories, I relived much of it.

I saw the house my grandfather had given my father and one of his sisters; but which was lost during the Great Depression because my father and his sister could not even pay the taxes. Many times, when I took food from the refrigerator to set the table, my mother remarked how, during "the Dirty Thirties," dust had worked its way past the rubber seals of the ice box and left little dust rows inside. In scrapbooks, I saw the young faces and yellowed headlines about the war that killed some but lifted others out of the hard times with jobs in defense plants and drafted-in military jobs.

The stories did not come from the end of the Earth ... but you can see it from there. It is the small farm town of Huron, South Dakota on the Northern Great Plains where I grew up.

The hard times started about six months after the 1929 crash in the stock market. As with many things, the Depression arrived in Huron several months after the crash in the east.

All the banks closed. My father, who was running a gas station had no place to deposit receipts for the end of the week so he nervously guarded several hundred dollars in a cigar box at home. A week later the worries about the gas-station money, along with the gas-station job, had gone away. There were new worries and no job.

America has had many booms and busts, but this bust stayed on and grew into the Great Depression. For two years my father had no regular job. He developed a circuit in his neighborhood of washing and waxing cars, putting on storm windows in the fall, shoveling winter snow, replacing storm windows with screens in the spring, and back to washing and waxing cars the next summer.

The Great Depression got worse when drought turned the fields to brown dust that lofted upward with the slightest wind, and South Dakota often has winds much stronger than slight. There were often dirt drifts instead of snow drifts, and there were sometimes soil drifts in town deep enough for shoveling and sweeping sidewalks. They called the area from South Dakota to North Texas the Dust Bowl. The farmers' crops often failed or produced very little, so they could only buy a few things in the stores when they came into town on Saturday evenings; so, the town grew poorer yet.

There was bitterness and despair among my parents' generation. They referred to pants pockets pulled inside out as "Hoover flags" because they blamed then president Herbert Hoover for people not having any money to put in their pockets. Hoover and others proposed many plans to fix the economy, but nothing worked.

A new president, Franklin Delano Roosevelt (FDR), came into office in 1933. President Roosevelt said that, "The only thing we have to fear is fear itself."

As a liberal democrat, Roosevelt had campaigned against Hoover's socialistic attempts to use big-government programs to cure the Great Depression. Then, when in office, he became what he called, "… a new kind of liberal," and went for government interventions even bigger than those of Hoover.

Roosevelt proposed an amalgam of programs he called the New Deal. It was a wild mixture of make-work jobs, subsidies for students to get schooling, and infrastructure projects such as rural electrification, cement sidewalks, and river dams. One of the most successful projects was planting quarter-mile-long strips of trees with the goal of slowing the winds that tore soil from farmer's sprouted crops. To his dying day, my father referred to such tree strips as Roosevelt's trees.

Still, the Dust Bowl went on, and the Great Depression continued on. "Doctor End-the-Depression" Roosevelt kept trying new programs, but the results were meager. Sometimes the crops or the economy did a little better. Sometimes they slipped and did worse. There was always a fear that much worse might come. Things only improved when outside events brought change.

In the 1930s a local lake had gone dry in the drought. Yet, the lake bed still had water only a few feet underground. Thus, the local farmers got several good wheat harvests from that lake bed. They finished harvesting in 1939 just as some promising fall rain came. Because the rain caused the ground to become mud, they had trouble moving the combine. They left it behind, with the intention of getting it when the ground dried sufficiently.

More rain was followed by snow. It was the end of the drought, and the dry lake filled back up. As far as we know, rusted pieces of the combine are still on that lake bed.

For the economy, there was another type of storm that ended the money drought. Wars in Europe and Asia started and grew fiercer. The next fear was that one of those wars might come to America. Yet the fear was mingled with opportunity for there were growing numbers of jobs in distant defense plants, and workers started sending some of their money back to their families at home.

The federal government instituted a draft of men for the military. Many better jobs started opening up. Still, nobody wanted to be in a war. Nobody wanted to, "die for Danzig," ... wherever that was.

Then, it became a real war. Japanese planes bombed Pearl Harbor ... wherever that was. Two years later, my father was at Pearl Harbor while his ship stocked up before sailing west. President Roosevelt no longer called himself "Doctor End-the-Depression." He became "Doctor Win-the-War," and everyone was for it because they saw a clear and present danger.

There were lots of jobs. Many men had guaranteed jobs in the military—in fact, they were not allowed to quit. They had mandatory savings programs, and the farmers were able to sell everything they produced. The war was followed by one of the longest and greatest periods of economic growth in history as people used their savings to build houses and families.

Yet, there was always a fear that it could all be taken away. The economy could slip back into hard times. There could be another super drought, a new Dust Bowl or worse.

Might there be another New Deal to combat different dangers? Would it work?

We have some theoretical dangers of major global warming that might affect my little former home town—and every place else. If those dangers are as severe as predicted, fixing them will indeed require some type of Green New Deal. If the threats are that severe, will such a new deal work? How could a Green New Deal be made better?

These are the questions that I will start to answer in this book with the following major themes:

- There is a serious danger of global warming, but the danger is not approaching quite as rapidly as declared by many Green New Deal partisans.

- This is fortunate because many of the Green New Deal proposals cannot be achieved as quickly as proposed, and attempting to do so might cause other problems.

- There is also a threat of short-term cooling events and other problems caused by volcanoes and other tectonic events. In this case, short term is only immediate events to as much as several years, but these events could be more severe than warming—such as the world cooling caused by a mega volcanic eruption.

- Three of the most promising ways to mitigate these threats are in the areas of (1) better agriculture; (2) relighting the fire of nuclear fission by adopting next-generation reactors; and (3) expanding agriculture and industry into the oceans.

This book contains my analyses and suggested solutions. You may have some better. If you do, more power to you. Add your contributions.

T

Before this century is over billions of us will die and the few breeding pairs of people that survive will be in the Arctic where the climate remains tolerable.

James Lovelock, 2006[1]

To his credit, Lovelock considered later data and decided that the global warming dangers were still significant, but were much less than his quote above. Then, he went on to research further and suggest mitigations for

This author, 2019[2]

1 Introduction to the Green New Deal

If human society is endangered by possible massive global warming coming in the future, what should we do? We could have a massive human effort to stop that warming.

The Green New Deal (GND) is a set of proposals and a budding political movement that was proposed to do just that. It is based on two key ideas:

(1) A belief that trace amounts of carbon dioxide and other materials in the air must be swiftly removed or terrible climate-related things will happen to humanity.

(2) A set of vast new government programs such as those in the 1930s depression and in World War II must be implemented to transcend those dangers.

This book is a quick summary of the global warming issues, the original New Deal, what that policy's descendent—the Green New Deal (GND, resolution (included in Chapter 11), what the GND got right, what the GND got wrong, and some new suggestions for a better GND.

1.1 The New Deal Background

The first New Deal has roots from a century ago. The Great War, or "The War to End all Wars," ended in 1918 after millions of soldiers had died in the first large war with mass-produced artillery shells and mass-produced machine guns. Many believed that the war would cause all

---

[1] Quoted from Mark Ballard, "Forget global warming - think boiling oceans, *The Register* (a United Kingdom newspaper), Jan. 16, 2006.
https://www.theregister.co.uk/2006/01/16/revenge_of_gaia/

[2] After the pace of global warming slowed noticeably in the early two-thousands through two-thousand-teens, Lovelock reduced his fear factor. He still believed that warming was serious, but decided that he and others had gotten much more worried than was needed.

civilization to collapse as political empires and kingdoms did collapse in Russia, Europe, and the present Middle East.

Yet, most of civilization did not collapse. Instead, there was an economic boom in the 1920s. They called that decade "The Roaring Twenties," as wild speculations generated unheard of profits selling new mass-produced marvels, including cars in the cities, tractors on the farms, and radios … everywhere.

It seemed that the boom would never end, and the best way to make a fortune was to invest in the stock market and double your money in a couple years. Even better, you could invest ten percent of the stock price and buy the stock on margin. Then, you could get twenty times your investment! It all worked as long as the stock market was booming upward.

Then, on Thursday, October 24, 1929, the stock market's rocketing rise tipped sharply downward. Later, that day would be called "Black Thursday." As the stock traders say, "It was a correction," or, "What goes up, must come down." Another term for these periodic corrections is "a panic." Over the centuries, speculators made or lost great fortunes from booms of wildly increasing prices of land, oil futures … or tulips in the fervent expectation that the prices would continue rising. Then eventually, the irrational exuberance overshot the actual market for such items. The investments collided with reality, and panic selling ensued.

This panic was one of the biggest. Many stocks dropped so far in value that it triggered margin calls. Many of the stock owners did not have enough cash to pay the margin call, so the banks took their stocks and immediately offered the stocks for sale in the market, which caused stock prices to fall farther. Others had money in bank accounts, so they ran to their banks and withdrew savings to pay their margin calls. This was called a run on the bank.

If a bank gave out all of its cash on hand to the depositors withdrawing savings, it would have to close for the day, which would look as if the bank might be unsound (and some were). When account holders began seeing lines at certain banks, they began to wonder if their bank might run out of money, so they ran to their banks to draw their money out. Consequently, all the banks worked even harder selling the foreclosed stocks, which drove the stock prices lower yet. There were more margin calls, and more people running to their banks.

More people saw lines at the banks and decided to join the runs. Eventually, many banks ran out of cash and closed their doors; some went out of business. That caused the stock market to drop even farther, causing more margin calls. After terrible losses for stock-market investors and some bank failures, the economy tipped into recession. President Herbert Hoover told businesses not to lay off any workers because the losses would be temporary. The losses were not temporary, and losses got even worse.

Then Congress decided to help by increasing tariffs 20 percent. That was supposed to keep business within the country. Instead, it caused other countries to jack up tariffs in return. Within four years, world trade had declined by about two-thirds, and there was a world depression with failed businesses and a quarter of all U.S. workers laid off. Many governments fell, and radical authoritarian ideas bloomed.

A great drought in the center of the country made matters worse as bankrupted farmers from the Great Plains states fled to more prosperous areas such as California.

Marxist socialists (communists) in present day Russia said that the depression showed the inherent failure of capitalism. They said that only a planned socialist economy could fix the problems. Their mutant National Socialist (Nazi) Party cousins in Germany and Fascist Party cousins in Italy said the same thing. Nazis, Fascists, and Communists all claimed that a planned socialist economy could solve what was already called the Great Depression. Many in America feared that American voters would turn to either Nazi or Communist socialists out of desperation.

To prevent a major socialistic revolution, in 1933 the new president Franklin Delano Roosevelt began what he called "the New Deal" to fight the depression. He had a large majority of his party in Congress, so with that majority he was able to start many new federal government programs. The programs addressed the "3 Rs" of fighting the depression: Relief, Recovery, and Reform—relief for the poor and jobless, recovery for the economy, and reform of the nation's banking system to prevent future depressions.

Roosevelt's New Deal had a number of socialistic aspects, but America remained a democratic republic while helping many people with immediate jobs and other forms of relief. Moreover, Roosevelt made great speeches and "fireside chats" over that new mass medium, radio. Even though the New Deal only produced a weak recovery (and had occasional backsliding into recession) it kept American hopes alive, especially in poorer communities. Roosevelt detractors said that we would have been better without the New Deal. Roosevelt supporters said that the New Deal had worked its economic cure and that it would have worked even better if there had just been more money for a bigger stimulus.

That economic cure of vastly more money came in the form of Japanese planes bombing the U.S. fleet in Pearl Harbor, Hawaii on December 7, 1941. The United States was in the second great war, World War II. President Roosevelt said that, "Doctor End the Depression has been replaced by Doctor Win the War!" and the Congress gave him incredible amounts of money to fight that war.

Vastly increased spending jolted the American economy into a boom. Yet, people were forced to save much of their earnings in bonds and later benefits. Those deferred income streams helped buy houses and start families. They fueled sustained growth with only mild recessions for two decades after World War II.

The New Deal and World War II often blur together in people's minds—and with good reason. They provide inspiration and precedents for struggling against seemingly insurmountable odds. The Green New Deal is one such quest.

1.2 The Green New Deal Political Background

The GND has been largely identified with brash new Congress woman Alexandria Ocasio-Cortez representing the Bronx (and part of Queens) areas of New York City and Senator Ed

1 Introduction

Markey of Massachusetts. Together, they proposed a resolution for the House and the Senate summarizing some of the ideas.

Just as candidate and then President Donald Trump used, "Make America Great Again," to further his political career, Congresswoman Ocasio-Cortez has been using the "Green New Deal" to further the movement she is seeking to develop.

Of course, both of these major idea themes (memes) had been originally used and developed by others.

For Donald Trump, "Make America great again," was a time-worn theme from Ronald Reagan's 1980 political campaign[3] when Reagan was opposing the problem-plagued President Jimmy Carter who had been facing a gasoline shortage, inflation, and lingering bitterness about the failed Vietnam War. President Carter made a July 15,[4] 1979 speech entitled "Energy and the National Goals - A Crisis of Confidence." However, both Democrat Senator Ted Kennedy and Republican presidential candidate Ronald Reagan both the malaise speech.'"[5] Hence, the verbal counter stroke was, "Make America Great Again."

President George Herbert Bush, who followed after Raegan's two terms, did not continue the term, so Arkansas Governor Bill Clinton appropriated the unused term for his presidential campaign in 1991 and 1992.[6]

After Clinton's two terms (1993–2001), the phrase languished under the two terms of George Walker Bush (not to be confused with his father) and the two terms of Barrack Obama.

Finally, Republican candidate Donald Trump invented "Make America Great Again" … (again!) in 2016. He added the acronym MAGA, and used it as the theme for his 2016 campaign and all campaigning since then.

Likewise, the term "Green New Deal" has had a long history since President Roosevelt's original New Deal in the 1930s. In 2006 the Green New Deal Group in Great Britain held

---

[3] Luke Kinsella, "The Little-Known History of 'Make America Great Again,'" *News.com*, Dec. 2, 2017. https://www.news.com.au/world/north-america/the-littleknown-history-of-make-america-great-again/news-story/fb8a09b40aa59defd39ef0bcdeaeb281 (accessed Aug. 22, 2020)

[4] Jimmy Carter, "Energy and the National Goals - A Crisis of Confidence," speech delivered July 15, 1979, *American Rhetoric Top 100 Speeches*. https://www.americanrhetoric.com/speeches/jimmycartercrisisofconfidence.htm (Aug. 22, 2020)

[5] Quin Hillyer, "Forty Years Ago, Jimmy Carter's Malignant 'Malaise' Speech," *Washington Examiner*, July 15, 2019. https://www.washingtonexaminer.com/opinion/columnists/forty-years-ago-jimmy-carters-malignant-malaise-speech (accessed Aug. 22, 2020)

[6] "Bill Clinton Said 'Make America Great Again' In 1991," *YouTube*, Sept. 14, 2016. https://www.youtube.com/watch?v=voMW-P9bU8I (accessed Aug. 22, 2020)

discussion panels of economists and journalists on the concept.[7] In 2007, columnist Thomas Friedman proposed the concept for the United States in a *New York Times Magazine* article.[8] A United Nations report in 2009 proposed A Global Green New Deal.[9] The American Green Party has argued for a Green New Deal since the twenty-teens.[10,11]

However, the Occasio-Cortez–Markey joint resolution was the first time it became a theme used by a major party in the United States. Candidate Occasio-Cortez had it in her 2018 election campaign, and Representative Occasio-Cortez continues to use it in speeches for fundraising and developing a movement throughout the country.

The joint resolution did add some of the more dire climate computer models and say that failure to implement a Green New Deal (GND) before a certain deadline might destroy civilization or even humanity itself. Our very survival might be at stake. Furthermore, early speeches on the topic seemed to suggest a deadline of eleven years (apparently counting down to 2030).

With that putative existential danger, it would seem that Trump's MAGAs and Occasio-Cortez's GNDs need to find some common ground for developing a Green New Deal that could actually prevent the threatened climate disaster … even if (hopefully) the worst dangers might be delayed some years past those ten years.

The following material summarizes the good, the bad, and some suggested improvements for the GND.

---

[7] Caroline Lucas, "Britain needs its own Green New Deal," *Financial Times*, June 8, 2019. https://www.ft.com/content/79a7b226-8928-11e9-b861-54ee436f9768 (accessed Aug. 22, 2020)

[8] Thomas L. Friedman, "A Warning from the Garden," *The New York Times*, Jan. 19, 2007. https://www.nytimes.com/2007/01/19/opinion/19friedman.html?module=inline (accessed June 6, 2020)

[9] Edward B. Barbier, *Rethinking the Economic Recovery: A Global Green New Deal*, Economics and Trade Branch, Division of Technology, Industry and Economics, United Nations Environment Programme, United Nations, April 2009. https://www.cbd.int/development/doc/UNEP-global-green-new-deal.pdf (accessed Sept. 28, 2020)
Barbier updated the report and released it in a book form as:
A Global Green New Deal: Rethinking the Economic Recovery, Cambridge University Press, 2010.

[10] Howie Hawkins, Mark Dunlea, and Jon Rynn, "Whatever Happened to the Green New Deal?" *Counterpunch*, Aug. 21, 2020. https://www.counterpunch.org/2020/08/21/whatever-happened-to-the-green-new-deal/ (accessed Aug. 22, 2020)

[11] "Green New Deal Timeline," Green Party US website, undated. https://www.gp.org/green_new_deal_timeline (accessed Aug. 22, 2020

1 Introduction

2 What Are the Global Warming Threats?

The threat that Representative Occasio-Cortez, former Vice-President Gore, Senator Ed Markey, and others have posited is that the possible warming could suddenly multiply beyond anything in recorded human history. Such unprecedented global warming could present several threats to humanity.

(1) Global warming could be more significant than anything experienced in recorded history.

(2) Rising sea levels could flood many coastal cities comprising a significant fraction of our infrastructure.

(3) Climate bands of moist areas and desert both moved farther from the Equator. For the United States, that means the desert band in northern Mexico would move farther north into the Northern Great Plains. The Sand Hills of West Nebraska might look like the Sahara Desert … again.[12] Even though decreased plant fertility in some temperate areas would be partially compensated by warmer temperatures farther poleward, adjusting to the different climate bands would lower production for many years.

(4) Decreased nutrient flow from the deeper ocean to the surface because of warmer surface waters could decrease oceanic food production and increase global-warming carbon dioxide in the air because the ocean would absorb less.

(5) (As a result of the first four threats) there could major losses in food supply with resulting famines and government collapses.

Note: The term climate change means nothing because the climate is always changing, either warming or cooling in (usually) moderate cycles of decades or centuries. Significant warming and cooling are the climate changes that are dangerous.

2.1 How Global Warming (and Cooling) Happens

For warmth, our planet has an insulating atmospheric blanket called the greenhouse effect, which is a good thing. Without the greenhouse effect, the average world temperature would be colder than the freezing point of water, and most of Earth would look like the ice sheets in Antarctica.

That insulating greenhouse blanket can vary between heating and cooling depending on a number of gases and particles in the atmosphere. These greenhouse agents allow visible sunlight

---

[12] Venkataramana Sridhar, David B. Loope, James B. Swinehart, et al., "Large Wind Shift on the Great Plains During the Medieval Warm Period," *Science*, vol. 313, pp. 345–347, July 21, 2006. http://eas2.unl.edu/~dloope/pdf/Sridhar%20et%20al.,%202006.pdf (accessed Sept. 30, 2020)

through, but when the ground absorbs that sunlight and sends some heat waves (infrared) back up into the sky, those agents absorb some of the heat waves. They warm up, and so they send infrared heat energy in all directions—including some going back down. The heat energy radiating back down toward the ground thus keeps more heat on Earth. The most important greenhouse warming agents are the following:

(1) Water vapor is by far the greatest mass of greenhouse warming gas in the atmosphere. Furthermore, when the climate warms up, more water is evaporated up into the atmosphere, causing more warming. This is a positive feedback loop for more warming. However, water vapor is not as simple as steam from a tea kettle. Some of the water vapor condenses into clouds of liquid or even ice. Some clouds cause even more warming. Other clouds reflect significant amounts of sunlight right back into space—causing cooling. Some clouds drop snow onto the ground, which makes a reflective cooling surface. Climate scientists are still trying to calculate how much warming or cooling water vapor causes and under what conditions. For example, it has been suggested that clouds at certain altitudes cause warming and that clouds at other altitudes cause cooling.

(2) Carbon dioxide is formed when carbon (C) in wood or fossil fuel is burned by two oxygens (O_2), making CO_2. That burning heats our homes, runs our cars, and powers our chemical processes. We, and all other animals, do a slow burn of food to power our organic lives, and we breath out carbon dioxide. To balance all the burning that emits CO_2, plants use sunlight energy to pull carbon out of the air, combine it with hydrogen (H_2) from water H_2O), and synthesize compounds of hydrogen and carbon (hydrocarbons) we use as food, fuel, chemical feedstocks, and structural materials. CO_2 also comes out of volcanoes. At last count, CO_2 made up about 415 parts per million in the atmosphere. That is less than half of one percent. Still, it is increasing at 2 or 3 parts per million each year, and it remains in the atmosphere two centuries or more. Carbon dioxide accounts for about 75% of the human-caused global warming.

(3) Methane (also called natural gas) is one carbon (C) and four hydrogens (H)— CH_4. Methane is produced when bacteria or heat partially break down the hydrocarbons of organic matter. The prematurely-released FAQ (frequently asked questions) document from Occasio-Cortez's office said it was from cow farts. Actually, it is released in burping as cattle throw up their cud of partially chewed grass and chew it some more to break down cellulose in the grass into more digestible sugars.[13] Termites are simply much smaller herbivores that digest wood—in houses and in

---

[13] Amy Quinton, "Cows and Climate Change," University of California Davis website, June 27, 2019. https://www.ucdavis.edu/food/news/making-cattle-more-sustainable (accessed Aug. 22, 2020)

forests around the world while also making methane.[14] Methane emissions have been considered a major concern because methane is 24 times as potent in causing global warming as carbon dioxide. Fortunately, methane decomposes chemically in the atmosphere within 12 years, and it only makes up 18 parts per million in the atmosphere versus 400+ parts per million for carbon dioxide. Still, methane accounts for about 14% of the human-caused global warming because is a stronger greenhouse agent than carbon dioxide.

(4) Nitrous oxide (two nitrogen and one oxygen, N_2O) is produced as a byproduct of hot combustion[15] and by the breakdown of artificial fertilizers. It is about 3 parts per million in the atmosphere, and it stays in the atmosphere for about 114 years. Nitrous oxide accounts for about 8% of the human-caused global warming.

(5) Soot or black carbon is a particulate aerosol rather than a gas. It is produced by incomplete combustion of fossil fuel and biomass. Soot increases global temperature by warming surrounding air (which then reradiates some of the heat toward the ground) and warming nearby clouds (causing those clouds to evaporate and allowing more sunlight through to warm the ground because there is less cloud reflection. Soot on snow and ice absorbs more heat, thus increasing their melting. Recent studies of soot may skew the warming percentages because soot may produce about 10% of global warming.[16] On average, soot only stays in the atmosphere a few weeks, but soot generated by jet airplanes in the stratosphere may stay much longer because there is very little water that high to wash the soot down to the ground.

The purpose of the GND is to stop global warming or at least slow the warming. Prevention or cooling happens because of reduced emission of greenhouse gases into the atmosphere, removal of greenhouse agents, and/or increases in anti-greenhouse agents. A review of these cooling mechanisms provides suggestions for reducing or even reversing global warming. A partial list follows.

- Increased efficiency of structures, vehicles, and processes: Better efficiency in car mileage, better insulation of homes, and hundreds of other improvements over the centuries have increased the amount of well-being (miles traveled, warmer houses, and so forth) per amount of fuel burned. Consequently, less carbon

---

[14] Philipp A. Nauer, Lindsay B. Hutley, and Stefan K. Arndt, "Termite Mounds Mitigate Half of Termite Methane Emissions," *Proceedings of the National Academies of Sciences of the United States of America*, vol. 115, no. 52, pp. 13306–13311; December 26, 2018. https://www.pnas.org/content/115/52/13306 (accessed Aug. 22, 2020)

[15] The worst smog in Los Angeles came from nitrous oxide. Catalytic converters in our cars were put there to reduce that smog. As a side benefit, petroleum refiners had to remove lead and sulfur (both of which could poison the nitrous catalysts) from their gasoline blends, so two other nasty pollutants decreased greatly.

[16] Richard A. Kerr, "Soot Is Warming the World Even More Than Thought," *Science*, vol. 339, issue 6118, p. 382, Jan. 25, 2013.

dioxide has gone up into the atmosphere than would have been the case, and efficiency can be increased more (at a cost).

- Carbon capture by reforestation: There are large areas that were forests in the past, and many of these areas are available for replanting as forests. Growing trees draw down carbon dioxide from the atmosphere to grow both the trees we see and the roots below, so less global-warming carbon dioxide is in the atmosphere. Even if the trees are later removed, they can make a major difference ... and the trees have other uses such as erosion prevention, wildlife habitat, tree agriculture (silviculture), watershed, and recreation.

- Carbon capture by regenerative agriculture: Roughly a quarter of carbon dioxide being emitted to the atmosphere comes from soil (particularly eroded soil) oxidizing into carbon dioxide. Better farming practices (such as reduced plowing and use of cover crops) decrease the global-warming-causing carbon dioxide, methane, and dust in the air, thus slowing global warming. This is regenerative agriculture. This is even more important than the soil building of trees because trees generally maintain soil or gain slowly, whereas agriculture is often actively degrading soils and releasing global-warming carbon dioxide into the atmosphere. In addition, poor soils require heavier nitrate fertilization, much of which gets into the atmosphere as greenhouse-warming oxides of nitrogen and gets into groundwater causing algae overgrowth. Representative Occasio-Cortez has mentioned regenerative agriculture in speeches.

- Carbon capture by increased ocean fertility: The oceans cover 71 percent of our planet's surface. Algae, coral, shellfish, and other organisms in the ocean draw down carbon dioxide from the air as part of their life processes. This is a major carbon-capture process, probably more than carbon-capture on land. It could be made much larger with improved habitat and fertilization. The other benefits would include increased production of food, chemicals, and even fuel.

- Noncarbon energy production from nuclear fission, solar, and wind: If energy production does not produce carbon dioxide, that is less carbon dioxide being emitted into the atmosphere. (The limitation on that is that you need to factor in the carbon dioxide produced in building your infrastructure for wind, nuclear fission, solar, or whatever.)

- Electrified transportation: **If** you have enough **affordable** noncarbon electricity, cars, trains, ships, and even airplanes can run electrically. That would decrease emissions of not just carbon dioxide, but other greenhouse agents as well. Of course, as Jeff Gibbs noted in *Planet of the Humans*, electrifying transportation using fossil-fuel power plants just transfers the place where burning is done.[17]

---

[17] Jeff Gibbs, *Planet of the Humans*, Michael Moore, producer, 2019.

2.2 The Pluses and Minuses of Increased Heat in General

The good/bad warming news is that it is getting better before it gets worse. More people are surviving the milder winters, and that increase in survival is greater than the increased mortality from heat waves in the summer.[18] For food, slightly warmer years (and more advanced agricultural techniques) have actually increased food production.

Another factor increasing food production is that increased carbon dioxide levels make photosynthesis in plants more efficient. Photosynthesis works when a plant catalyst manages to grab a carbon dioxide molecule in the air and makes plant sugar. The more carbon dioxide molecules in the air, the better it works. As a further benefit, the efficiency means that plants do not need to circulate as much air through their leaves, so they do not need as much water. Thus, there has been a noticeable greening of about 5% in desert areas.

Despite news stories about increasing storm intensities, the severity of storms has decreased. The number of tornadoes per year has decreased. There was a nearly 12-year pause in hurricanes striking the lower 48 states; this was the longest such pause in recorded American meteorological history.[19]

Conversely, warming may be slower, coming in rises, pauses, and temporary retreats. In 2019, the U.S. Parks Service discreetly removed a sign at the entrance to Glacier National Park predicting that all the park's glaciers would be gone by 2020; the glaciers in question appeared to be growing.[20]

Scandinavia, Russia, Alaska, and Canada might see a 50% increase in forests at the expense of polar low shrubs and grassland tundra by the mid-21st Century, croplands in those areas might comparably increase, and the Arctic Ocean would be navigable all year long, which would shorten trade routes between East Asia and ports on both sides of the Atlantic.[21]

However, the good times might turn bad if warming continues beyond a certain point. There are historical analogs.

---

[18] Antonio Gasparrini, "Mortality Risk Attributable to High and Low Ambient Temperature: A Multicountry Observational Study," *Lancet*, vol. 386, pp. 369–375, July 25, 2005. https://www.thelancet.com/journals/lancet/article/PIIS0140-6736(14)62114-0/fulltext (accessed Sept. 28, 2020)

[19] Susan Jones, "U.S. Has Gone a Record 142 Months Without a Major Hurricane Strike," *CNS News*, Aug. 24, 2017. https://www.cnsnews.com/news/article/susan-jones/record-142-months-without-major-hurricane-may-be-coming-end (accessed July 23, 2020)

[20] Roger I. Roots, "Glacier National Park Quietly Removes Its 'Gone by 2020' Signs," *Watts Up with That?* June 7, 2019. https://wattsupwiththat.com/2019/06/07/glacier-national-park-quietly-removes-its-gone-by-2020-signs/ (accessed July 23, 2020)

[21] Richard G. Pearson, Steven J. Phillips, et al., "Shifts in Arctic Vegetation and Associated Feedbacks under Climate Change," *Nature Climate Change*, doe: 10.1038/nclimate1858, March 31, 2013.

2 What Are the Global Warming Threats?

The Medieval Warm Period (roughly 950 to 1300 A.D.) was great for Europe—consistently warmer and not much dryer, but it brought devastating droughts to the Americas.[22] The Mayan civilization in southern Mexico and Central America collapsed.[23,24] The cliff-dwelling Anasazi in Colorado faced not the typical droughts of 2 or 3 years, but droughts as long as a dozen years.[25] The Anasazi abandoned their cities and hopefully reached help with relatives farther south. About this time, the culture that archeologists call the Mound-Builders in the present corn belt of the Midwest also collapsed.[26,27]

The changes of the Holocene Thermal Maximum were much more intense. That was the warmest time between the melting of the glacial ice sheets and now. It was roughly 8000 years before present when the Arctic was roughly 4°Celsius (9°Fahrenheit) warmer. (A worrisome thing for planning future climate policies is that there are no firmly established explanations as to what caused the Holocene Thermal Maximum or why temperatures generally decreased since that maximum.)[28]

The warming was less severe in the middle latitudes going away from the poles, but there was a general ratcheting of climate bands away from the Equator.

As a good thing, the equatorial jungle band expanded poleward both to the north and the south.[29] Similarly, the savannah grasslands adjoining jungle also expanded farther from the Equator. For example, the Sahara in Africa became a lush green prairie with lakes and rivers. Likewise, desert areas in the Saudi Peninsula, the Indian Subcontinent, and northern Mexico became greener.

---

[22] William K. Stevens, "Severe Ancient Droughts: A Warning to California," *The New York Times,* July 19, 1994. https://www.nytimes.com/1994/07/19/science/severe-ancient-droughts-a-warning-to-california.html

[23] G. H. Haug, D. Gunther, et al., "Climate and the Collapse of Maya Civilization," *Science*, vol. 299, pp. 1731–1735, 2003.

[24] G. H. Haug, K. A. Hughen, et al., "Southward Migration of the Intertropical Convergence zone through the Holocene, *Science*, vol. 293, pp. 1304–1308.

[25] Meko, David M., Christopher A. Baisan, et al., "Medieval Drought in the Upper Colorado River Basin, *Geophysical Research Letters*, vol. 34, paper doi:10.1029/2007GL029988, 2007.

Also, "Colorado River Streamflow History Reveals Megadrought Before 1490," *Science Daily*, May 18, 2007. http:/www.sciencedaily.com/releases/2007/05/070517152428.htm on Dec. 2, 2007 (accessed Sept. 28, 2020)

[26] Robert Silverberg, *The Mound Builders: The Archaeology of a Myth*, Second Edition, Ohio University Press; May 1, 1986.

[27] Timothy R. Pauketat, *Cahokia: Ancient America's Great City on the Mississippi*, Penguin Group, July 1, 2009.

[28] Zhengyu Liu, Jiang Zhua, Yair Rosenthal, et al., "The Holocene Temperature Conundrum," *Proceedings of the National Academy of Sciences of the United States*, pp. E3501–E3505, Aug. 11, 2014. https://www.pnas.org/content/pnas/111/34/E3501.full.pdf (accessed Sept. 28, 2020)

[29] Dian J Seidel, Quiang Fu, et al., "Widening of the Tropical Belt in Changing Climate," *Nature Geoscience*, vol. 1, pp. 21–24, 2008.

However, the desert bands did not disappear. They also just moved farther from the Equator. The continental United States may have been a couple degrees Celsius (three degrees Fahrenheit) warmer on average than today, but with much less rainfall in a band from the Great Basin (still largely desert today),[30] through the northern Great Plains, and extending eastward to the present Corn Belt of Iowa to Indiana. Dry areas used only for grazing now were desert. Western Nebraska has an area of low rolling hills called the Sand Hills. These Hills have the exact contours of the sand dunes called ergs in the Sahara Desert because that area and much of eastern Colorado were definitely desert.[31] The corn belt of Iowa to Indiana would probably have only been fit for more dry-land crops such as the wheat and sorghum in Kansas.[32] Areas producing wheat now would probably been only suitable for grazing land.

Barring significant farming advances, a return to the Holocene Thermal Maximum would cause a major decline in American food production.

Wildly more warming did happen in the distant geological past. The Paleocene-Eocene Thermal Maximum (PETM) at roughly 56 million years before present is the most extreme ancient analog of modern climate change carried through to the worst-case scenario of several more centuries. The carbon dioxide in the atmosphere was probably about 1000 to 2000 parts per million (or two-and-one-half to four times that of today), and there were also much higher levels of methane (another greenhouse gas) in the atmosphere. Where the carbon dioxide and methane came from has not established, but theories include one or more of the following:

- Major regional volcanic eruptions in an area with coal deposits
- Impact from a large carbon-filled (carbonaceous) meteorite
- Release of methane from frozen methane–ice (clathrates) from ocean floor deposits
- Release of methane from melting permafrost
- A slight increase in solar radiation

Whatever the cause or causes, the PETM results on Earth were major.[33] Comparable results on Earth would probably come with two centuries of increased greenhouse gasses in the atmosphere. Some have used the Paleocene-Eocene Thermal Maximum as an example of the

---

[30] Matthew S. Lachniet, Yemane Asmerom, Victor Polyak, and Rhawn Denniston, "Great Basin Paleoclimate and Aridity Linked to Arctic Warming and Tropical Pacific Sea Surface Temperatures," *Paleooceanography and Paleoclimatology*, June 12, 2020.

[31] Stephen Stokes and Richard B. Swinehart, "Middle- and late-Holocene Dune Reactivation in the Nebraska Sand Hills, USA," *The Holocene*, vol. 7, issue 3, pp. 263–272, Sept. 1, 1997.

[32] P. J. Bartlein, T. Webb, and E. Fleri "Holocene Climatic Change in the Northern Midwest: Pollen-Derived Estimates," *Quaternary Research*, vol. 22, issue 3, pp. 361–364, Nov. 1984.

[33] Phil Jardine, "The Paleocene-Eocene Thermal Maximum," vol. 1, article 5, *www.palaeontologyonline.com*, https://pdf.palaeontologyonline.com/articles-2011/The_Paleocene-Eocene_Thermal_Maximum-Jardine_P-Oct2011.pdf (accessed July 23, 2020)

ultimate global warming could actually do the results were (and would likely repeat) as the following:

- Warming and other resulting changes came quickly in GEOLOGICAL terms—over just a few thousand years.

- Great extinctions in sea life, although land life flourished.

- Global temperatures averaged about 6° Fahrenheit (4° Celsius) warmer.

- That warmer weather made more water evaporate from the oceans making more water in the atmosphere … probably making even more heating and definitely making more rain, especially in the higher latitudes.

- The greatest temperature increases were the farthest distances from the Equator, and equatorial areas were largely unchanged. No evidence has been found of equatorial dead zones in the Eocene caused by excess heat, so the tropics would still be livable even if more uncomfortable.

- Climate bands expanded and shifted poleward north and south away from the Equator with more equatorial jungle, savannah grasslands pushing out desert, and desert invading the next grassland area farther away. However, much life flourished—just in different places. The Earth became mostly ice free. Greenland and Antarctica had forests. Antarctica had a cool rainforest that was dark half of the year.[34] (Think of Seattle winters, only more so.)

- Sea levels were as much as 520 feet (160 meters) higher than today. All the river deltas were submerged. Broad slow-rising flat lands were under water.

- Significantly warmer surface waters in the ocean suppressed vertical circulation of nutrients to the surface. This caused oxygen poor (anoxic) conditions below the surface in many areas, sea-life die offs, and probably even higher carbon dioxide levels in the atmosphere.

- The Arctic region, which is now a frozen desert with little precipitation, had land and sea comparable to today's Florida. There was so much rain that the Arctic Ocean was freshwater on the surface.[35] At that point, a freshwater fern (azolla) often covered much of the Arctic Ocean in free-floating mats. Algae and the azolla ferns helped create the Arctic's vast deposits of coal, oil, and natural gas that are eagerly sought today. As a side effect, the fertile Arctic may have pulled enough carbon dioxide from the atmosphere to drop from 3500 parts per million to a measly

---

[34] Tibi Puiu, "Antarctica was home to a rainforest some 50 million years ago," *ZME Science*, Aug. 2, 2012. https://www.zmescience.com/research/antarctica-rainforest-eocene-warming-321432/ (accessed July 23, 2020)

[35] H. Brinkhuis, S. Schouten, M.E. Collinson, et al., "Episodic Fresh Surface Waters in the Eocene Arctic Ocean," *Nature*, vol. 441, no. 7093, pp. 606–609, June 2006. https://www.ncbi.nlm.nih.gov/pubmed/16752440 (July 23, 2020)

600 parts per million (still much more than today), and that lower carbon dioxide put the world into a danger zone for possible cooling.[36]

The key lessons from a much bigger warming than humanity is likely to cause are that

1. Global warming will not destroy humanity.

2. Any problems will probably not arrive suddenly.

3. Still, those problems could be very large expensive problems that could threaten many lives.

4. All things considered, we had better work on some solutions for the most serious of those problems.

2.3 The Bottom Line: Food Supply Is the Only Crucial Danger

Large sea-level rises would be horrendously expensive for real estate markets, but land a few miles away would become prime real estate.

Greater heat levels a la the Paleocene-Eocene Thermal Maximum would increase the discomfort and inconvenience of being outside in the summer, in daylight, and in lower latitudes; but there is no evidence of "dead zones" in the vicinity of the Equator during that time period. Remember, the climate changes are greatest near the poles and the amount of change is progressively less toward the Equator. Consequently, the equatorial regions of a new hothouse Earth would probably not be significantly hotter than today.

Moreover, people can and do adopt with fans, siestas, nocturnal activities, subsurface living (terratecture), and other responses. This is the opposite of human adoptions farther from the Equator where people often die from the cold. People there adopt with furnaces, heavy clothes, and sometimes building underground. At present, they are more susceptible to disease in the winter when they are crammed into confined quarters with heating that is often minimal. Their death rate would radically decline in a warmer world. The net effect of warming to date has been an increase in average worldwide lifespan.

Only decreased food supplies are an existential threat to a large percentage of humanity. Although, many areas would become more fertile, there would be many costs.

- Exact movement of desert bands is guesswork, but there would definitely still be deserts.

- Every existing river delta—all of them rich agricultural areas—would become shallow sea. A major part of Bangladesh is delta, which would retreat significantly with rising sea level.

---

[36] *The Azolla* Foundation, web page, http://theazollafoundation.org/azolla/the-arctic-azolla-event-2/, accessed August 28, 2018. *ZME Science*, Au. 2, 2012. https://www.zmescience.com/research/antarctica-rainforest-eocene-warming-321432/ (accessed June 30, 2020)

- Many of the rich river valleys for many miles upstream would become long inlets. This would cover much rich bottom land along these rivers.

- Many rich coastal plains would become ocean shallows. For example, the North European Plain includes Berlin at 34 meters (112 feet). The elevation of Modesto in the California Central Valley is 89 feet (27 meters), so San Francisco Bay would extend north and south into the Central Valley to form a significant inland sea. The Florida peninsula, pointing down toward Cuba is very flat for a reason—higher sea levels during warmer times planed that land down to just sea level so that it was an area of shoals, and a significant seal-level rise would do so again.

- All areas outside of the tropics would change to different temperature regimes with major difficult adjustments of farm practices. This might be a happy change for places such as permafrost areas changing to temperate growing of corn and wheat. It might be more painful in corn growing areas switching to crops such as sorghum, millet, dates, and palms—all much less productive per unit of land area. There would most likely be decades to make the transitions, but there are no guarantees.

The bottom line is that there are no guarantees with sufficient food supply being one of the biggest unknowns. Thus, this book is heavily weighted toward food production advances, although many other possibilities are covered.

3 A Quick Summary of the Green New Deal?

The Green New Deal (GND) was described in a House Resolution[37] and in a frequently asked questions (FAQ) paper that was released and then withdrawn. These two items are included in Chapter 11.

The GND is still not clearly formed in many ways. It is a partial set of directions rather than a detailed prescription. As a first cut, we can examine it and make suggestions.

The climate-fixing proposals are as shown in the following bullet list. The following list does not include the measures that the GND writers apparently believe would make society better (such as guaranteed good housing and Medicare for all). These items may be good or bad, but they do not apply to the existential threat of global warming. The proposal items that seem to be directed toward mitigating or stopping global warming are as follows:

- In the first list of (1) through (4), (4) (A): global reductions in greenhouse gas emissions from human sources of 40 to 60 percent from 2010 levels by 2030

- In the first list of (1) through (4), (4) (B): net-zero global emissions by 2050

- [There are two additional parenthesized number lists before the next items dealing with climate.]

- In (1)(A): achieve net-zero greenhouse gas emissions

- In (1)(C): invest in the infrastructure and industry of the United States to sustainably meet the challenges of the 21st Century

- In (2): [The goals of (1)] should be accomplished through a 10-year national mobilization

- In (2)(A): building resiliency against climate change-related disasters, such as extreme weather

- In (2)(C): meeting 100 percent of the power demand in the United States through clean, renewable, and zero-emission energy sources

- In (2)(D): building or upgrading to energy-efficient, distributed, and "smart" power grids, and ensuring affordable access to electricity

- In (2)(E): upgrading all existing buildings in the United States and building new buildings to achieve maximum energy efficiency, water efficiency, safety, affordability, comfort, and durability, including through electrification

---

[37] H.Res.109 — 116th Congress (2019-2020), Feb. 7, 2019. https://www.congress.gov/bill/116-congress/house-resolution/109/text (accessed June 20,2-19)

- In (2)(F): spurring massive growth in clean manufacturing in the United States and removing pollution and greenhouse gas emissions from manufacturing and industry as much as is technologically feasible, including by expanding renewable energy manufacturing and investing in existing manufacturing and industry

- In (2)(G): working collaboratively with farmers and ranchers in the United States to remove pollution and greenhouse gas emissions from the agricultural sector as much as is technologically feasible, including … (iii) investing in sustainable farming and land use practices that increase soil health

- In (2)(H): overhauling transportation systems in the United States to remove pollution and greenhouse gas emissions from the transportation sector as much as is technologically feasible, including through investment in—(i) zero-emission vehicle infrastructure and manufacturing; (ii) clean, affordable, and accessible public transit; and (iii) high-speed rail

- In (2)(J) [overlaps with (2)G)]: removing greenhouse gases from the atmosphere and reducing pollution by restoring natural ecosystems through proven low-tech solutions that increase soil carbon storage, such as land preservation and afforestation

- In (2)(M): identifying other emission and pollution sources and creating solutions to remove them

- In (2)(N): promoting the international exchange of technology, expertise, products, funding, and services, with the aim of making the United States the international leader on climate action, and to help other countries achieve a Green New Deal

The following chapters describe the good, the bad, and some suggested better upgrades for the Green New Deal.

4 The Good of the Green New Deal

4.1 Take a First Cut at an Overall Plan and Work from There

The first plus of the Green New Deal (GND) is that it combines all the global warming issues in one overall set of ideas. This has been proposed before in essays or studies, but this is a government proposal. Even though it is only a resolution (and one that was voted down in the Senate at that), it drew a great deal of interest.

Most often, ideas for or against a particular issue have been considered in isolation. This GND, or some future version of it, will allow consideration of a more complete matrix of issues related to global warming.

4.2 Regenerative Agriculture Is Huge

Regenerative agriculture [(2) (G) and (H) in the GND Resolution] could stop or reverse a major part of global warming. If worked just by itself. More about that in Chapter 7 Food, Fear, and Climate Fixes from the Land.

4.3 Zero-Sum Accounting Works—What Is the Net Warming?

Many have argued that zero-sum accounting was as important as the discovery of the Americas in building the world's economy to today's prosperity. People could quickly determine if they were making money or losing money, so that they could adjust their actions accordingly

Likewise, the first action statement of the GND resolution [(1)(A)] calls for net-zero greenhouse gas emissions. One major implication of this is that you could continue running a coal-fired power plant with its production of global warming carbon dioxide if you had some other compensating process (or processes) that drew carbon dioxide from the atmosphere.

That is the principle of carbon trading done on an individual or corporate scale; however, it could be done on an international scale to allow phase-out of long-term investments. A side issue is the danger of fraudulent carbon trading or (worse) carbon trades that make the world worse off.

4.4 Considering Resiliency as Part of the Picture

The concept of "building resiliency against climate change-related disasters" [(2)(A)] is an excellent addition to the issue of climate change. It does not reduce global warming, but it mitigates potential damage from rising sea levels or other extraordinary conditions.

In fact, it would help under ordinary conditions. A large percentage of the prime urban real estate looks out onto ocean fronts, streams, or lakes. Much of that property is referred to as flood plain. That means that they periodically flood NOW. Some of the angriest people on Earth are economic geographers who repeatedly say, "Don't build on flood plains!"

Environmentalists have often failed to address this issue because realtors are a powerful political force. How ironic it is that environmentalists can stop a billion-dollar power plant, but a ten-million-dollar housing tract—with that beautiful riverfront view—whips them every time ... or could it be bribes them every time.

4.5 Yes, Commercial Aviation Contributes to Global Warming

Those giant buses in the sky place the global warming agents of soot and nitrous oxide above most rain clouds, so they are not washed out of the sky. Consequently, these global-warming agents stay up much longer than they would otherwise, and they are more harmful than the carbon dioxide, which airlines also produce. And yes, commercial aircraft also emit a significant amount of that best known global-warming gas, carbon dioxide.

Surprisingly, a significant percentage of air transportation could conceivably be made electric or hybrid-electric. More on that later too (see 8.9 Electric Airplanes ... Really!).

4.6 The Power of Positive Insanity

Giant government initiatives are often failures.

Large government programs do best when they copy existing successful technologies on a large scale rather than developing entirely new technologies. The technology set is known to work, there are already functioning systems to learn from, and there are already workers and managers experienced with the technology. Examples include:

- The United States initiative to build the Transcontinental Railroad across North America in the 1860s was expensive but relatively low-risk because there were already major rail lines in Europe and in the eastern United States.[38] Furthermore, each stretch of line extending into new territory had immediate use.

- The Czarist Russian initiative to build the Trans-Siberian Railway had even more precedents including the American Transcontinental Railroad.[39] Also, as with the Transcontinental Railroad in North America, it had profitable uses even before it was complete.

---

[38] "Transcontinental Railroad," *History*, Sept. 11, 2019. (accessed Oct. 5, 2020) https://www.history.com/topics/inventions/transcontinental-railroad (accessed Oct. 5, 2020)

[39] Zack Beauchamp, "The Trans-Siberian Railway Reshaped World History," *Vox*, Oct. 5, 2016. https://www.vox.com/world/2016/10/5/13167966/100th-anniversary-trans-siberian-railway-google-doodle (accessed Oct. 5, 2020)

- The United States' construction of the Panama Canal in the early 1900s was on the edge of established technology and new science. Canal-building technology was well established, but American doctors had to find what caused yellow fever, and prevent the worst of its outbreaks. Without success in that endeavor, the United States would have failed as the earlier French canal-building effort had failed.[40]
- The 1956-through-earl-1990s construction of the United States Interstate Highway System was started because President Dwight Eisenhower had been impressed by a similar highway system when he directed troops in the invasion of Germany. The Germans had used their nationwide Autobahn system to quickly reposition troops and munitions while defending against Eisenhower's invasion in 1944 and 1945.[41]

When giant government crash programs invest vast sums, there is often no visible return on the investments. Phrases such as "boondoggle" and down a rat hole" often apply,

Yet, many important things just would not happen without giant government (and large-business) initiatives that would not have happened without dreamy and/or desperate thinking.[42, 43] Those government research and development initiatives often only had government uses at the time (usually wars). Most did not return any profits for years or even decades. Some had no use at all when started or even years or even decades later.

Yet, these boondoggles provided entirely new and technologies that eventually became technological revolutions.

- Agriculture: The Hatch Act of 1887 provided federal funding for agricultural experiment stations, most of them set up in proximity to land-grant colleges, to engage in state-of-the-art research that could increase the productivity of the nation's farms. The Smith-Lever Act of 1914 funded cooperative extension services, including the employment of thousands of "county agents," to diffuse the latest knowledge to farmers.
- Heavier-Than-Air Aircraft/Airplanes: The Wright brothers made their first heavier-than-air flights with their own money, but eventually they got some federal money. Then, in World War I, many governments developed many kinds of

---

[40] David McCullough, *The Path Between the Seas: The Creation of the Panama Canal, 1870-1914*, Simon & Schuster, New York, 1977.

[41] Lee Lacy, "Dwight D. Eisenhower and the birth of the Interstate Highway System," United States Army, Feb. 20, 2018. https://www.army.mil/article/198095/dwight_d_eisenhower_and_the_birth_of_the_interstate_highway_system (accessed Oct. 6, 2020)

[42] Fred L. Block and Mathew Keller, State of Innovation: The U.S. Government's Role in Technology Development, Routledge, Nov. 17, 205. https://www.huffpost.com/entry/nine-government-investmen_b_954185

[43] William Lazonick, "Nine Government Investments That Made Us an Industrial Economic Leader," *HuffPost*, Nov. 8, 2011. https://www.huffpost.com/entry/nine-government-investmen_b_954185 (accessed Oct. 6, 2020)

airplanes and used them under wartime conditions. One of the development goals was to make airplanes that could fly high enough and fast enough to shoot down another wild-eyed initiative, the giant lighter-than-air airships that carried out the first aerial bombardments on England. After the war, there was an existing infrastructure of airplane manufacturers, mechanics, and pilots. With just a little more investment, some of the individuals in that infrastructure became the entrepreneurs who started commercial airlines … and looked for a few more government contracts. World War II (1939–1945 was the same type of massive investments again, only more so. This time, it led to a massive and usually profitable industry—by 2017 there were 4 billion passengers in a single year.[44]

- Jet engines: Frank Whittle in Great Britain and Hans-Joachim von Ohain in Germany developed competing turbojet engines just before and during World War II, and both countries had crash programs to build jet fighters and bombers. Jets did little in World War II because of poor efficiency, hence limited range. However, jet efficiency quickly improved, and jet commercial passenger jets were a major part of that revolution in commercial airlines just described.

- Computers and the Internet: Before World War II, computers were mathematical workers trained to do laborious calculations as quickly as possible. For code breaking and developing better artillery firing tables, governments needed much more computing than people could do. The result was several massive programs to develop mechanical and electronic computers. A 1999 study, *Funding a Revolution: Government Support of for Computing Research*, stated, "Federal funding not only financed development of most of the nation's early digital computers, but also has continued to enable breakthroughs in areas as wide ranging as computer time-sharing, the Internet, artificial intelligence, and virtual reality as the industry has matured." Among other things, the study details the now well-known role of the U.S. government's three-decade development effort with precursors that eventually became the commercially profitable revolution of the Internet.[45] That government research paid off. Moreover, many later applications on the Internet could be developed by small software companies—entrepreneurs.

- Solid-State Electronics: The switch from vacuum tubes to solid-state transistors started largely with the nongovernment American Telephone and Telegraph Bell Labs. However, their interest started with attempts to make germanium "crystal" mixer diodes for microwave radar receivers in World War II, and governments were soon funding many projects to get lighter and more capable equipment. Transistors

---

[44] "Traveler Numbers Reach New Heights: IATA World Air Transport Statistics Released," Press Release 51, International Air Travel Association, Sept. 6, 2018. https://www.iata.org/en/pressroom/pr/2018-09-06-01/ (accessed Oct. 6, 2020)

[45] *Funding a Revolution: Government Support for Computer Research*, National Academies Press, Washington, D.C., 1999. https://www.nap.edu/download/6323 (accessed Oct. 5, 2020)

got steadily smaller until it became hard to connect them all, so designers Built more than one circuit together—an integrated circuit. Putting many integrated circuits and different kinds of circuits together on a chip became a microcircuit, and the transistors in microcircuits have shrunk exponentially in size down to the size of nanometers (billionths of a meter. 25 millionths of an inch) with a resulting exponential growth in capabilities. This growth in capabilities is often called the Electronics Revolution, and it is unquestionably as great a change as the Industrial Revolution of steam power. This is one case where wild-eyed government investments paid off handsomely.

- Nuclear fission: Nuclear fission was a crazy idea in World War II. It was only a wild theoretical possibility because two physicists had actually achieved a tiny fission reaction in 1938. The United States spent a preposterous amount of money creating whole industries with the capability of fabricating fission bombs more powerful than chemical explosives. It was insane ... except that the other side (Germany) might have been making a similar effort. (They were, but their effort was far behind, partly because of Allied commando attacks but mostly because the task was simply too massive for German resources.) The war effort pushed nuclear technology decades ahead of where it would have been in peaceful times. Nuclear technology evolved to the point in the 1970s that fission power reactors produced a significant fraction of electrical power at a profit, not counting the giant government research input that had preceded them. Nuclear fission reactors went through hard times due to spectacular accidents, but it could now reemerge as the only commercially practical replacement for much of present fossil-fuel use.

- Staged Rockets and the Space Race: The Versailles Treaty after World War I prohibited defeated Germany from building cannons larger than a certain size, but the treaty did not mention the obsolete technology of rockets, which were more expensive and less accurate than cannons. The German government responded by spending large sums on improved rockets. They developed three rocket improvements: (1) guidance systems for improved accuracy, (2) liquid fuels and oxidizers that provided greater range, and (3) Russian theoretician Constantin Tsiolkovsky's system of rail cars (stages) that provided still more range. The most notable result was the V-2 guided ballistic missile with a 200-mile (320-kilometer) range carrying a bomb weight of 1 metric ton (2,200 pounds) into the stratosphere and crashing down at supersonic speed. It was much less effective than a bomber, but it was unstoppable. Its accuracy was still poor, but that would not matter if it carried a nuclear bomb. That led to a 1950s and 1960s arms race of several countries developing intercontinental ballistic missiles (ICBMs). Those rockets allowed launches of artificial satellites that, in turn, touched off the Space Race. Arthur C. Clarke said that the Space Race probably advanced space technology three quarters of a century beyond what it would have been in ordinary times. Then, development by the governments atrophied in many ways for decades until Elon Musk became the James Watt of the 21st Century by introducing profit-making improvements including reusable stages and redesigned stages with far fewer parts. The baton was passed to (or grabbed by) an entrepreneur.

- Solar, Wind, and Almost Everything Else Alternative Energy: The two Energy Crises of the 1970s led to a scattergun set of research initiatives on nearly every possible alternative energy, plus efficiency and improvements in nuclear and fossil fuels. Many did not get past the stage of studies and proposals. Many development projects did not achieve enough progress to get further funding. Subsidized commercial ventures often failed spectacularly—wind-power sites often looked like war zones with flying pieces of propeller blades and blazing gear boxes at the propeller hubs. Large solar power ventures generally failed (although small solar at remote sites steadily pushed their competitive zone closer to the low-cost-price areas of the central power grids). After a pause of a couple decades due to low oil prices, the high prices of the early 2000s led to a similar government energy frenzy in the 2010s; only this time, solar, wind, and a number of technologies did much better.

5 The Bad of the Green New Deal

Analyses condemning the Green New Deal (GND) were not long in coming, especially expressing reservations about cost estimates being too low, entirely new costs that would result, and accusations that GND costs would fall disproportionately on the poor,[46,47] and this skepticism has continued.[48]

5.1 A Perceived Socialist Bait and Switch Leads to Loss of Credibility

The GND was proposed as a way to save humanity from the new threat of runaway global warming. However, a quick reading of the Green New Deal resolution shows dozens of other expensive unrelated goals.

Within hours of the release of the Green New Deal (GND) resolution, political conservatives noted that it was mostly not a new deal. Instead, it had every old socialist wish list item dredged from the 1890 progressives, through Franklin Roosevelt's original New Deal, to the socialist proposals of today. The resolution calls for socialized medicine for all, guaranteed incomes for all, universal access to healthy food, clean and affordable public transit, cleaning up of hazardous waste sites, strengthened labor-union-organizing rights, reduced transfer of jobs and pollution overseas, guaranteed access to adequate housing, counteracted systemic injustices (whatever that means), guaranteed economic security for all, and undoubtedly many others not listed here.

Many of these proposed goals may be laudable, but they have nothing to do with climate, which is said to be an existential threat, a death threat, for humanity. At best, these many additional goals and associated items of legislation would slow implementation of the climate aspects. At worst, they would stop the whole process. The fact that the resolution has the complexity of myriad socialist dream goodies leads to a strong suspicion of a bait and switch scam.

---

[46] Benjamin Zycher, *The Green New Deal: Economics and Policy Analytics*, American Enterprise Institute, April 2019.

[47] Jude Clementa, "Five Practical Problems for the 'Green New Deal'," *Forbes*, April 29, 2019. https://www.forbes.com/sites/judeclemente/2019/04/29/five-practical-problems-for-the-green-new-deal/#75889ee53e8a (accessed (Oct. 24, 2020)

[48] Valerie Richardson, "'Forcing Americans in the Dark': Green Energy Push Blamed in California's Rolling Blackouts," *The Washington Times*, Aug. 18, 2020. https://www.washingtontimes.com/news/2020/aug/18/california-rolling-blackouts-caused-green-energy-p/ (accessed Oct. 24, 2020)

5 The Bad of the Green New Deal

The cynicism about the GND grew stronger when Representative Ocasio-Cortez's chief of staff at the time, Saikat Chakrabarti, said in an interview that,[49,50]

"The interesting thing about the Green New Deal, is it wasn't originally a climate thing at all. Do you guys think of it as a climate thing? Because we really think of it as a how-do-you-change-the-entire-economy thing.

The bait-and-switch suspicion, reinforced by that quote, would make it very difficult to pass various pieces of legislation needed to implement a GND.

5.2 The Short Deadlines Are Unworkable and Undermine Credibility

Even if all the proposed social-justice proposals were acceptable to a majority of American voters, the proposed speed of implementing the infrastructure of the Green New Deal is problematic.

Many people have questioned the 12- and 32-year deadlines (2030 and 2050) for implementing the GND. With the mantra that might be summed up as, "We're all going to die if we do not reduce greenhouse gas emissions by 40 to 60 percent by 2030," it would not matter whether or not we had guaranteed incomes and socialized medicine for all. What good is an excellent health-care plan if you are dead.? The suspicion again grows that the resolution writers did not truly expect to reach their stated goals.

There is a plethora of items that would be impractical to implement by 2030, and many would be impractical even by the 2050 deadline. For instance,

- Upgrading all buildings becomes tremendously complex and expensive for old buildings. The building plans are often not available or too faded for detailed redesign. Furthermore, every old building is different, and the renovation may require drawing entirely new plans and getting them approved. Every building upgrade is a negotiation, a contractor estimate, and monitoring of the contract performance. The contractor on each building will likely bid high because workers often find asbestos, black mold, or other toxic substances needing remediation, often find damage from termites or dry rot, and (most of all) often find basic structural components that are not compliant with current code.

---

[49] "How Saikat Chakrabarti Became AOC's Chief of Change," *Washington Post*, July 10, 2019. www.washingtonpost.com/news/magazine/wp/2019/07/ (accessed Feb. 19, 2020)

[50] Jack Crowe, "AOC's Chief of Staff Admits the Green New Deal Is Not about Climate Change," *National Review*, July 12, 2019. https://www.nationalreview.com/news/aocs-chief-of-staff-admits-the-green-new-deal-is-not-about-climate-change/ (accessed Sept. 20, 2020)

- Moreover, making buildings more airtight to save energy increases indoor air pollution. This increases the health hazards for asthma, allergies, and other breathing problems, particularly for children.[51]

- Making the U.S. electrical grid 100-percent renewable would have many advantages. However, it would be an immense undertaking, especially considering that the resolution proposes electrification of not just transportation in electric cars and trains but also all building processes including air conditioning, water heating, and cooking stoves that are often gas-fired at present. Try to imagine an electrical capacity three or four times larger—nearly all provided by solar or wind (after shutting down coal, gas, and nuclear fission). Because most hydroelectric power sites have already been taken, the only major way renewables could increase would be solar and wind. Thus, they would need to increase at least tenfold. An additional investment in time and money would be that these renewable power sites tend to sprawl out to remote sites, so additional high-power transmission lines would be needed along with the transformer stations at each end.

- Furthermore, renewable power without fossil-fuel backup would require major investments in industrial-scale electrical-storage backup for nights, clouds, and windless periods. This storage would probably be giant industrial batteries. They would be cheaper per kilowatt hour than the lithium cells of cars, but they would still be expensive, and they would be large facilities. An entire industry would have to develop for supplying those batteries or some even more exotic storage methods. Another alternative is to develop better ways to use excess electricity generated for splitting water, then develop mass storage for the hydrogen generated and finally hydrogen-burning generating capacity for use when needed,

- A smart grid with higher quality power and variable metering for time of day is another worthy investment. However, utilities are slow-moving organizations. They still use transformers that are fifty or sixty years old. Conversion to a smart grid is long overdue, but totally rebuilding the entire electric utility industry in a shorter time period would radically increase costs.

- Switching to an electric transportation network would require not only new vehicles, but an infrastructure of charging stations and the construction and installation of those facilities across the country. For electrified trains and trolleys, there would need to be electrical transmission facilities either as a third rail at ground level or as above-ground power supplies for trolleys. Plus, that charging infrastructure would be part of that much-larger electrical generation capacity.

---

[51] *The Inside Story: Health Effects of Indoor Air Quality on Children and Young People*, Royal College of Pediatrics and Health Care, London, 2020.
https://www.rcpch.ac.uk/sites/default/files/2020-01/the-inside-story-report_january-2020.pdf
(accessed Jan. 30, 2020)

- Likewise, chemical processing for paints, plastic, paper, etc. would have a vast expense of conversion from fossil chemical inputs (such as coking coal for smelting iron or natural gas to make plastics). Chemical plants typically evolve over decades as investments are paid off. Moreover, there are thousands of individual processes, many of which are developed over years of trial and error.

- Removing fossil fuels from cement and steel production would require changing massive facilities that have been built over decades. Furthermore, as with many of the other processes mentioned, changing cement and steel production from fossil fuel production to electrical production, or production using hydrogen from electrolysis would require additional major increases in renewable electrical power production.

- Regenerative agriculture has great possibilities, but it requires different technologies and different sets of knowledge for the farmers and ranchers in different climate areas. One solution cannot be tested once and then applied to all. Instead, different solutions in different areas must be refined over years, and some experimental solutions will fail in some areas. Thus, it would be unwise to have all farms experimenting at the same time. Most importantly, bad practices degraded soil over many decades; drawing down global-warming carbon dioxide from the sky by regenerating soil is a similarly slow process.

- Just gearing up to run a giant multi-billion-dollar space bureaucracy, the National Aeronautics and Space Administration (NASA) for the Space Race took several years. That was for a narrowly focused goal of building upgraded German V-2 rockets from World War II and flying a few excursions to the Moon and back. The GND would require thousands of entirely new technologies that would be produced in millions of units.

- The most dangerous thing about short deadlines is that it might drive policy makers to do dangerous unproven policies. During the new-ice-age scare of the 1950s through early 1970s,[52] there was a serious proposal to (expensively) air drop black coal dust onto the Arctic ice to increase sunlight absorption, melt the ice, and thus warm the planet. The resulting increased heat was expected to melt large areas of pack ice and create more open water that would absorb more sunlight **to warm the planet**. The net effect was supposed be reversing the expected dangerous chilling

---

[52] Peter Gwynne, "The Cooling World," *Newsweek*, p. 64, April 28, 1975.
Peter Gwynne, "My 1975 'Cooling World' Story Doesn't Make Today's Climate Scientists Wrong," *Inside Science*, May 21, 2014. https://insidescience.org/news/my-1975-cooling-world-story-doesnt-make-todays-climate-scientists-wrong (accessed Nov. 8, 2020)

of the planet.[53,54] Conversely, there are now serious proposals to loft millions of tons of reflective aerosols into the stratosphere **to cool the planet**.

The optimistic view is that perhaps the GND resolution framers did not really expect to fit within their stated deadlines. If they did not expect things to happen that quickly, then we should not require such desperate measures. Then there would be time to evolve those exotic technologies that could transcend global warming—but not use them unless they are definitely needed.

As a practical matter, former president Barack Obama purchased a house on Martha's Vineyard, Massachusetts in 2019. That house is 3 feet (1 meter) above sea level[55] with a view across Edgartown Pond to a narrow barrier island that is typically breached during the stormy season. Obviously, someone as smart as a former president would not buy such a house if he believed that global warming would be severe enough to cause major sea-level rise in the near future.

5.3 Governments and Big Business Do Stunts and Disasters; Entrepreneurs and Outsiders Are More Likely to Make Technology Revolutions

The GND resolution suggests that great profits will accrue from the GND, but entrepreneurs are never mentioned. The Frequently Asked Questions document explains that American society would cease using energy from both fossil fuels and nuclear energy within ten years because it would be so much more affordable, but the alternative energies of solar and wind are only marginally competitive now. Considering that lack of competitive advantage, replacing existing fossil and nuclear facilities would entail massive compensations to the present owners and higher prices for the customers. Suitable rewards to entrepreneurs might elicit greatly improved solar and wind energy production. However, nothing like that is mentioned, nor are carbon credits mentioned except to say that they should not be used.

These are a major failing because technological revolutions almost invariably grow from the efforts of entrepreneurs. In fact, former Vice President Al Gore, the major iconic figure railing against the dangers of global warming said:

---

[53] P. M. Borisov, "Can we Control the Arctic Climate?", Bulletin of the Atomic Scientists, March, 1969, pp. 43-48.

[54] James Rodger Fleming, *Fixing the Sky: The Checkered History of Weather and Climate Control*, "The Climate Engineers," Columbia University Press, New York, pp. 236–237, 2010.

[55] Daniel Greenfield, "Obama Ocean Mansion 3-10 Feet Above Sea Level Extremely Vulnerable to Global Warming," Frontpagemag, Aug. 23, 2019.
https://www.frontpagemag.com/point/2019/08/obama-ocean-mansion-3-10-feet-above-sea-level-daniel-greenfield/ (accessed Oct. 24, 2020)

5 The Bad of the Green New Deal

The businesses will be the driver for change because if the businesses do not make new products like cars or planes or buildings, then nothing is going to change.[56]

Governments see a bright shiny new technology and want to push it (technology push). If they lose interest, they move on to another bright shiny object. In contrast, Coen van Oostrom, the person who recounted Gore's quote above, is a Dutch entrepreneur. He has built more efficient buildings and pioneered procedures to run them more efficiently.

Entrepreneurs need to make a profit; hence, they find a market for a new technology or opportunity. They then develop a product to sell in that market so they can make a profit (demand pull). Governments, and even large industrial concerns, do not jump at new opportunities. If you doubt, look at history; innovations have lowered costs

- From 1405 to 1433, Admiral Zhen He of China commanded a fleet with hundreds of ships and as many as 28,000 men that sailed throughout the South China Sea and the Indian Ocean. The Jade Throne of Imperial China sent him on seven voyages of discovery and trade. They lost money, so they took their ball and went home.[57,58] In 1497, Vasco de Gama sailed from Portugal commanding just three caravel sailing ships crewed by 170 men total. They sailed for profit, and their country built an empire, the first of several European empires throughout the world.

- The United States telegraph industry did not invest in Alexander Graham Bell's telephones. Likewise, the British Post Office was not interested in telephones during the 1890s. As their chief engineer said, "The Americans have need of the telephone, but we ... have plenty of messenger boys."[59]

- In 1903, one of America's leading scientists, Samuel Langley, made his final attempt in a government development effort to fly a steam-powered heavier-than-air craft. He failed ignominiously. That same year, a pair of bicycle mechanics from Dayton Ohio used an internal-combustion engine to successfully fly. The Curtis-Wright Corporation they helped found is still operating today.

- In the 1920s, the United Kingdom had a government-directed research initiative to build commercial passenger airships (dirigibles) for long-distance travel. Interestingly, they had a two competing prongs of the development. One (called the R100) was built by a private concern, and the other (called the R101) was built

---

[56] Al Gore, quoted by Coen van Oostrom, "Smart Cities: How Technology Will Change Our Buildings, TEDx Talks, Oct. 28, 2016. https://www.youtube.com/watch?v=hT4ZsaZsEgc (accessed Feb. 14, 2020)

[57] Edward K, Dreyer, *Zheng He: China and the Oceans in the Early Ming Dynasty, 1405-1433*, Pearson; 1 edition, Library of World Biography Series) 1st Edition, May 13, 2006.

[58] Louise Levathes, *When China Ruled the Seas: The Treasure Fleet of the Dragon Throne, 1405-1433*, Oxford University Press; Revised edition Jan. 9, 1997.

[59] Mathew Dunn, "20 Times in History People Were Really, Really Wrong about Technology," *New York Post*, March 8, 2016. https://nypost.com/2016/03/08/20-times-in-history-people-were-really-really-wrong-about-technology/ (accessed Aug. 7, 2020)

by a government entity. The privately-developed R100 airship met its design goals while the government-built R101 was underpowered and overweight. Despite the R101's problems the British government sent it on a maiden flight toward India. When it crashed in France, it killed the program … and most of the passengers.[60]

- During World War II in 1944 and 1945, the German government developed a staged liquid fueled rocket to lob long-range artillery warheads at Britain. Shortly before the German surrender, teams from the communist world (led by the USSR)[61] and the western powers (chiefly the United States) raced each other to capture as many of the German rocket team members and as much of the equipment used as possible, and the two sides commenced a rocket missile race. The USSR launched the first artificial satellite in 1957, and the space race was on. However, the artillery-shell paradigm (use once and crash) was fabulously expensive. Surprisingly, that paradigm remained unchanged until 2015 (more than 60 years!). During all that time, there was one attempt at a reusable booster, the Space Shuttle, that ended up costing more per mass of payload than the expendable Saturn V rockets that it replaced. Finally, the SpaceX Corporation made a fresh start building boosters that could return and fly again. That was the paradigm shift that made space flight practical, and it had been suggested many times by experts within and outside of the U.S. National Aeronautics and Space Administration. Still for six decades the space agency bureaucracy and their civilian managers in Congress and all the way to the White House could not imagine a change … until there were successful business launches with returning reflyable boosters.

- Also, during World War II, the United States developed nuclear fission (atom splitting) reactors as part of the effort to build fission bombs. The U.S. government then applied the same technology into powering submarines with fission reactors so that they could cruise for months without resurfacing. In the 1950s, the U.S. government encouraged large businesses to apply that technology to make electrical power. They succeeded somewhat, but their large reactors were touchy—without careful supervision, they could fail catastrophically. Worst of all, they were behemoths that had to be built on site. Only in the twenty-teens did entrepreneurs proposing multiple compact reactors of a few tens of megawatts (a megawatt is a thousand kilowatts) get serious consideration. Such compact reactors (or mini-reactors) could be built on an assembly line, were small enough to be trucked to a site, and were small enough to be shipped back for disassembly (and recycling of materials) after operational life.

---

[60] Nevil Shute, *Slide Rule: Autobiography of an Engineer*, House of Stratus, Jan. 12, 2008 (first published by William Heinemann Ltd., London, 1954).

[61] The Union of Soviet Socialist Republics (USSR) was a Marxist-socialist continuation of the Russian Empire that had been run by the czars. It began in 1917 and included present-day Russia, Ukraine, Belarus, Kazakhstan, Georgia, Azerbaijan, and several other present-day countries. The USSR collapsed and fragmented into the present independent states in 1992.

5 The Bad of the Green New Deal

- From 1953 until 1963, Nikita Khrushchev's USSR ran the Virgin Lands campaign to farm semi-arid lands in the province (now country) of Kazakhstan. As with the Great Plains in the United States, initial high production eventually transitioned into a dust bowl.

- From 1958 through 1960, Mao Zedong's Marxist-socialist China attempted "The Great Leap Forward." The goal was to leap from being a poor agricultural economy to a rich industrial one. The Great Leap had a number of ambitious component programs. Farming was done by communal work brigades rather than farmers working private plots. Trees were cut down to fuel backyard blast furnaces. The program was cancelled as a result of a great famine that killed between thirty and ninety million people.[62] As with casualties from the corona virus (or Wuhan virus as it is known in China), the true number of Chinese casualties is still a state secret.[63] The number of lives cost are still a secret because the Great Leap was part of a Marxist dream of a socialist utopia for which the ends justified the means. Admitting that the disaster happened might slow achievement of that dream; therefore, lies to hide the famine were justified.

- The first world energy crisis came in 1973. An Arab reduction in oil exports demonstrated that the United States was vulnerable to interruptions in oil imports. President Nixon responded with Operation Independence for more U.S. production, particularly by expanding use of coal and nuclear fission. In 1979, the fall of the Iranian government in Iran caused the second energy crisis. President Carter doubled down with his New Foundation initiative that included even more coal use to increase available energy while sabotaging the nuclear industry. Both Nixon and Carter had other, much smaller, research initiatives to release the oil and gas in shale. George Mitchell of the (relatively small!) Mitchell Energy led a team working to find the right mixtures of water, sand, and solvents at pressure to fracture oil- and gas-bearing shales so that he could continue servicing his customers in Chicago and elsewhere. He succeeded, and others followed his lead. Between 2008 and 2016, United States petroleum production increased 85 percent, and United States natural gas production increased 25 percent, and the revolution was just getting started. The United States is once again the world's largest energy producer thanks to the Shale Oil and Gas Revolution and thanks to all the relatively small production companies that helped make it happen.

- The Soviet Union (Russia plus a number of presently independent countries was the largest experiment in government-directed research and development. That

---

[62] Yang Jisheng (author), Edward Friedman (editor and introduction), Stacy Mosher (editor and Translator), et al., *Tombstone: The Great Chinese Famine, 1958-1962*, Farrar, Straus and Giroux; Nov. 19, 2013.

[63] Frank Dikötter, *Mao's Great Famine: The History of China's Most Devastating Catastrophe*, Walker & Company, 2010.

experiment tended to develop good weapons innovations.[64] Meanwhile, development of consumer good lagged. This had dire results for the Soviet economy. For instance, Personal computers were rare and expensive for Soviet citizens.[65,66] (In fairness, another large monolithic organization, IBM, also failed to develop the personal computer even though IBM marketed the major advance in personal computers that actually carried the name "PC".) Even if the Soviet Union had survived the crises of the early 1990s, they would have totally missed the personal computer and Internet revolution.

There is a pattern here. Governments develop or latch onto new technologies, but they produce large expensive products. Those products work poorly or not at all. Years later, entrepreneurs improve and market those products to make a profit, and then those products grow major industries.

Sad to say, businesses also sometimes make the wrong choices. When they do, small businesses either succeed or are quickly reminded of reality when they check their account balances. The bigs—big government and big companies—can continue burning fortunes for years.

- The early market for Otto cycle four-stroke spark-plug-fired horseless carriages had strong competition from electric-battery cars and steam-powered cars that were playthings of the rich. What we now know as the "automobile" was considered an unlikely contender. Henry Ford decided that the Otto cycle (four-stroke internal-combustion engine) had the best chance of success because it delivered the most power per unit of weight. His Ford Motor company developed the modern assembly line to build cheap Otto-cycle cars so cheaply that his own workers could afford them. Within a decade steam and electric cars were curiosities.

- In the 1880s and 90s, Thomas Alva Edison, had become fabulously wealthy selling phonographs, light bulbs, electric power plants to run the light bulbs, and dozens of other inventions. He attempted another major innovation of using magnetic separation for a vast (and yes, revolutionary) mining system. He spent what would be billions of dollars in today's money and transformed a vast fortune into a smaller one.

- In the late 1950s, the by-then massive, Ford Motor Company introduced a new larger-sized model car called the Edsel. It had a number of technological advances,

---

[64] Matt Ridley, "Chapter 9 The Economics of Innovation," *How Innovation Works: And Why It Flourishes in Freedom*, Harper, May 19, 2020.

[65] Benj Edwards, "The Lost World of Soviet PCs," *PC Magazine*, Nov. 5, 2015. https://www.pcmag.com/news/the-lost-world-of-soviet-pcs (accessed Dec. 13, 2020)

[66] Daniel Seligman, "The Great Soviet Computer Screw-up, *CNN Money*, July 8, 1985. https://money.cnn.com/magazines/fortune/fortune_archive/1985/07/08/66122/index.htm (accessed Dec. 13, 2020)

but people were going to more compact cars. The market had changed, and Ford lost $250 million, the equivalent of several billions of dollars today.[67]

- After the price of oil collapsed (again) in the 1980s, several of the largest United States oil companies continued spending hundreds of millions of dollars on trying to profitably extract oil-like bitumen from tar sands in Utah. It was an idea that might succeed ... eventually, but they lost more than a billion dollars. Meanwhile, like the U.S. federal government, the oil majors had fracturing of shale deposits way down on their list of possible solutions.

- The Kodak camera and film company that dominated the photography market had experts who warned them about the approaching disruptive technology of digital pictures. The company that had been built on innovation ignored the warnings, and Kodak is gone today.[68,69]

5.4 The Long-Term Perils of National Mobilizations

The GND resolution proposals include a 10-year national mobilization to deal with global warming comparable to the United States mobilization of World War II. That mobilization included vast government research and development projects, price controls, rationing and an excess profits tax. but those policies successfully fused with an entrepreneurial spirit to win the war.[70] The mobilization could essentially defy the laws of economic gravity for a limited time. The mobilization controls were only in effect during 4 years from 1942 through 1946.

Even during that short period, problems were beginning to show. There was a drift toward corruption. Roughly 20% of all the meat, and one gallon out of every twenty, were bought on the back market. Informal trades of ration stamps, services, and favors–a gray economy—became steadily more common. Political concerns became more increasingly important in locations of defense plants, defense research monies, and military bases.

What would have happened in ten years?

---

[67] Glenn Arlt, "Whatever Happened To: Edsel?" National Historic Vehicle Association web page, Nov. 10, 2013. https://www.historicvehicle.org/whatever-happened-to-edsel/ (accessed June 7, 2020)

[68] John Kotter, "Barriers to Change: The Real Reason Behind the Kodak Downfall," *Forbes*, May 2, 2012. https://www.forbes.com/sites/johnkotter/2012/05/02/barriers-to-change-the-real-reason-behind-the-kodak-downfall/#4132862e69ef (accessed June 7, 2020)

[69] David Gann, "Kodak Invented the Digital Camera - Then Killed It. Why Innovation Often Fails," *World Economic Forum*, June 23, 2016. https://www.weforum.org/agenda/2016/06/leading-innovation-through-the-chicanes (accessed June 7, 2020)

[70] Eric Lofgren, "How the US won World War II with economic policy," *Acquisition Talk* blog web page, Nov. 4, 2020. https://acquisitiontalk.com/2020/11/how-the-us-won-world-war-ii-with-economic-policy/ (accessed Nov. 8, 2020)

Consider the Moon Race. The American psyche had been shocked, shamed, and threatened by the demonstration of an entirely new concept, an orbiting artificial satellite—*Sputnik*. This new technology was demonstrated by America's arch rival, the Union of Soviet Socialist Republics, whose leader had bragged, "We will bury you." Worst of all, this new technology had an obvious military application in that a satellite carrying nuclear bombs could drop from the sky with very little warning. Early American launch failures made the national shame even worse.

The response was President John F. Kennedy's 1962 speech saying that America would put a man on the Moon in that decade.[71] America succeeded in putting a man on the Moon within 7 years, but then things began to change.

The urgency slowly drained away over later years. The plans for a Moon base and expeditions to Mars gathered dust. The first space station, Skylab, was hardly used and was left to crash to Earth.[72] The second space station went through name changes and design changes. First, it was an American space station, but then it became an international space station at great cost so that rocket scientists from the politically disintegrating Soviet Union would be kept employed and not go work for rogue states in building missile programs for those countries.[73] Then it became part of a fully international program coordinating and negotiating with many space programs—an expensive diplomatic process.[74] Missions were added of increasing science and engineering education, particularly for minorities. Along the way, space contracts have been spread throughout all fifty states to maintain political support—politically expedient but no way to run a space enterprise.

Yet, the space program is still a narrowly focused set of technical fields, and it is tiny compared to the national economy. Multiple controls in a full-blown national economy that feeds, cloths, and employs the populace—while generating new technologies—would be much harder: To paraphrase Nobel-prize-winning economist Milton Friedman, economists may not know much, but they do know how to produce a shortage or surplus.

Price controls lead to shortages.[75] In an ordinary market, an increased demand leads to higher prices. Those higher prices decrease demand but also encourage manufacturers to produce more or even new manufacturers to begin production. In contrast, price controls on gasoline

---

[71] "John F. Kennedy Moon Speech - Rice Stadium," web page, National Aeronautics and Space Administration Johnson Space Center, Sept. 12, 1962. https://er.jsc.nasa.gov/seh/ricetalk.htm (accessed Nov. 8, 2020)

[72] Nike Wall, "40 Years Ago, NASA's Skylab Space Station Fell to Earth," *Space.com*, July 11, 2019. https://www.space.com/skylab-space-station-fall-40-years.html

[73] Robin McKie, "Twenty Years of the International Space Station – But Was It Worth It?" *The Guardian*, Oct. 25, 2020. https://www.theguardian.com/science/2020/oct/25/twenty-years-of-the-international-space-station-but-was-it-worth-it?ref=hvper.com

[74] Jeff Foust, "The Trouble with Space Stations," *The Space Review*, Sept. 12, 2005. https://thespacereview.com/article/453/1 (accessed Nov. 8, 2020)

[75] Fiona M. Scott Morton, "The Problems of Price Controls," Cato Institute, June 20, 2001. https://www.cato.org/publications/commentary/problems-price-controls (accessed Nov. 7, 2020)

benefit neither consumers nor the overall economy because the apparent price savings to consumers are transformed into costs of waiting in lines, purchasing only allowed, shorter sale hours during certain days, or other forms nonmarket rationing that exceed the monetary savings. Price controls also worsen shortages by reducing the incentive to provide additional supplies.[76,77]

Rent control often causes the price for rental units to fall below the market price. Consequently, more people want to rent, and the resulting shortage of rental units arises as landlords become less interested in providing rental units.

When government adopts a price control, it forces a large percentage of transactions to happen at that price instead of the equilibrium price set through the interaction between supply and demand. Unfortunately, the government price only changes after a lengthy political process, so the government price will usually be either too high or too low. When the price is too high, as with farm-price supports in Europe and the United States, there is excess food ("butter mountains" in Europe and excess millions of bushels of grain in the United States). These excesses are dealt with by either dumping the food as "aid" to poor countries or by using them as feedstock for biofuels.

In 1971, President Richard Nixon enacted comprehensive wage-and-price controls.[78] The results were almost immediate complaints. Labor leader Walter Reuther caustically observed that workers' wages were not allowed to rise, but the number of matzo balls in a can of soup had decreased. More immediately noticeable, there were suddenly shortages of things such as toilette paper, just like in the command-and-control economy of the Soviet Union.

With government-mandated wage-and-price control, the costs of most items eventually become disconnected from demand. In the old Soviet Union, a guitar pick was sometimes more expensive than a barrel of jet fuel. That is part of why the Soviet Union collapsed in 1992.

Climate change might be an existential threat after all, just not in the way that the proponents of the GND imagined.

---

[76] W. David Montgomery, Robert A. Baron, and Mary K. Weisskopf, "Potential Effects of Proposed Price Gouging Legislation on The Cost and Severity of Gasoline Supply Interruptions," *Journal of Competition Law & Economics*, vol. 3, issue 3, pp. 357–397, Sept. 1, 2007. https://academic.oup.com/jcle/article-abstract/3/3/357/775279 (accessed Nov. 8, 2020)

[77] John Hirschauer, "Why the Government Should Not Impose Price Controls," *National Review*, March 16, 2020. https://www.nationalreview.com/2020/03/coronavirus-outbreak-government-should-not-set-price-controls/ (accessed Nov. 8, 2020)

[78] Robert P. Murphy, "Removing the 1970s Crude Oil Price Controls: Lessons for Free-Market Reform," *The Journal of Private Enterprise*, vol. 33, no. 1, pp. 63–78, spring 2018.

5.5 Only Nuclear Fission Can Provide Enough Noncarbon Power

The notion that renewable energies and batteries alone will provide all needed energy is fantastical. It is also a grotesque idea, because of the staggering environmental pollution from mining and material disposal, if all energy was derived from renewables and batteries.

James Hansen, Jeremiah of global warming, 2018[79]

Nowhere in the GND resolution is there any mention of nuclear fission (splitting atoms), an entire class of energy-producing devices. The frequently asked questions (FAQ) document simply says that nuclear fission will be rendered unnecessary by the GND energy changes. Considering the backtracking and/or outright failures of some other government-run attempts at technological revolution—The Chinese Communist Great Leap Forward, the Soviet Russian Virgin Lands, United States President Nixon's Operation Independence, and others—this claim could be seriously over optimistic.

If fossil fuel electrical power generation were to be eliminated, we would need a lot more energy from elsewhere. If electrical cars and trucks were to replace vehicles burning gasoline and diesel fuel, the electrical power demand would grow again. Finally, replacing natural gas cooking, space heating, and industrial process heating would cause another giant increase in demand for electricity. The needed electrical generating capacity would need to double, triple, or increase even more.

Besides nuclear fission, the only other electrical generation sources suggested as major replacements and/or additions to burning fossil fuels are solar, wind, and some backup from hydroelectric dams. Unfortunately, solar and wind have the following issues.

1. They are intermittent; the Sun goes down, and the winds go calm—meaning that there must be backup power plants or expensive backup electrical storage. As muttered darkly by solar experts, "He who cannot store has no power after four." In 2020, the state of California experienced rolling blackouts (much as developing world countries experience) because there was too much intermittency of the solar and wind power without adequate energy storage.[80] Sufficient backup power to prevent those blackouts is expensive.

2. They are diffuse, not concentrated, so generation facilities require large land areas.

---

[79] James Hansen, "Thirty Years Later, What Needs to Change in Our Approach to Climate Change," *Boston Globe*, June 26, 2018. https://www.bostonglobe.com/opinion/2018/06/26/thirty-years-later-what-needs-change-our-approach-climate-change/dUhizA5ubUSzJLJVZqv6GP/story.html (accessed Jan. 1, 2021)

[80] Charlotte Whelan, "California Blackouts Result from Intermittent Renewable Energy Sources," *Townhall*, Aug. 27, 2020. https://townhall.com/columnists/charlottewhelan/2020/08/27/california-blackouts-result-from-intermittent-renewable-energy-sources-n2575104 (accessed Aug. 27, 2020)

3. Covering every roof in a city with solar panels is possible but expensive because there would be many small builds with the added complexity of developing electrical grids that can receive and integrate power from thousands of these small power sources.

4. Giant wind mills are definitely not acceptable in urban areas due to real-estate costs, safety concerns, and noise complaints.

5. Because of 1 through 4, major new solar and wind power production must be largely in distant areas. This requires another major expense of long-distance power-transmission lines.

6. Although solar and wind electricity production facilities have only limited environmental costs, the materials and energy investments to produce those facilities do. Significant increases in solar and wind would require a major increase in the environmental effects and energy use for steel in windmill towers, foundations for cement in concrete foundations, rare-earth elements for the magnets in windmill generators, and various high-tech materials in solar panels.[81]

7. Except for the Great Plains and the Southwest, large areas of the United States have much greater limitations on solar due to cloudiness.

8. The Southwest is ideal for solar energy, but backing it up with hydroelectric dams is limited because there are no more major rivers to dam, and this area is already experiencing water shortages due to competing uses such as farming and municipal water supplies. Furthermore, reservoirs are already experiencing shortages, possibly associated with a drier climate due to global warming, and significant global warming would probably make the water shortages worse.[82,83]

Germany has already demonstrated many of the problems resulting from attempting to replace nuclear fission with solar and wind energy. Prime Minister Angela Merkel started an energy turning point (or *Energiewende*) to replace nuclear fission with solar and wind—they were not even attempting to phase out coal and gas. Within two years, Germany's electricity prices were the highest in Europe with prospects of going higher, emissions of global-warming carbon

---

[81] Mark P. Mills, Mines, Minerals, and "Green" Energy: A Reality Check," Manhattan Institute, New York, July 2020. https://media4.manhattan-institute.org/sites/default/files/mines-minerals-green-energy-reality-checkMM.pdf (accessed Nov. 1, 2020)

[82] Ian James, "Risk of Colorado River Shortage Is on the Rise, Could Hit Within 5 Years, Officials Say," *AzCentral*, Sept. 16, 2020. https://www.azcentral.com/story/news/local/arizona-environment/2020/09/16/risks-water-shortage-loom-along-colorado-river-officials-warn/5804337002/ (accessed Oct. 24, 2020)

[83] Sarah Zielinski, "The Colorado River Runs Dry," *Smithsonian Magazine*, Oct. 2010. https://www.smithsonianmag.com/science-nature/the-colorado-river-runs-dry-61427169/ (accessed Oct. 24, 2020)

dioxide had increased, and the costs were falling disproportionately on the poor.[84] By 2017, the *Energiewende* was causing "Germany's Green Energy Meltdown" with intense voter dissatisfaction.[85] By 2019, many were calling the *Energiewende* a failure.[86] By 2020, the average price of German electricity had risen to 35 U.S. cents (30.03 euro cents).[87] (This is about twice the price of electricity in California and nearly three times the overall American average.)

Bottom line: solar and wind are more expensive than nuclear fission and much more expensive than burning most fossil fuel … except when utilities are required to accept and pay premium rates for all wind and solar electricity while other sources become backup—more expensive because they only run part time.

The real price contrasts have sparked an ongoing debate among energy economists. The 2015 Jacobson et al. study proposed that all fossil fuels and nuclear fission power could be replaced by wind and solar. A key item in the study was that hydroelectric power dams could be switched on and off to back up the times when the sun did not shine and/or the winds did not blow. Hence, the United States could be entirely powered by alternative energy by 2055.[88]

Then, the 2017 Clack et al. paper criticized the 2015 paper as using invalid modeling tools, containing modeling errors, and making implausible and inadequately supported assumptions.[89]

---

[84] Frank Dohmen, Michael Fröhlingsdorf, Alexander Neubacher, et al., "Germany's Energy Poverty: How Electricity Became a Luxury Good," *Der Spiegel*, April 9, 2013. https://www.spiegel.de/international/germany/high-costs-and-errors-of-german-transition-to-renewable-energy-a-920288.html (accessed Sept. 18, 2020)

[85] "Germany's Green Energy Meltdown: Voters Promised a Virtuous Revolution Get Coal and High Prices Instead," *Wall Street Journal*, Nov. 17, 2017. https://www.wsj.com/articles/germanys-green-energy-revoltgermanys-green-energy-revolt-1510848988 (accessed Sept. 18, 2020)

[86] Michael Shellenberger, "The Reason Renewables Can't Power Modern Civilization Is Because They Were Never Meant to," *Forbes*, May 6, 2019. https://www.forbes.com/sites/michaelshellenberger/2019/05/06/the-reason-renewables-cant-power-modern-civilization-is-because-they-were-never-meant-to/#3f5b13c7ea2b (accessed Sept. 18, 2020)

[87] P. Gosselin and Holger Douglas, "Green Energy: German Electricity Prices Skyrocket to Record Highs," *Climate Change Dispatch*, Jan. 27, 2020. https://climatechangedispatch.com/german-electricity-prices-record-highs/ (accessed Oct. 3, 2020)

[88] Mark Z. Jacobson, Mark A. Delucchi, Mary A. Cameron, and Bethany A. Frew, "Low-Cost Solution to the Grid Reliability Problem with 100% Penetration of Intermittent Wind, Water, and Solar for All Purposes," *Proceedings of the National Academy of Sciences of the United States of America*, vol. 8, pp. 15060–15065, 2015. https://www.pnas.org/content/112/49/15060 (accessed June 7, 2020)

[89] Christopher T. M. Clack, Staffan A. Qvist, Jay Apt, et al., "Evaluation of a Proposal for Reliable Low-Cost Grid Power with 100% Wind, Water, and Solar," *Proceedings of the National Academy of Sciences of the United States of America*, June 27, 2017. https://www.pnas.org/content/114/26/6722 (accessed June 7, 2020)

5 The Bad of the Green New Deal

Also, the critiquing paper said that the available hydroelectric backup capacity is much less than had been claimed; hence the hydroelectric backup would be insufficient.[90]

The two sides are still arguing.

5.6 Change the World or Forget It

A major failing of the GND is that—even if totally successful—it would not be enough to save the world from its alleged doom. If the United States implemented all the policies proposed in the GND, the result by the year 2100 would still be a world temperature change of less than 0.2° Celsius (0.4° Fahrenheit) worldwide. That difference is smaller than the margin of error in calculating worldwide temperature. The result is disappointingly weak because many other countries throughout the world, particularly developing countries, are building their economies and burning more fossil fuels.

A successful GND must convince, cajole, or force a significant number of other countries in two places,[91] the GND resolution states that:

> ... *promoting the international exchange of technology, expertise, products, funding, and services, with the aim of making the United States the international leader on climate action, and to help other countries achieve a Green New Deal.*

The implications of, "... help other countries achieve a Green New Deal," has not been factored into the GND costs. However, Stan Cox's *The Green New Deal and Beyond* said that converting the world to "green energy" would be larger than converting America and that it "will require a large U.S. contribution."[92] Ka-ching—more taxes!

Obviously, the GND will not matter unless technologies and policies developed are so massively successful that other countries will desperately work to copy them.

Has that happened yet? No. Western governments and environmentalist organizations demanded that developing countries in Africa and Asia stop building coal-fired power plants. These countries pointed out that coal power is cheaper and that western countries built their prosperity on coal.

In 2020, the Chinese government announced that the country will be carbon neutral by 2060. However, China already burns more coal than any other country, and China has another 150 Gigawatts (150,000 megawatts or 150 million kilowatts) of coal-fired power under development. That is roughly equal to the European Union's total electrical generating capacity

---

[90] Chris Mooney, "A Bitter Scientific Debate Just Erupted over the Future of America's Power Grid," *The Washington Post*, June 19, 2017. https://www.washingtonpost.com/news/energy-environment/wp/2017/06/19/a-bitter-scientific-debate-just-erupted-over-the-future-of-the-u-s-electric-grid/ (accessed March 30, 2020)

[91] House Resolution 109, item (2) Feb. 17, 2019.

[92] Stan Cox, *The Green New Deal and Beyond: Ending the Climate Emergency While We Still Can*, City Lights Books, San Francisco California, 2020.

now.[93] When challenged on their coal-centered policy, the Chinese government refused to change, and they called western pressure to do so neocolonialism.

Old King Coal is still king because coal is still the cheapest source of heat; and from that, heat, electricity and chemical processing energy. The virtue-signaling of converting the United States' economy to noncarbon (or even low-carbon) will draw few converts unless industry throughout the world can demonstrate low-carbon and no-carbon technologies that can be more profitable than coal and/or demonstrate profitable operations that remove global-warming agents from the atmosphere.

5.7 People Will Only Support a Green New Deal If There Is a Visible Threat

The same issues of cost versus long-term environmental worries apply in developed countries.

The American people widely supported Franklin Roosevelt's New Deal because there was a visible Great Depression. Even more so, the American people later supported World War II because Japanese planes had attacked the American fleet at Pearl Harbor in Hawaii and their German allies had brutally invaded countries in Europe.

The warnings about global warming have been based on slight increases in temperature for a quarter century and computer models that predict cataclysmic heat waves, famines, and other baneful effects decades in the future. However, the immediate visible results have been increased food production and milder winters—balanced against a few more heat waves—in temperate climates. Because the continental United States (the lower 48 states) is mostly in temperate climate, most Americans have actually benefitted from global warming so far.

Consequently, the people have not been willing to make major sacrifices, although they say they want things to change. By their actions, from local to national levels, they have not worked for major changes. Warnings to cease building on flood plains oceanside and elsewhere have been largely ignored. World coal burning continues to increase. Recycling of paper and plastics has been declining since China (previously the world's largest recycling destination) became prosperous enough to only take the highest quality trash. (Who knew there was high-quality trash?)

The Green New Deal House of Representatives resolution was defeated 0–57 in a 2019 Senate vote with not a single democrat courageous enough to vote yes.[94] In 2018, a French increase in fuel prices that was enacted to slow global warming led to months of "yellow-vest"

---

[93] James Temple, "If China Plans to Go Carbon Neutral by 2060, Why Are They Building So Many Coal Plants," *MIT Technology Review*, Sept. 23, 2020. https://www.technologyreview.com/2020/09/23/1008786/china-carbon-neutral-coal-plants-un/ (accessed Sept. 24, 2020)

[94] Penny Starr, "Senate Votes Down Green New Deal Resolution 0-57," *Breitbart*, March 26, 2019. https://www.breitbart.com/politics/2019/03/26/senate-votes-down-green-new-deal-resolution-0-57/ (accessed Oct. 16, 2020

protests and rioting[95] that have continued into 2020.[96] Such events do not suggest popular support for a national mobilization, much less a 10-year national mobilization.

A 2019 Associated Press poll surveyed the depth of American support for sacrifices to combat climate change:

> *Suppose a proposal was on the ballot next year to add a monthly fee to consumers' monthly electricity bill to combat climate change. If this proposal passes, it would cost your household $____ every month. Would you vote in favor of this monthly fee to combat climate change, or would you vote against this monthly fee?*

In response, 57% said they would be willing to pay one additional dollar per month. However, if the monthly assessment rose to ten dollars per month, the yes responses dropped to only 28%.[97] That is hardly a cry of support for national mobilization.

This is important because many of the climate-mitigation investments are up-front costs that would only repay the financial debt and the global-warming debt over many years. For instance, the steel and concrete for giant wind turbines require heat from burning gas or coal, which emits global-warming carbon dioxide. Furthermore, shipping those materials to the construction sites emits more carbon dioxide, and every construction aspect costs money. Only after construction is complete does a solar or wind plant start to repay those investments, and then several years later, they break-even and start paying returns in energy and money

Similarly, a $50,000 rebuild on a house to make it more energy efficient might yield a yearly savings of $1,000—hence, the payback would be fifty years.

Such tradeoffs have led to one study conclusion that expenses of the mitigation investments might be more than the return at least until 2100.[98]

---

[95] Jake Cigainero, "Who Are France's Yellow Vest Protesters, And What Do They Want?" *NPR*, Dec. 3, 2018. https://www.npr.org/2018/12/03/672862353/who-are-frances-yellow-vest-protesters-and-what-do-they-want (accessed Oct. 2, 2020)

[96] Niamh Kennedy and Rory Sullivan, "Yellow Vest Protesters Return to Paris for the First Time since Lockdown," *CNN*, https://www.cnn.com/2020/09/12/europe/yellow-vest-france-protests-intl/index.html (accessed Oct. 2, 2020)

[97] Emily Ekins, "68% of Americans Wouldn't Pay $10 a Month in Higher Electric Bills to Combat Climate Change," *Cato at Liberty*, March 8, 2019. https://www.cato.org/blog/68-americans-wouldnt-pay-10-month-higher-electric-bills-combat-climate-change (accessed Oct. 16, 2020)

[98] Eric Worrall, "Study: 'Mitigation Costs of Limiting Global Warming … Are Higher Than … Avoided Damages This Century,'" *Watts up with That?* Oct. 10, 2020. https://wattsupwiththat.com/2020/10/10/study-mitigation-costs-of-limiting-global-warming-are-higher-than-avoided-damages-this-century/ (accessed Nov. 8, 2020)

> *We find that even under heightened damage estimates, the additional mitigation costs of limiting global warming to 1.5 °C (relative to 2.0 °C) are higher than the additional avoided damages this century under most parameter combinations considered.*[99]

Furthermore, the GND resolution presupposes that the support for a national mobilization would continue for a number of years. This is also a questionable assumption. The Moon Race was a major effort spurred on by the embarrassment of the Soviet Union being the first to orbit a satellite, a Cold War competition to land men on the Moon, and the primal drive to explore new territory. At one point, space exploration was 1-percent of the U.S. federal budget. Then, the public interest waned; consequently, government funding waned also. The wild dreams of space hotels and Moon settlements drifted from plans back into science fiction.

Energizing the public (the masses as progressives tend to say) would require something real and something threatening. What have the purported threats been so far?

- More hurricanes? There have been five centuries of documented hurricanes since the Europeans arrived in the Caribbean waters, and probably many more before that.[100] One of the first things Christopher Columbus learned from the native Caribs was the word *Huricán*, a very large storm named after their god of evil. Columbus learned about hurricanes and that he should avoid them. As for increasing hurricanes, there was a recent history-making 11.8-year (4324-day) period without a single hurricane striking the lower 48 states.[101] Convincing the public about a hurricane epidemic would require several years of a more-than-average number of hurricanes.

- More powerful hurricanes? Not so far. The most powerful hurricane, with a measured minimum pressure of 892 millibars (26.4 inches of mercury), was the Florida Keys hurricane of 1935.[102,103] Superstorm Sandy had declined from being a hurricane to only being a big slow-moving tropical storm. It caused extensive

---

[99] Patrick T. Brown and Harry Saunders, "Approximate calculations of the net economic impact of global warming mitigation targets under heightened damage estimates," *PLOS One*, Oct. 7, 2020. https://journals.plos.org/plosone/article?id=10.1371/journal.pone.0239520 (accessed Oct. 17, 2020)

[100] Eric Jay Dolin, *A Furious Sky: The Five-Hundred-Year History of America's Hurricanes*, Liveright, New York, Aug. 4, 2020.

[101] Anthony Watts, "It's Over – 4324 Day Major Hurricane Drought Ends as Harvey Makes Landfall at Cat 4," *WUWT Watts up with That? Website,* Aug. 25, 2017. https://wattsupwiththat.com/2017/08/25/its-over-4324-day-major-hurricane-drought-ends-harvey-lands-as-cat4/ (accessed Oct. 2, 2020)

[102] Jonathan Allen "The Most Intense Hurricanes to Hit the United States," *Reuters*, Oct. 10, 2018. https://www.reuters.com/article/us-storm-michael-strength-factbox-idUSKCN1MK2KS (accessed Oct. 4, 2020)

[103] William Mansell, "The Strongest Hurricanes to Hit the US Mainland and Other Tropical Cyclone Records," *ABC News*, Aug. 30, 2019. https://abcnews.go.com/US/strongest-hurricanes-hit-us-mainland-tropical-cyclone-records/story?id=65277296 (accessed Oct. 4, 2020)

damage because it struck at high tide and because it moved so slow that it dropped a great deal of rain in one area—that might become a concern about global warming.

- Hurricanes striking farther north in the United States? That is not a new thing. There is a long history of hurricanes hitting the New York area[104] and on up into New England.[105,106]

- More floods? Well, maybe. Taking floods along the Mississippi River as an indicator, the 2018–2019 flooding around New Orleans, Louisiana was the greatest in history. However, the 1927 flood on the Mississippi was a close second, and it did the most flood damage of any flood in United States' history.[107,108] A flood bigger than the 2018–2019 flood just might break through the U.S. Army Corps of Engineers' flood defenses enough to break through the levees (as in 1927) and wreak havoc all the way to New Orleans. That might be a major start on support for a mobilization.

- More tornadoes and more deadly? Not so far. Data from 1954 through 2016 show no change in the number of tornadoes.[109] As for deadliness, the Tri-State Tornado of March 1925 had the largest toll and the longest-known ground path of any tornado.[110]

- Giant forest fires in North America? The worst recorded forest wildfires happened in the late 1800s and early 1900s. Skeptics can look up "America's Most Devastating Wildfires,"[111] and point out that millions of acres burned and thousands died in North American wildfires dating back to the dawn of the Industrial Revolution.

---

[104] "List of New York Hurricanes," *Wikipedia*. https://en.wikipedia.org/wiki/List_of_New_York_hurricanes (accessed Oct. 4, 2020)

[105] "List of New England Hurricanes," *Wikipedia*. https://en.wikipedia.org/wiki/List_of_New_England_hurricanes (accessed Oct. 4, 2020)

[106] R. A. Scott, *Sudden Sea: The Great Hurricane of 1938*, Little, Brown and Company, 2003.

[107] "Mississippi River Flood History 1543-Present: New Orleans/Baton Rouge," National Weather Service web page, updated Aug. 10, 2019. https://www.weather.gov/lix/ms_flood_history

[108] John M. Barry, *Rising Tide: The Great Mississippi Flood of 1927 and How it Changed America*, Touchstone, New York, NY, p. 40, 1997 (Touchstone edition 1998).

[109] Todd W. Moore, "Annual and seasonal tornado trends in the contiguous United States and its regions," *International Journal of Climatology*, vol. 38, issue 3, pp. 1582–1594, March 15, 2018. https://doi.org/10.1002/joc.5285 (accessed Oct. 4, 2020).

[110] Christopher C. Burt, "The Five Deadliest F/EF5 Tornadoes on Record," *The Weather Channel*, April 10, 2020. (accessed Oct. 4, 2020)

[111] "America's Most Devastating Wildfires," *American Experience*, Public Broadcasting System web page, undated. https://www.pbs.org/wgbh/americanexperience/features/burn-worst-fires/ (accessed Sept. 28, 2020)

- An ice-free Arctic Ocean in summer? In 2009, Former Vice-President Al Gore semi-predicted that the Arctic Ocean might be totally ice free during some summer months within 5 to 7 years (2014 to 2016). He said a 75% chance of it happening to cover himself,[112] but the credibility damage was done. By September 2020, the lowest ice volume for that year, there was still pack ice in the Arctic ocean.[113] The credibility of global warming threats has decreased with every year since the predicted ice-free year passed. Worse, the Arctic Ocean is far away from most people, and like an ice cube in a glass, the floating pack ice would not cause any change in water volume if it melted. If and when an ice-free summer were to happen in the Arctic Ocean, it would be a mere abstraction for most people but a rich new frontier for shipping lines sailing shorter trade routes through those newly opened waters.

- Severe ongoing oceanic flooding in famous cities such as Venice, New Orleans, Miami, and Djakarta? No, these cities have been sinking for decades because of groundwater withdrawals (all of them), oil withdrawal (Djakarta and New Orleans), building on sand ("… like a foolish man, who built his house on the sand," (Miami)[114], and building on a naturally sinking delta no longer being replenished with sediment because levies channel sediments to the ocean rather than replenishing the delta (New Orleans again).[115]

- A noticeable rise in sea level in a several-day period? Well, maybe. a global sea-level rise in a few days averaging at least a meter (3 feet) or more would be an obvious and impressive change because it would immediately endanger some of the most impressive seaside real estate in the world. There have been no such rapid rises documented in historical times. However, the geological record shows major rises in seal level during the last roughly 20,000 years as the great ice sheets largely melted back toward the poles. The total rise was about 120 meters (390 feet) with significant amounts of rise coming in pulses from catastrophic drainage of giant glacial lakes that were held back by ice dams. These ice dams each collapsed as the Earth slowly warmed out of the worst of our current ice age, and the resulting

---

[112] "Al Gore Warns Polar Ice May Be Gone in Five Years," quoting comments at Copenhagen, Denmark United Nations Climate Change Conference, Dec. 14, 2009., YouTube, Dec. 16, 2006. https://www.youtube.com/watch?v=Msiolw4bvzl (accessed Sept. 28, 2020)

[113] "Arctic sea ice decline stalls out at second lowest minimum," *Arctic Sea Ice News & Analysis*, Sept. 21, 20028. https://nsidc.org/arcticseaicenews/2020/09/arctic-sea-ice-decline-stalls-out-at-second-lowest-minimum/ (accessed Sept. 28, 2020)

[114] *Mathew 7:26, King James Version of The Bible*.

[115] Richard Campanella, "Above-Sea-Level New Orleans The Residential Capacity of Orleans Parish's Higher Ground," Center for Bioenvironmental Research, Tulane University, New Orleans, Louisiana, April 2007. http://richcampanella.com/wp-content/uploads/2020/02/study_Campanella%20analysis%20on%20Above-Sea-Level%20New%20Orleans.pdf (accessed Dec. 2, 2020)

floods were enough to quickly raise sea levels.[116] Plato's story of Atlantis sinking in a day and a night,[117] as well as many other ancient flood legends, would fit perfectly with the emptying of a major glacial lake. There are similar lakes today under the ice sheets of Greenland and Antarctica, but they are much fewer than when the great ice sheets covered much of Eurasia. Today, there are only ice sheets on Antarctica and Greenland, so the potential meltwater pulses are much fewer. Minor pulses of sea-level will probably happen, but there is a good chance that it will not be for decades.

- A major drought on the Great Plains? Yes, but it already happened for 7 years. Severe drought started in 1931 and lasted until 1938 when rains broke the drought in most of the affected areas.[118] This was in the 1930s, long before the doubling and quadrupling of fossil-fuel burning that happened in the great economic booms after World War II. Dust Bowl warnings would only be accepted after 8 years or more.

- A major drought in the U.S. Southwest? Yes, but the ancient droughts were not 2 or 3 years, but 10 or 12, to apparently devastate agriculture for the First Nations peoples from the American Midwest through the Southwest.[119] Again, it would only be appreciated as a threat after a number of years.

- Major droughts extending from California to the Midwest? Yes, that would definitely get national attention. Still, it would need to happen for at least several years to be a noticeable trend.

---

[116] Vivien Gornitz, *Rising Seas: Past, Present, Future*, Columbia University Press, New York, 2013. (In particular, see Chapter 7, "Sea Level Rise on a Warming Planet," pp. 166–187.)

[117] Plato, *Dialogues of Timaeus and Critias*, about 360 B.C. (The story discussed the sinking of the great island of Atlantis supposedly about nine-thousand years earlier still from Plato's time.)

[118] "Timeline: The Dust Bowl," *Surviving the Dust Bowl*, Public Broadcasting System, Jan. 29, 2019. https://www.pbs.org/wgbh/americanexperience/films/dustbowl/ (accessed Oct. 3, 2020)

[119] Larry V. Benson, Michael S. Berry, Edward A. Jolie, et al., "Possible Impacts of Early-11th-, Middle-12th-, and Late-13th-Century Droughts on Western Native Americans and the Mississippian Cahokians," *Quaternary Science Reviews*, vol. 26, issues 3–4, pp. 336–350, Feb. 2007. https://www.sciencedirect.com/science/article/abs/pii/S0277379106002447 (accessed Oct. 2, 2020)

5.8 The Perils of Trend Extrapolation Amid Human Responses

Trend extrapolation is a term from mathematics whereby a straight line, or some type of known curve, is extended in exactly the same way that it was going. This always works in mathematics but not necessarily in the real world.

Geology/climatology has major swings of hothouse warming and ice ages, but there are many more minor cycles of warming and cooling that could boost or cover human inputs for decades. These include the Roman Warm Period,[120,121] the cold temperatures of the Dark Ages, the Medieval Warm Period, the Little Ice Age,[122] and today's era of warming.

These were swings of several hundred years. There have also been smaller swings. The 1910s through 1930s had noticeable warming. Climate warming happened in the 1930s (along with a great drought on the Great Plains of North America). That led to fears of a continued warming and drying. Climate cooling happened from the 1940s through the early 1970s. That was also about the time when geologists got more accurate dating for ice advances and retreats. Some of the data at that time suggested that our present low-ice period (an interstadial in geologist jargon) had already lasted about ten-thousand years, and that length of time was thought to be a typical time limit before new ice advances—there were serious concerns.[123,124,125]

The two primary conclusions from examining the global warming (and cooling) threats above is that they were largely maybes and ifs. They are still heavily weighted toward theories and allegations.

It gets even more ambiguous when the predictors begin including so-called tipping points that might cause the world climate to change into a different stable climate regime. Cynics have noted that the United Nations IPCC has not subscribed to the tipping point theories, and that

---

[120] Jan Esper, David C. Frank, Mauri Timonen, et al., "Orbital Forcing of Tree-Ring Data," *Nature Climate Change*, vol. 2, pp. 862–866, July 8, 2012. https://www.nature.com/articles/nclimate1589 (accessed Oct. 16, 2020)

[121] Tree-Rings Prove Climate Was WARMER in Roman and Medieval Times Than It Is Now - and World Has Been Cooling for 2,000 years," *Mail Online*, July 11, 2012. https://www.dailymail.co.uk/sciencetech/article-2171973/Tree-ring-study-proves-climate-WARMER-Roman-Medieval-times-modern-industrial-age.html (accessed Oct. 16, 2020)

[122] Brian Fagan *The Little Ice Age: How Climate Made History 1300–1850*, Basic Books, New York, 2000.

[123] "The 1970s Ice Age Scare," *Real Science*, May 21, 2013. https://stevengoddard.wordpress.com/2013/05/21/the-1970s-ice-age-scare/ (accessed Oct. 15, 2020)

[124] Lowell Ponte, *The Cooling: Has the Next Ice Age Already Begun?* Prentice-Hall; 1st Printing edition, 1976.

[125] Anthony Watts, "The 1970's Global Cooling Compilation – Looks Much Like Today," *Watts Up with That?* March 1, 2013. https://wattsupwiththat.com/2013/03/01/global-cooling-compilation/ (accessed Oct. 15, 2020)

simplified models (read less expensive computer time) tend to create greater numbers of such tipping points.[126]

Extending trend extrapolation to responses by people multiplies the unknowns, and new technologies multiply the odds again.

- In the 1790s, Robert Malthus began to write about an unchanging law that had applied throughout history: population increased geometrically while land under production only increased linearly. The result was that people eventually bred themselves down to poverty until famine, plague, or war reduced the population again. However, that was already changing. Steam engines brought wealth to buy food and provided powered trains and ships to transport it. Then, artificial fertilizers multiplied food production everywhere. Those two technological revolutions undid the Malthusian paradigm in the last two centuries and suggest that continued technological revolutions can do so even more into the future.[127,128]

- The prosperity of the 1920s ("the Roaring Twenties") was expected to continue indefinitely. Then, in the 1930s, many feared that the economic depression would continue, and some thought it would lead to a general collapse.

- The world oil price briefly reached $145 per barrel in 2008, and it was undoubtedly a contributing factor to the economic crash that year. However, the fearful expectation was that the eventual economic recovery would lead to even higher prices, and the greatest fear was that the world might slide into a long period of economic decline comparable to the Great Depression of the 1930s. Hence, the proposed Global Green New Deal would have a more resilient economy with much less fossil fuel moving toward the 2100s; however, that report and the related book did not mention any existential dangers within the next several decades.[129]

---

[126] P. Gosselin, "Little Evidence Showing Climate "Tipping Points" Are Tipping at All! Artefacts of Simplistic Models," *NoTricksZone* web page (translation of article by Sebastion Luning and Fritz Vahrenholt), April 27, 2016. https://notrickszone.com/2016/04/27/little-evidence-showing-climate-tipping-points-are-tipping-at-all-artefacts-of-simplistic-models/ (accessed Oct. 17, 2020)

[127] Matt Ridley, Chapter 4, "The Feeding of the Nine Billion: Farming after 10,000 Years," Chapter 6, "Escaping Malthus' Trap: Population after 1200," *The Rational Optimist: How Prosperity Evolves*, Harper Collins, June 10, 2010.

[128] Charles C. Mann, Chapter 4, "Earth: Food," *The Wizard and the Prophet: Two Remarkable Scientists and Their Dueling Visions to Shape Tomorrow's World*, Alfred A. Knopf, New York, 2018.

[129] Edward B. Barbier, *A Global Green New Deal: Rethinking the Economic Recovery*, Chapter 1, "Introduction: Opportunity from Crisis," Cambridge University Press, 2010.

A classic example of unexpected changes in technology and human behavior is the "The Great Manure Crisis of 1894."[130,131] In the late 1800s, cabs, trolleys, and wagons for material deliveries were primarily horse drawn. One estimate was that London had 50,000 horses and New York City twice as many. Each of those horses left a trail of manure and urine behind them, leading to a comment in *The Times* fretted that, "In 50 years, every street in London will be buried under nine feet of manure."

That manure burial did not happen because horses were replaced by cars and trucks powered by internal-combustion engines. Henry Ford managed to start manufacturing these vehicles cheaply enough to outcompete horses, and by 1912, The Great Manure Crisis of 1894 was a fading memory.

The pace of innovation has increased considerably since 1912, so the likelihood of such revolutions is even greater. Green-New-Dealers, and others, may be pleasantly surprised by something new. Of course, they might also be unpleasantly surprised so it would be wise to continue looking for solutions.

5.9 Patience Should Be a Green New Deal Virtue

Another technological hope (but also a concern) is that the GND proposal set depends on a number of technologies needing further development to be profitable enough to save the world from the potential dangers of global warming. Waiting just a few years will make many of these technologies more profitable in terms of both money and reduction of global warming. For instance, the cheaper natural gas provided by the Fracking Revolution has decreased American use of coal, and hence emissions of global-warming carbon dioxide, more than solar and wind power so far.[132]

Conversely, rushing development programs could cause disasters. For instance, during the 1930s, a number of governments attempted to build fleets of giant commercial lighter-than-air dirigibles. This led to a string of disasters ending with the explosion of the Hindenburg in May 1938 over Lakehurst, New Jersey. That was the end for commercial passenger airships.[133],[134] In

---

[130] Ben Johnson, "The Great Horse Manure Crisis of 1894," *Historic UK,* undated. https://www.historic-uk.com/HistoryUK/HistoryofBritain/Great-Horse-Manure-Crisis-of-1894/ (accessed Oct. 13, 2020)

[131] Stephen Davies, "The Great Horse-Manure Crisis of 1894" *Foundation for Economic Education*, Sept. 1, 2004. https://fee.org/articles/the-great-horse-manure-crisis-of-1894/ (accessed Oct. 14, 2020)

[132] Zeke Hausfather and Alex Trembath, "Could Fracking Actually Help the Climate?" *Politico*, Oct. 19, 2020. https://www.politico.com/news/agenda/2020/10/19/could-fracking-help-the-climate-429929 (accessed Oct. 20, 2020)

[133] List of airship accidents," *Wikipedia*, undated. https://en.wikipedia.org/wiki/List_of_airship_accidents (accessed Oct. 20, 2020)

[134] Jessie Szalay, "Hindenburg Crash: The End of Airship Travel," *LiveScience*, May 4, 2017. https://www.livescience.com/58959-hindenburg-crash.html (accessed Nov. 26, 2020)

5 The Bad of the Green New Deal

another instance, the communist Chinese Great Leap Forward in 1958–1962 attempted to extend the experimental concept of making high-quality steel in backyard blast furnaces into a national program with the result that farm implements and kitchen pans were miraculously changed into cheap pig iron, and millions died.[135,136,137]

Similarly, solar and wind were very expensive and very unreliable when governments and entrepreneurs attempted to introduce them in the 1970s, but they became more cost effective over time. Similar improvements applicable to GND concerns are coming, but more years will be needed for many of them.

The following items are the most important items that will need patience.

- Regenerative agriculture first includes rebuilding soils (which will draw down global-warming carbon dioxide) that have dwindled over decades, centuries, or even millennia. The rebuilding can be quicker, but it would still be a slow process of decades. Second, plants can be bred with more extensive root systems, but just the breeding process would require years, followed buildup of seed production, growing in test fields, and finally widespread use

- More efficient solar- and wind-power generation: The millions of dollars per kilowatt of solar panel capacity were literally astronomical when space programs began using them in the late 1950s, and wind was a faded technology for small wind chargers. The federal government pushed both technologies during the two Energy Crises of the 1970s, but the efforts were largely failures. Prices have dropped, and quality has improved since then. However, they both need even more improvements because of the intermittency problem—getting power during times of no sunlight and/or no wind.

- Energy storage or regeneration of energy: There are two ways around the issue of solar and wind intermittency: (1) electricity storage (batteries or mechanical storage) and (2) using harvested energy to make some type of synthetic fuel and then using that fuel to generate electricity when needed (dispatchable power).

- Nuclear fission: The most abundant and least polluting of energy sources may be the largely discredited splitting of atoms. However, as with the cheaper and reusable rocket boosters of SpaceX, designers of nuclear fission reactors have new approaches that can reduce reactor prices, increase reliability, and even make

---

[135] Martina Petkova, "China's "Great Leap Forward" That Killed Over 30 Million People," *History of Yesterday*, July 30, 2020. https://medium.com/history-of-yesterday/chinas-great-leap-forward-that-killed-over-30-million-people-c2f23cb231fd (accessed Oct. 20, 2020)

[136] Yang Jisheng (author), Edward Friedman (editor and introduction), Stacy Mosher (editor and Translator), et al., *Tombstone: The Great Chinese Famine, 1958-1962*, Farrar, Straus and Giroux; Nov. 19, 2013.

[137] Vaclav Smil, *Energy Myths and Realities: Bringing Science to the Energy Policy Debate*, Chapter 3, "Soft Energy Illusions," the AEI Press, Washington, D.C., Aug. 16, 2010.

generation of synthetic fuels practical. The missing link for energy has been cheaper and more efficient catalysts. Those catalysts are needed for both separating hydrogen from water and for reassembling it either back to water while delivering energy or building up into valuable chemicals. It has only been a few years since the lack was identified, but the catalysts are coming.

- The oceans have tremendous capacity to provide energy, provide food, draw down greenhouse-warming gases, and cool the planet directly. There were wild-eyed proposals and attempts from the 1920s through the 1970s. Those ideas, and many more

As will be described in the following chapters, all these technological initiatives (and possibly others yet unknown) will make it progressively easier and more affordable to meet the climate-mitigation goals of the GND. Yet, making them work will require decades for development.

Patience!

5 The Bad of the Green New Deal

6 Ground Rules for a Better Green New Deal

6.1 Invent, Develop, and Only Then Mass Produce

Humans have always used innovation to get around new threats. Roughly ten or twelve thousand years ago, people bumped up against limitations on food supply, and they (gradually) shifted from nomadic hunter-gather living to farming the land. When wood became scarce, people built better stoves and got serious about mining coal. To better mine the coal, they developed steam engines to pump water from the mines. When whale oil for lamps grew scarce, they started drilling for rock (*petro*) oil (*oleum*)—petroleum.

Innovation must be applied to the double threat of too much warming and possible resulting reduced world food supplies. There are no limits to the technologies and suggested technologies for using less energy, for running our machinery without adding greenhouse agents to the atmosphere, and for drawing down greenhouse agents from the atmosphere.

However, economics (and accounting) is the art of balancing infinite wants but only limited means. Innovations generally start with something less than working machinery. They may be little more than theoretical ideas, dreams. They may be just inventions that have working prototypes. Eighty percent of the cost and risk of innovation is transforming an invention into a product, building it in quantity, making it affordably, and convincing customers to buy the new item. Then it may work better than anything else and dominate the market ... or not.

It does not always happen even after individuals, companies, and governments have spent vast sums of money.

- Giant lighter-than-air cylinders called dirigibles seemed to be the future of aviation in 1900. Dirigibles using hydrogen for lift tended to explode, and even those using helium were prone to weather-related crashes. Today, we have just a few craft such as the Goodyear and Fuji blimps that provide aerial camera views and advertising billboards on their sides.

- 1950s through 1960s nuclear fission reactor power could not compete with the price of coal-fired electricity, and it had serious safety concerns. Worse, the companies and countries building reactors kept redesigning to ever more giant facilities in an effort to better compete. Each had to be tested for safety, and all required expensive on-site construction. Many entities lost billions of dollars, and the losses got worse when the competing innovation of fracking made natural-gas-fired power cheaper than fission.

- The energy crisis of the 1970s led the United States government to institute a crash program of tax subsidies for windmills generating electrical power. It was said that the tax subsidies were enough that a wind machine could produce a tax-write-off profit even if it never generated a single kilowatt of power. This was fortunate for the entrepreneurs who rushed into the market because the windmills were not ready for commercial use. There was a trail of burned-out bearings and broken blades in the prime wind sites. Then the energy crisis ended with a collapse in oil prices; worse, there was often insufficient power-line capacity to carry the generated electric power from the distant wind sites to urban power users. Years later, windmills and their supporting transmission lines got much better—although still suffering from those pesky wind calms.

- During the first energy crisis in the early 1970s, the Mazda company attempted to innovate with the rotary Wankel engine that simply rotated rather than having pistons that rise, stop, and fall. Mazda produced a high-performance sports car, but the large area of seals was hard to maintain. The major seal replacement in just a few thousand miles made the rotary engine too expensive to compete; it remained a sports car novelty.[138]

The lesson is that instead of rushing into some desperate venture, companies and governments should experiment with a number of them on a small scale and allow the best to thrive. Then expand to larger scale production. Edison did that with phonographs; Ford did it with cars.

6.2 Always Build—*Semper Facilitat*— Rather Than Tear Down

Trying to start a green revolution like this is like shooting all the horses to start the industrial revolution.

Al Watt. 2020[139]

Dear reader, you may be excused for thinking that this comment should be in the preceding negative chapter. However, the positive comment is so important that it must be placed here.

Much of the activism proposing activities to reduce the dangers of global warming seem to aimed at destroying the present energy systems: These measures include shutting down coal and nuclear power plants, preventing construction of oil and gas pipelines, preventing construction of electrical transmission lines, forbidding new natural gas connections for homes and businesses, blocking new mining ventures, and declaring that Henry Ford's internal-combustion engine will be forbidden at some future set date. All these negative activities come at the cost of positive activities not done.

---

[138] "14 Biggest Pros and Cons of Rotary Engines," Green Garage web page, Nov. 13, 2019. https://greengarageblog.org/14-biggest-pros-and-cons-of-rotary-engines (accessed June 7, 2019)

[139] Al Watt, Twitter, Dec. 11, 2020. https://twitter.com/almwatt/status/1337541761837756419 (accessed Dec. 12, 2020)

Consider the following positive activities that come from these politically incorrect industries:

- Better electrical grids with long transmission lines will be needed for bringing solar-generated and wind-generated electricity to the users. Moreover, a better electrical grid can be more secure lest a massive solar storm or a deliberate electromagnetic attack cause a long-term blackout worse than global warming.

- Natural gas is a much better energy carrier than hydrogen. Thus, if society were to use solar and wind to pull hydrogen out of water, it would be much more practical to combine that hydrogen with a little carbon to make synthetic methane (CH_4) gas that could be transported in the existing and working network of natural/synthetic gas lines. Besides, large-scale production of hydrogen from breaking water does not exist yet. (Presently, hydrogen is produced by half burning natural gas with a byproduct of global-warming carbon dioxide that is vented to the atmosphere.)

- Preventing mining in the United states means that platinum-family elements and rare-earth elements (both needed for fuel-cell catalysts, windmill super magnets, and many other alternative-energy initiatives) will only be available from countries willing to do that mining. One country in particular has a name spelled C-H-I-N-A, where Russian Stalinist state terror has fused with robber-baron capitalism. If fools give robber-baron capitalists a monopoly, what should robber barons do? They would charge much more because fools obviously have more money than they know what to do with. Indeed, there have already been episodes in which China has jacked up prices or demanded proprietary information or control in return for rare-earth elements.[140,141,142]

- Coal-fired steel production provides the structural steel and steel rebar for building windmills and solar-power structures. Eventually, the "coking" of carbon in steel production in steel making will be replaced by hydrogen. However, this is still a theoretical advance at present. It still needs years of development. Meanwhile, coke from coal is a requirement for steel production. If well-meaning idealogues choke off American coal production, they also choke off American steel production needed for solar and wind energy production.

---

[140] Tim Treadgold, "China's Rare Earth Threat Sparks an International Backlash," *Forbes*, Aug. 7, 2020. https://www.forbes.com/sites/timtreadgold/2020/08/07/chinas-rare-earth-threat-sparks-an-international-backlash/?sh=7202e2064394 (accessed Aug. 13, 2020)

[141] Alexandra Ma, "From iPhones to fighter Jets: Here's a List of American Products That Could Be Affected If China Banned Rare-Earth Metal Exports to the US as a Trade-War Weapon," *Business Insider*, May 31, 2019. https://www.businessinsider.com/china-rare-earth-list-of-us-products-could-affected-2019-5 (accessed Dec. 13, 2020)

[142] Robert Maginnis, *Human* Events, "China's 'Rare Earth' Monopoly," April 20, 2010. https://humanevents.com/2010/04/20/chinas-rare-earth-monopoly/ (accessed Dec. 13, 2020)

- Meanwhile, until better systems are designed, built, and operating, coal, natural gas, and nuclear power plants are the electrical energy that keeps our society moving. Build a working alternative first before destroying something that works now.

- Most importantly, seeking to destroy existing energy systems to provide the more-moral energy system of the ecological moral future, sucks energy from dozens, or even hundreds, of good initiatives that could use political or economic support.

6.3 Green New Deal Innovation Do's and Don'ts

The following list summarizes the overall opportunities and pitfalls any proposed GND will need to overcome the pitfalls.

1. **Work for Reduction of the Net Amount of Greenhouse Agents:** It does not matter if there is a coal-burning power plant in Idaho, an increased number of trees in Maine can balance it—or many other things you can read about in the following three chapters. This is one key point that the GND has right—analyze in terms of the overall sum of things.

2. **Work for Fun and Profit:** All policies must provide fun and/or profit for somebody who can make things happen. Recycling of containers was always a laudable goal, but the streets always had empty beer and soda containers. I was in Pasadena, California the morning that empty containers could first be redeemed for the 5-cent container fee. Suddenly, the empty containers were gone (except for "health drink" Gatorade bottles, which were not charged a container fee and were thus not redeemable). Similarly, although planting a billion trees is possible and can capture tons of carbon both above and below ground, they can fall prey to pests, fires, or firewood thieves unless they are protected by someone who stands to benefit from the fruits, lumber, tourist income, and/or firewood trimmings from those trees. In a larger case, the fracking revolution happened first in the United States because landowners have mineral rights—landowners can negotiate profitable deals with drilling companies wanting to produce oil and gas from their property. Most important of all, if an endeavor provides a notable benefit (that is profit) to one group of people, others will want to do the same thing. This is another aspect of "the invisible hand" that economist Adam Smith spoke about when he described people all working for their personal benefit serving the common good.

3. **First, Do No Harm:** More than two thousand years ago, Hippocrates wrote what is called the Hippocratic oath of physicians. One of its cardinal precepts is, "First, do no harm." Likewise, for the world's energy and environment, doing no harm means refraining from efforts to destroy existing industries (such as coal and nuclear fission) in the expectation that people will then be forced to build something more suitable to one's views. This does not make a profit; nor is there a guarantee that a replacement will be built. For instance, Hernando Cortez ordered all his ships burned after his troops had landed on the shores of the Aztec empire in present-day

Mexico. That way, they would have to win or die in their invasion. It worked for him. The Xhosa people in present-day South Africa slaughtered their cattle and burned their grain so that they would have to win or die fighting the encroaching Europeans—it didn't work out for them.[143]

4. **Do Not Cry Wolf until and Unless There Is a Visible Wolf:** In one of Aesop's fables, a shepherd boy was bored, so he cried an alarm of wolves to see the villagers come running to help. That was funny to him, but later a wolf did come and attack the sheep. The shepherd boy cried wolf, but nobody came to help because, as one villager said, "Liars are not believed even when they speak the truth." With climate fears, there have already been predictions that fell short. In 2008, former Vice President Al Gore said that some studies predicted a totally ice-free Arctic Ocean by the summer of 2013. There was at least one such computer simulation, but the reality is that summer ice still prevails in much of the Arctic Ocean. Climate skeptics have been making snide remarks ever since. There could even be a ten- or-twenty-year period of cooling—part of a natural cycle we don't know about— despite the overall trend of increasing greenhouse warming, so "twelve years fix or we die" might court open derision. Far better it would be to use modest improvements that are profitable and prototype more ambitious techniques that can be expanded if and when a real crisis appears. As noted earlier, people will support a crash program against a visible threat.

5. **Modestly Encourage Climate-Fix Technologies:** Overly ambitious government or big business ventures attempt to pick technology winners and push those technologies onto the world. Rushed big projects often produce the least per dollar expended. For example, nuclear fusion, "the power of the Sun on Earth," has received tens of billions of dollars since it was first declared to be practical within twenty years … in 1962. Conversely, the research grants on fracturing shale to get oil was one of the smaller efforts, but it helped start the fracking revolution that now yields tens of billions of dollars yearly. Relatively small contest prizes have jump-started revolutions. The Emperor Napoleon Bonaparte held a contest to find some way to can food, and food canning was born. Business people in St. Louis, Missouri offered a prize for the first solo flight across the Atlantic, which was won by "Lucky Lindy," Charles Lindbergh. That contest inspired the Ansari X Prize of $10 million for the first non-government organization to launch a reusable manned spacecraft into space twice within two weeks, and that contest broke the paradigm of disposable space launchers.

6. **Continue Pursuing the Interlinked Visions of Efficiency, Noncarbon Energy, and Technologies that Reduce Global Warming in the Atmosphere:** In 1852, a heroic French inventor, Henri Giffard, exhausted his health constructing the

---

[143] Rael Jean Isaac, *Roosters of the Apocalypse: How the Junk Science of Global Warming Nearly Bankrupted the Western World*, Revised and Expanded Edition, CreateSpace Independent Publishing Platform; 2 edition, Nov. 25, 2013.

lightest steam engine per unit of power produced up until that time. He used it to power a guidable balloon (a dirigible), a lighter-than-air craft that flew 27 kilometers (17 miles) from Paris to a nearby town. but it still did not have enough power to return against the wind. A half century later, the wonderful Brazilian dilettante, Alberto Santos-Dumont, simply cobbled together two small French automobile engines to power his dirigible in a course that handily rounded the Eiffel Tower to win an aviation prize ... even against the wind.[144] Evolving technology had made the heroic merely clever. This is another example showing the power of allowing the maximum amount of time to capture evolving technology before tackling the thermostat adjustment for a planetary climate.

7. **While preparing to Adapt to Bad Climate Changes, Also Plan to Use the Advantages of Climate Change:** The Russian government has announced the beginnings of such a plan particularly since areas near the North and South Poles are expected to be more strongly affected by global warming than countries in lower latitudes, and Russian territory is largely within or near the North Polar region. It lists preventive measures such as dam building, improved forest-fire control, switching to more drought-resistant crops, and preparations for emergency vaccinations and/or evacuations. However, it also proposes to explore possible "positive" effects including decreased heating costs, increased good farming land, better access to and navigational opportunities in the Arctic Ocean.[145]

8. **Rest Assured:** God's universe has more abundance than humanity will ever need. Just a small part of that abundance is human ingenuity enough to surmount threats of global warming and resulting secondary effects.

The following three chapters provide a quick list of methods that are the most promising.

---

[144] Paul Hoffman, *Wings of Madness: Alberto Santos-Dumont and the Invention of Flight*, Hachette Books, June 11, 2003.

[145] "Russia Announces Plan to 'Use the Advantages' of Climate Change," *The Guardian, Jan. 5, 2010.* https://www.theguardian.com/world/2020/jan/05/russia-announces-plan-to-use-the-advantages-of-climate-change?utm_source=Daily%20on%20Energy%20120620_01/06/2020&utm_medium=email&utm_campaign=WEX_Daily%20on%20Energy&rid=9580 *(accessed Jan. 6, 2020)*

7 Food, Fear, and Climate Fixes from the Land

The battle to feed all of humanity is over. In the 1970s hundreds of millions of people will starve to death in spite of any crash programs embarked upon now. At this late date nothing can prevent a substantial increase in the world death rate.

Paul Ehrlich, *The Population Bomb*, 1968[146]

Getting enough food to eat has always been a major human concern. Throughout history, there have been uncounted famines when the crops failed or blights and insects came.[147] Thus, it was natural to be concerned about possible famines in the 1960s and 70s. The population was growing, especially in poorer countries that were having increasing difficulties raising enough food. Thus, Ehrlich's prediction was plausible. If the prediction was correct, his proposed solution of abandoning countries such as India and Mexico to famine might have been the best course of action.

Fortunately, the prediction was wrong—in part because countries like India and Mexico had other ideas. From better spear points, to new ways to farm, to steam engines, to computer chips—humanity has continually found ways around the so-called limits to growth, including getting enough food supplies but not too much global-warming in the atmosphere.

Of course, those enough-of, but not too-much-of, factors apply particularly to food. The high food productivity in advanced economies, such as the United States, requires a great deal of energy, requires significant mineral resources, and often contributes significantly to global warming. In fact, agriculture makes 30% of the yearly contribution to global warming.[148]

However, more advanced farming can do just the opposite. It can decrease greenhouse warming emissions while producing more food and even some resources, as we shall see in this chapter.

Along the way, note that much of the world is seeking to emulate the farm productivity of the United States, Canada, and Europe, especially if global warming makes growing conditions more difficult.

Meanwhile, those productive areas really have the opposite concern of having too much food and the farm economy suffering because of collapsing food prices. For that reason, President Franklin Roosevelt's original New Deal in the 1930s included shooting cattle and burning grain.

---

[146] Paul R. Ehrlich, *The Population Bomb*, Ballantine Books, 1968.

[147] Ralph A. Graves, "Fearful Famines of the Past," *The National Geographic Magazine*, vol. 32, no. 1, pp. 69–90, July 1917.

[148] Jonathan Foley, "The other inconvenient truth," *TEDxTC*, Sept. 2, 2011. https://www.youtube.com/watch?v=uJhgGbRA6Hk (accessed July 17, 2020)

7 Food, Fear, and Climate Fixes from the Land

Planting tree strips, paying to fallow land in the "Soil Bank," "Food for Peace," and other programs did a number of good things, ... but their primary purpose was reducing production, or removing some food from the market, to maintain price levels.

This chapter describes some of the ways that humanity can get more needed food while decreasing factors that cause global warming. Many of the subsections are derived from the wonderful 1978 book, *Future Food: Alternate Protein for the Year 2000*, which is still the best introduction to the endless assortment of existing and possible food sources.[149] The text describes existing and potential human food sources from alfalfa, to earthworms, to mushrooms, to processed feathers, to dozens more—the book particularly describes immense sources of protein. And of course, every weight unit of new protein can free up three to eight times the grain that would have been used for livestock feed.

In 2020, Mosby, Roz, and Evans provided a few new and updated angles of food production ... including questions such as will there even be grocery stores in the future.[150]

In *Meals to Come: A History of the Future of Food*, Warren Belasco quotes the poet Homer as saying, "A hungry stomach will not allow its owner to forget it, whatever his cares and sorrows." Thus, Belasco focuses quite entertainingly on the story of food from famines of Malthusian doom, to "cornucopian" optimism, to the present and on to future possibilities.[151]

In *Our Livable World*, Marc Schaus provides an excellent and entertaining survey of many climate-related technologies, including regenerative agriculture, culturing foods, corals, kelp, and energy.[152]

7.1 Beyond the Plow: Regenerative Agriculture

What if they gave a revolution and nobody noticed? That is just what has been happening in agriculture. There are ongoing trends of reduced use for plowing, artificial fertilizers, and pesticides. These advances are helping to both increase food production and help mitigate global warming.

Plows have been the mainstay of farming since the beginning of agriculture. (The fame of our third president, Thomas Jefferson, was further burnished after he invented a better plow.) However, when plowing, a farmer cuts into the soil and turns it over, thus exposing the carbon in the plant material of rich soils to oxygen in the air. Some of the plant material oxidizes into carbon dioxide, which goes up into the atmosphere just like when coal burns in a power plant. Worse, plowing exposes the soil to erosion—it can be washed away by rains or blown away by winds.

---

[149] Barbara Ford, *Future Food: Alternate Protein for the Year 2000*, Barbara Ford, William Morrow & Co., New York, 1978.

[150] Ian Mosby, Sarah Rotz, and Evan D. G. Fraser, *Uncertain Harvest: The Future of Food on a Warming Planet*, University of Regina Press, Saskatchewan Canada, 2020.

[151] Warren Belasco, *Meals to Come: A History of the Future of Food*, University of California Press, October 2006.

[152] Marc Schaus, *Our Livable World: Creating the Clean Earth of Tomorrow*, Diversion Books, October 13, 2020.

That eroded soil oxidizes into even more carbon dioxide. Roughly a quarter of the global-warming carbon dioxide entering the atmosphere has been coming from formerly rich soils oxidizing into increasingly carbon-poor soils.

But farmers have been learning some new tricks. Now, plows are becoming increasingly less common because the farm implement makers developed machinery to kill plants at the surface without plowing them under and to insert seeds for the next crop, again without plowing.

This reduced or avoided plowing (low-till or no-till) method is the start of regenerative agriculture.[153] The minimal disturbance of the surface keeps an insulating layer on the surface to prevent too much hot or cold for the sprouting crop. It allows water to better soak in while shielding it from wind and water erosion. Just as important, reduced plowing allows development of a soil community including pest predators, mineral-nutrient-dissolving, bacteria, and the mycelium (fungus) tendrils that allow transfer of nutrients throughout the soil—nutrients and deeper supplies from the fungi and sugar from the plants in payment.

As a result, the soil loses less carbon in the form of carbon dioxide. Instead, the plants are actually transforming carbon dioxide from the air and banking it as organic matter in their roots.

Still, there is more. Regenerative (or conservation) farmers can go to the extra trouble of planting what they call a "cover crop" after the harvest of the cash crop (corn, wheat, or whatever). A cover crop often does not have enough time before winter to produce much, but it covers the ground and protects the soil from those bare-ground problems mentioned earlier. If the cover crop is alfalfa or some other "nitrogen-fixing" plant, that cover crop pulls nitrogen out of the air and combines it into chemicals that the next season's cash crop can use so that the resulting nitrogen compounds are available next year spring for the next cash crop … all without an expensive stop at the fertilizer store. Furthermore, reduced use of artificial fertilizer lessens runoff of nitrates that contribute to excess algae in nearby water bodies (eutrophication and stagnant dead waters) and to oxides of nitrogen evaporating into the atmosphere and contributing to global warming.

With some extra effort, a farmer can plant not just a single crop in a field—a monoculture; that farmer can plant several crops—a polyculture. Such polycultures make it much harder for pests to zero in on any one particular crop so the crops are protected, and helpful predator species have more habitat. This more-diverse-targets strategy can help reduce another expense, pesticides. Just as importantly, working several crops at once provides insurance against problems (such as pests, a late frost, drought, too much rain, too much heat, not enough heat, an early frost, and more) that strike one crop but not the other crops. Farmers are the last of the big-time gamblers!

Once again, there is more. Two-thirds of world farm land is grazed by cattle, sheep, and other large grass eaters. This grazing land is generally dryer and/or hillier than the prime farm land. Because of those features, it's generally not suitable for plowed crops.

Here is the disagreement with one initial detail of the GND. Yes, it is true that cattle produce methane (mostly from burping as they chew their cuds—NOT cow farts!). Also, cattle

---

[153] Charles Massy, "How Regenerative Farming Can Help Heal the Planet and Human Health," *TEDxCanberra*, Nov. 13, 2018. https://www.youtube.com/watch?v=Et8YKBivhaE (accessed Dec. 4, 2020)

raised on grain require a lot of grain for each pound of beef. However, cattle that are grass-fed rather than grain-fed do not use grain; they use output from marginal lands (remember, hilly and/or dry) that often should not be risked for row-crop production in the first place.

Moreover, production from those cattle pays for maintaining grassland that can draw down carbon dioxide from the air for building soil. That draw-down of carbon dioxide from the atmosphere more than compensates for the methane burps.

An additional spin from regenerative agriculture can make grazing much more efficient for production and for protecting us from global warming. That spin comes from analyzing the dynamics of natural grazing herds.

For millions of years, large grazing animals evolved along with Earth's grasslands ... and they may have even prevented global warming. They always moved in herd formation to get the best tall grasses. That kept them steadily migrating—rarely to the same spot twice in a year. Each year, the decaying roots of those grasses steadily increased the depth of the grassland soils.

In their migrations, the herds spread distributed fertilizing manure over the grasslands ... even on the hilltops where rains would overwise leach away and erode away nutrients. Also, their hooves made imprints that allowed water to soak in better and thus minimize water erosion.

The result was giant areas with some of the richest and deepest soils in the world. One byproduct of that incredible soil building was that it drew down billions of tons of carbon dioxide from the atmosphere to make it happen. Without those tons becoming part of the grassland soils, Earth might have experienced greenhouse warming eons ago. Wild cattle, bison, buffalo, wildebeest, caribou, horses, moose, elk, kangaroos, camels, and even fur-clad elephants were the climate defense. We could call that the green old deal.

Then, human technology became powerful enough to cause a new brown deal. Fire drives and kill pits were the start when hunters destroyed the northern elephants, but the main apocalypse came with railroads and repeating rifles. In North America, within a half century, hunters (who took only the hides) and sport hunters (who often took nothing) killed nearly all the buffalo in North America.

Then it got worse with plowing. Again, North America is the prime example. The land-sale promoters of the day proclaimed that "Rain follows the plow." They seemed to be right for a few decades.

Then, In the 1930s, it was time to pay the piper. A sustained drought temporarily transformed parts of the Great Plains from North Texas to South Dakota into semi-desert. The area was called "The Dust Bowl," and those years were called "The Dirty Thirties."[154] Giant dust storms transported millions of tons of top soil all the way to the Atlantic Ocean, and of course, much of the organic matter in that dust oxidized into climate-warming carbon dioxide.

---

[154] Timothy Egan, *The Worst Hard Time: The Untold Story of Those Who Survived the Great American Dust Bowl*, Houghton Mifflin Harcourt, 2006.

The practical lesson of today is that grazing of perennial grasses, rather yearly planted crops, is a safer way to get food from marginal lands. The lesson will probably be even more important with a hotter drier climate.

Even in grazing, there is a better way—mob grazing or short-term high-intensity grazing. Stock growers fence off a number of smaller pastures, and move the herd regularly so (as with the ancient wild herds) the cattle move along just getting the tops of the grasses—no lollygagging close to the water available at the stock pond, chewing up everything around that area down to the dirt, and turning that area into a little dust bowl.

Just as with the natural herds, the cattle graze through everywhere. Then they leave and allow that area to recover. The production of meat per acre increases, and soil carbon content increases.[155]

In fact, the cooling from drawing carbon dioxide down from the air into the soil is greater than the warming from the methane. That is especially so considering that methane breaks down within 8 years while carbon dioxide may stay in the air for 20 to 200 years.

Regenerative agriculture of both crops and grazing can slow or reverse soil exhaustion.[156] Healthy soils with adequate organic carbon levels can absorb rainfall faster, hold moisture longer, and provide an environment for fungi and microbes to make minerals in the deeper ground accessible to plants. This is important for food production, particularly if global warming were to cause more droughts and heat stress on crops.

Bottom line: Regenerative agriculture can help reduce global warming and maintain food production even in a hotter dryer climate.

Major Things for Governments and Businesses to Do

- Of course, more farmers, ranchers, and farm-implement suppliers should increase applications of regenerative agriculture, but that may entail business life-or-death investments for them. Governments should modify the regulations in favor of subsidizing increased use of regenerative agriculture. For instance, there might be payments for increased percentage of carbon in the soil.

---

[155] Tony Lovell, "Soil carbon -- Putting carbon back where it belongs -- In the Earth," *TEDxDubbo*, Sept. 10, 2011. https://www.youtube.com/watch?v=wgmssrVlnP0 (accessed Dec. 4, 2020)

[156] D. L. Evans, J. N. Quinton, J. A. C. Davies, et al., "Soil Lifespans and How They Can Be Extended by Land Use and Management Change," *Environmental Research Letters*, Sept. 15, 2020. https://iopscience.iop.org/article/10.1088/1748-9326/aba2fd (accessed Sept. 16, 2020)

What You Can Do

- Find the nearest agriculture extension station and ask the agents about programs in regenerative agriculture that you might learn about and possibly participate in. Depending on the ag extension station there may be lectures, literature, classes, ... maybe even wine tasting. (The county ag extension stations are everywhere. Even New York City has the Cornell Cooperative Extension Station near the Hell Gate Bridge.)

- Using the knowledge that you've gained, contact your local and national government representatives expressing support for regenerative agriculture.

- If you have yard space, plant a garden so that you can appreciate the work and the rewards of farming.

- For health and climate reasons, eat less meat than is in the typical American diet.

- However, treat yourself to a nice grass-fed steak now and then. (If you are a vegetarian for whatever reason, that is also fine. However, be sure to get a full assortment of nutrients, including those from animal products that exist but are sometimes harder to find in the plant kingdom.

7.2 Try a Little Variety: Food, Fiber, and Materials

Of Earth's estimated 400,000 plant species, we could eat some 300,000 armed with the right imagination, boldness and preparation. Yet humans, possibly the supreme generalists, eat a mere 200 species globally, and half our plant-sourced protein and calories come from just three: maize, rice and wheat.

Adrian Barnett, 2015[157,158]

In September 1945, a 20-year-old private, Keith Harrigle, my future junior high school science teacher, was on a lush green Pacific Island facing a daunting challenge. World War II had just ended, and he was one of thousands of U.S. military helping manage hundreds of thousands of Japanese prisoners of war.

The problem: his prisoners were starving. Vengeful Americans were feeding them vile disgusting things such as ground beef... potatoes ... wheat bread ... and corn. Men who had gone toe-to-toe with U.S. Marines during 4 years of war were near death because they feared those strange nasty things on the food trays. American privates like my future teacher were sent into the mess halls, forks and food trays in hand, to demonstrate that wheat bread does not kill.

---

[157] Barnett was reviewing *The Nature of Crops: How We Came to Eat the Plants we Do* by John Warren. The review was in the *New Scientist*, p. 43, July 18, 2015.

[158] John Warren, The Nature of Crops: How We Came to Eat the Plants We Do, CABI, June 2015.

"It's good! Um yum!"

The prisoners finally ate the strange food, … and they did not die. Indeed, thanks to later western subversives, such as Ronald McDonald of Golden Arches, all Asians in general changed their diet.

There were two results: One, they did not die. Two, the average height of Asian people increased dramatically. (Later, those richer diets contributed to increased heart and artery conditions in later years leading to earlier deaths, but that is an entirely different issue.)

There is more food everywhere if people would just look around. Ever hear of the poison apple? It grows from a plant much like its cousin the deadly nightshade. Ah, hah! People KNEW just how DEADLY that related plant must be.

Then, in 1830, American Robert Gibb Johnson gave notice that he would eat the evil fruit on a certain day. On that appointed day, he went to the steps of his local court house where a crowd had gathered, waiting to see. He bit into the soft juicy flesh, and he did eat. But he did NOT die when he ate … a tomato!

That rhymes with pot-A-TO (from the faraway Andes Mountains of South America). Spanish conquistadores brought potatoes back from South America, but that alien root vegetable was only used as pig food in France until Antoine-Augustin Parmentier waged a decades-long struggle to popularize potatoes including the presentation of a potato blossom bouquet to the French queen. (Now, the French have many excellent potato dishes with Parmentier in the title … but that too is another story.) Thanks to potatoes, available calories in Europe roughly doubled, and famine was averted for some decades.

The exception is that in the early 1840s. several wet cool summers allowed a blight to attack the lone variety of potato that had reached Europe. Unfortunately, that variety was very sensitive to a fungus blight during cold wet years. Historians call that event the Irish Potato famine.[159] It was another painful lesson in the story of farming … and why we should want crops resistant to greater heat that might be coming.

More broadly, a society depending on too limited a set of crops becomes vulnerable to disease or bad weather. For decades, French farmers tried to stay with grains during cold wet years of what has been called the Little Ice Age, (roughly late 1200s to mid-1800s), and famines resulted until they switched to more moisture-tolerant practices.[160]

Still, the examples from the past tell us that we can add to our menu again … or just borrow from foods already developed in certain areas. The limitation or barrier is, as always, that people get by with the least amount of change as they possibly can.

---

[159] Charles C. Mann, "How the Potato Changed the World," *Smithsonian Magazine*, Nov. 2011. https://www.smithsonianmag.com/history/how-the-potato-changed-the-world-108470605/ (accessed Dec. 25, 2018)

[160] Brian Fagan *The Little Ice Age: How Climate Made History 1300–1850*, Basic Books, New York, 2000.

Whatever happens, the following material gives just an inkling of those available possibilities ... if people get hungry enough. There are more vegetables, nuts, and grains than can be imagined. Likewise, there is a surprising array of protein sources, but those are some additional subsections.

Consider that roughly 300,000 of the estimated 400,000 plant species are edible while we only use a couple hundred as major crops. Broadening our palette repertoires would allow more productivity by use of plants that grow better in different conditions and would give more defense in depth against any plant diseases that might strike, such as the potato blight in the early 1840s and wheat rust throughout history.

There are potential crops that have not yet been discovered, there are existing crops that have not been put into widespread production, and there are crops that have drifted into obscurity. That leads to the question, "Are Forgotten Crops the Future?"[161] Many of these plants are tropical or subtropical, so they could become much more important in a hotter world. Some examples of both follow.

- The moringa tree's edible leaves are high in protein, iron, potassium, calcium, nine essential amino acids, and vitamins A, B, and C. Moringa seeds are rich in protein and omega-3 fatty acids.[162,163,164]

- The winged bean promises to become the soybean of the tropics because of its high production and high nutritional value. It has been called "a supermarket on a stalk" because it combines the desirable characteristics of the green bean, garden pea,

---

[161] Preeti Jha, "Are Forgotten Crops the Future of Food?" *BBC Future*, Aug. 22, 2018. http://www.bbc.com/future/story/20180821-are-forgotten-crops-the-future-of-food (accessed Dec. 19, 2018)

[162] Greg Rienzi, "Is the Moringa Tree the Next Superfood?" *Johns Hopkins Magazine*, winter 2016. https://hub.jhu.edu/magazine/2016/winter/moringa-the-next-superfood/ (accessed Dec. 19, 2018)

[163] Lloyd Phillipps, "Moringa: Superfood for the Hungry," *Farmers Weekly*, June 24, 2014. https://www.farmersweekly.co.za/crops/field-crops/moringa-superfood-for-the-hungry/ (accessed Dec. 19, 2018)

[164] Amy Quinton, "Moringa — the next superfood?" University of California Davis website, Oct. 18, 2018. https://www.universityofcalifornia.edu/news/moringa-next-superfood (accessed Dec. 19, 2018)

spinach, mushroom, soybean, bean sprout and potato. People can eat winged bean flowers, leaves, seeds, and tuberous roots. [165,166,167]

- Amaranth was a major crop of the Aztecs in present-day Central America and southern Mexico. Amaranth leaves can be eaten as a green vegetable, and its seed has a high lysine level that compensates for the lack of lysine in corn. However, the Spanish *conquistadores* banned the plant because it was associated with pagan blood rituals.[168] There is archeological evidence that amaranth as a crop extended at least as far as the northern Great Plains of South Dakota.[169]

- Quinoa is one of the few "neglected crops that has recently found its way to a strong beginning of market acceptance.[170] People of the high Andes Mountains in Ecuador and Peru have been growing quinoa since before the Spanish arrived. It is a high-protein low-gluten grain with a pleasant nutty taste. Getting quinoa developed as a crop with development for pest resistance is another issue, and this is a serious issue for any of these crops.[171] That issue will come up again because adding more major crops to the production list would greatly increase food security.

- Kudzu, or Japanese arrow root, is considered a weed in the southeastern United States because it often simply outgrows and smothers other plants. However, it is a versatile forage crop and food source. Its leaves can be cooked, its young shoots have a taste similar to snow peas, and it has nutritious potato-like roots. (Caution the berries and seeds are not edible for humans.) It is sometimes the only plant that can perform erosion control on steep land, and its fiber (called co-hemp) can be

---

[165] Jane E. Brody, "Winged Bean Hailed as a Potent Weapon Against Malnutrition," *The New York Times*, Feb. 23, 1982. https://www.nytimes.com/1982/02/23/science/winged-bean-hailed-as-a-potent-weapon-against-malnutrition.html (accessed Dec. 19, 2018)

[166] *The Winged Bean: A High Protein Crop for the Tropics*, Second Edition, National Academy Press, 1981. https://www.nap.edu/catalog/19754/winged-bean-a-high-protein-crop-for-the-tropics (accessed Dec. 19, 2018)

[167] P. Lepcha, A. N. Egan, J. J. Doyle, and N. Sathayanarayana, "A Review on Current Status and Future Prospects of Winged Bean (Psophocarpus tetragonolobus) in Tropical Agriculture, *Plant Foods for Human Nutrition*, vol. 72, no. 3, pp. 225–236, Sept. 2017.

[168] Jonathan B. Tucker, "Amaranth: The Once and Future Crop," *BioScience*, vol. 36, issue 1, pp. 9–13, Jan. 1, 1986.

[169] Ron Robinson, "Amaranth: The Once and Future Crop? revised from the March/April 2013 *South Dakota Magazine*. https://www.southdakotamagazine.com/amaranth (accessed Dec. 19, 2018)

[170] Johanne Uhrenholt Kusnitzoff, "Ancient Crops Are the Future for Our Dinner Plate," *ScienceNordic*, Oct. 11, 2016. http://sciencenordic.com/ancient-crops-are-future-our-dinner-plate (accessed Dec. 19, 2018)

[171] Lydia DePillis, "Quinoa Should Be Taking over the World: This Is Why It Isn't," *The Washington Post*, July 11, 2013. https://www.washingtonpost.com/news/wonk/wp/2013/07/11/quinoa-should-be-taking-over-the-world-this-is-why-it-isnt/?noredirect=on&utm_term=.438094afd8e1 (accessed June 11, 2020)

made into cloth or paper.[172] Only in a land of abundant food would it be considered a pest.[173] If global warming were to become serious, the Canadian and Russian kudzu harvests might feed billions.

- Trees in general could be a chapter by themselves. First, the hundreds of tree species available are part of that wondrous variety that can be applied for food, fiber, fuel, and structural material. Second, harvesting plantation trees rather than wild is much more productive, so land not needed for lumbering can be allocated to food production. Finally, trees can be part of a polyculture of multiple crops jammed together (intercropped) to increase productivity and to ward off pests. Third, (but not least), trees pull down large amounts of carbon for their above-ground structure and even more for their roots. By one account, growing trees could pull back a quarter of the carbon dioxide that humans emit each year. The proposal is that seedlings could be air dropped worldwide to reforest an area excluding existing trees and agricultural and urban areas that would be about 900 million hectares (about 3.5 million square miles, half the U.S. lower 48 states.[174] The caveat is that air-dropped, or otherwise untended forests, do not do well. Untended sprouting trees are browsed by large herbivores, lumber from grown trees is stolen by firewood thieves (and out-and-out loggers), and both are destroyed by drug farmers using the land for their own production. Most importantly, untended forests become overgrown with low-level brush and forest litter that provide kindling for giant firestorms[175] (as demonstrated recently in the United States Northwest and in Australia). Thus, forest mitigation of climate change requires significant costs for managing those forests.[176] From ancient Britain, to modern Madagascar, forests work best when the locals are getting some amount of return from the forest: some intermingling of trees and crops; controlled amounts of firewood; tourist revenues; and/or watershed fees.[177] (For GND climate concerns,

---

[172] "Kudzu," World Crops Database (website), undated. https://world-crops.com/kudzu/ (accessed June 29, 2020)

[173] David Denkenberger and Joshua M. Pearce, *Feeding Everyone No Matter What: Managing Food Security after Global Catastrophe*, Academic Press, Elsevier, 2015.

[174] Jean-Francois Bastin, Yelena Finegold, Claude Garcia, et al., "The Global Tree Restoration Potential," *Science*, vol. 365, issue 6448, pp. 76–79, July 5, 2019.

[175] Paul Hessburg, *USFS: Era of Megafires - Catastrophic Wildfires*, U.S. Forest Service Pacific Northwest Research station and North Forty Adventures in Creativity, (recorded May 31, 2018) June 1, 2018. [1 hour 20 minutes] https://www.youtube.com/watch?v=Ot_4NQVRVHs (accessed Dec. 20, 2020)

[176] K. G. Austin, J. S. Baker, B. L. Sohngen, et al., The Economic Costs of Planting, Preserving, and Managing the World's Forests to Mitigate Climate Change," *Nature Communications*, vol. 11, Article number: 5946, 2020. https://www.nature.com/articles/s41467-020-19578-z (accessed Dec. 2, 2020)

[177] Stewart Brand, *Whole Earth Discipline: An Ecopragmatist Manifesto*, Viking Penguin Group, New York, New York, 2009.

the smaller regular burns in managed forest would reduce the amount of carbon sequestered.)

- Kenaf, is a tall annual tropical and subtropical stalk plant that grows up to about 18 feet (5.5 meters) high. Kenaf can provide greens in its upper leaves or huge amounts of fiber for use in cloth, paper, and construction. A typical yearly crop is about 9 to 11 tons of dry weight per acre. That is roughly 3-to-5 times greater than the yield for southern pine trees, which may need 7 years to reach harvest size. Kenaf can sequester eight times as much carbon as per unit area as evergreen trees. Kenaf's short growing season, minimal water needs, and minimal fertilizer requirements make it among the most environmentally friendly and sustainable fiber crops on Earth.[178]

- The breadfruit tree has a tremendous history both lost and known. The lost history is that many people over centuries throughout the South Pacific bred a wonderfully productive and hardy low-maintenance tree that is highly productive. It provides both complex carbohydrates and a decent level of protein. For those reasons, Captain William Bligh of ("Mutiny on the Bounty" fame) brought breadfruit to the Caribbean to help feed slaves in British colonies there. That slavery history caused many Caribbean people to shun breadfruit, but breadfruit boosters are leading a resurgence in its cultivation.[179,180] Breadfruit has been suggested as a healthy option for food security.[181] Significant warming might make breadfruit a major crop in the American Southeast.

- Duckweed species are small floating aquatic plants found worldwide. They often grow in thick, blanket-like mats on still or slow moving, nutrient-rich fresh or brackish waters. They do not need to support themselves, so they have very little fiber. Hence, they can be digested by people and other one-stomach creatures. They can be used as livestock feed for chickens, pigs, rabbits, and tilapia. Duckweeds can double their mass in less than 2 days under optimal nutrient

---

[178] "Kenaf A Renewable Super Plant," *Ecotech Alliance newsletter*, vol. 365, issue 6448, pp. 76-79 http://www.ecotechalliance.com/kenaf-a-renewable-super-plant/#1448610320922-08325060-cb81 (accessed Dec. 19, 2018)

[179] Kelsey Nowakowski, "Can Breadfruit Overcome Its Past to Be a Superfood of the Future?" *National Geographic*, July 22, 2015. https://www.nationalgeographic.com/culture/food/the-plate/2015/07/22/can-breadfruit-overcome-its-past-to-be-a-superfood-of-the-future/ (accessed Sept. 17, 2020)

[180] Rebecca Rupp, "Breadfruit and 'The Bounty' That Brought It Across the Ocean," *National Geographic*, April 28, 2016. https://www.nationalgeographic.com/culture/food/the-plate/2016/04/28/breadfruit-and-the-bounty-that-brought-it-across-the-ocean/ (accessed Sept. 17, 2020)

[181] Ying Liu, Paula N. Brown, Diane Ragone, et al., "Breadfruit flour is a healthy option for modern foods and food security," *PLOS One*, July 23, 2020. https://journals.plos.org/plosone/article?id=10.1371/journal.pone.0236300 (accessed Sept. 18, 2020)

availability, sunlight, and water temperature. This is faster than almost any other higher plants.[182]

- When the Spanish invaders arrived in Mexico in the 1500s, the Aztecs and other Mesoamerican peoples were already making cakes out of dried freshwater microalgae, pond scum.[183] Today, people around Lake Chad in Africa still make similar cakes.[184] The microalgaes, such as *spirulina* and *chlorella pyrenoidosa*, are more productive than soy beans per unit of land—or in this case, surface area. They yield high protein, and they thrive in any water salinity level from fresh to that of the Dead Sea that is saltier than the ocean. A number of small firms produce algae as nutritional supplements. However, supplements are a boutique (expensive) commodity compared to other high-protein foods. the market model has been sales of an expensive nutritional food supplement, so the established producers have no interest in high-volume low-cost production. The potential is there if ever needed, but it is not developing now.

- Hemp is another old crop that, like Amaranth, went into eclipse for political reasons. Hemp went into cloth, paper, and rope for centuries because it has good fiber and because it generally outgrows weeds with little need for water or fertilizer. Unfortunately, certain varieties of the related marijuana plant contain a psychoactive intoxicant. Consequently, United States laws in 1937 and 1970 made hemp growing illegal (except when needed in World War II). Finally, the Agriculture Improvement Act of 2018 largely legalized hemp production at the federal level[185] (although state regulations are another complexity). Hemp could replace a sizeable fraction of the fossil fuels in plastics, and thus provide a crop replacement for some of that grain/feedlot beef being replaced by vegetarian food.

---

[182] R.A. Leng, J. H. Stambolie, and R. Bell, "Duckweed - a potential high-protein feed resource for domestic animals and fish," *Livestock Research for Rural Development*, vol. 7, no. 1, 1995. http://www.lrrd.org/lrrd7/1/3.htm (accessed June 17, 2019)

[183] Imen Hamed, "The Evolution and Versatility of Microalgal Biotechnology: A Review," *Comprehensive Reviews in Food Science and Food Safety*, vol.15, no. 6, Sept. 2016. https://www.researchgate.net/publication/308669168_The_Evolution_and_Versatility_of_Microalgal_Biotechnology_A_Review (accessed Dec. 22, 2018)

[184] "Nutrient-Rich Algae from Chad Could Help Fight Malnutrition," Food and Agriculture Administration of the United Nations website, July29, 2010. http://www.fao.org/news/story/en/item/44388/icode/ (accessed Dec. 22, 2018)

[185] Marian J. Lee, "The Legalization of Hemp," Food and Drug Law Institute web page,

Moreover, the plastic could replace some metal parts that are produced at the expense of significant amounts of coal burning in the smelting process.[186]

- Bamboo species, the largest type of plant in the grass family, are one of the fastest growing families of plants. They have provided food, fiber, and construction material in Asia for thousands of years.[187] Back at food, young bamboo shoots are a tasty part many Asian dishes.[188] They do well in the Southeastern United States, and their range may extend toward Canada if there were to be a significant warming. Another possibility is that advanced genetic engineering could extend both the climatic range and the usefulness of bamboo.

Uhm, yum! Good!

Major Things for Governments and Businesses to Do

- Governments should implement regulations for protecting and growing the wild ancestor species of existing crops. For instance, Mexico protects teosinte, the wild ancestor of maize.[189] The United States could protect prairie species.

- Farmers, ranchers, and farm-implement suppliers should increase use of more varieties of crops.

- Seed manufacturers: For each regional area, have an "Adopt a Forgotten Crop or Flower" for people to plant in small areas and contests to find plants that have been largely lost.

---

[186] There is an often-quoted story about Henry Ford's 1941 demonstration of bouncing an axe off a hemp automobile body. In a website entitled "The Hemp Car – Myth Busted," in Oct. 24, 2010 (https://theangryhistorian.blogspot.com/2010/10/hemp-car-myth-busted.html), an individual identifying him or herself as the Angry Historian tracked back to the original newspaper results and found that the car plastic was 30% resin binder and 70% cellulosic material. Of that cellulosic material, half was southern slash pine, 30% was straw, 10% was remie (which was used for cloth to wrap mummies), and only 10% was hemp. Still however, this history suggests that farm production could replace a great deal of fossil-fuel hydrocarbons in plastics manufacturing.

[187] Nirmala Chongtham, Madho Singh Bisht, and Sheena Haorongbam, "Nutritional Properties of Bamboo Shoots: Potential and Prospects for Utilization as a Health Food," *Comprehensive Reviews in Food Science and Food Safety*, vol. 10, issue 3, pp. 153–158, May 2011. https://onlinelibrary.wiley.com/doi/full/10.1111/j.1541-4337.2011.00147.x (accessed June 8, 2019)

[188] P. Nongdam and Leimapokpam Tikendra, "The Nutritional Facts of Bamboo Shoots and Their Usage as Important Traditional Foods of Northeast India," *International Scholarly Research Notices*, 2014: 679073, July 20, 2014 https://www.ncbi.nlm.nih.gov/pmc/articles/PMC4897250/ (accessed June 8, 2019)

[189] Tom Standage, *An Edible History of Humanity*, Bloomsbury USA, May 3, 2010.

What You Can Do

- If you have a garden or pot space, check with your nearest agriculture extension station, nursery and/or garden club about both native plants and possible new plants for gardening such as amaranth.

- ask your local nursery if they can get you a few exotic plants, such as amaranth, to add to your plantings. (Warning: Be sure the exotic plant is native to your area rather than a foreign plant that might become what local authorities might call "an invasive weed.")

- Treat yourself to a vegetable or fruit you have never had before.

7.3 Plant Breeding Miracles the Neo-Prairie, and the Ten-Foot Moon Shot

You don't need to be limited to the plants we have. People can breed better plants, and humans have been doing just that for thousands of years.

Few people know the name of a great historical biologist, Nikolai Vavilov. He developed a theory that breeding more productive plants is something that people have been doing from at least 9000 years before present (7000 B.C.). Furthermore, he identified centers of origin for domesticated plant breeding by correlating crop species with wild varieties still growing in the nearby regions. These "Vavilov centers" seemed to have been the source of domestic varieties from roughly 7000 to 3000 B.C. (9000–5000 years before present). The people's names and details of their plant-breeding revolutions are lost to history, but their crop species remain. (Vice President Al Gore provided an excellent introduction to Vavilov circles in his 1992 book, *Earth in the Balance*.[190])

Wild maize (which we Americans call corn) yields only tiny ears roughly as large as a fingertip. In the Vavilov center of Mexico and Central America, the locals had already bred maize into significantly larger ears of an established crop (now referred to as Indian corn) when the Spanish conquistadores arrived in 1519. By the mid-1900s, North American corn ears were a foot long and a major portion of the diets around the world as well as being feed for livestock.

Original rice varieties were just grasses with a few tiny seeds almost small enough to blow away. As early as ten thousand years ago, farmers in the Pearl River of China bred rice into a staple food crop. Surviving records in China describe rice production for at least the last four thousand years. Later, West Africans independently domesticated and improved local wild strains. There were almost certainly more rice innovators elsewhere.

Likewise, Pacific plants such as pineapple and breadfruit are far more productive than would be expected in the evolution of wild plants. One of the most prolific Vavilov centers—the

---

[190] Al Gore, Earth in the Balance: Ecology and the Human Spirit, Houghton Mifflin, Boston, MA, 1992.

birthplace of domestic almonds, apples, flax, and lentils—was in presently cold and dry Central Asia.

Here is a warning for those seeking to implement anti-warming policies: As the Central Asian history shows, some areas benefitted from a warmer and wetter climate as a result of global warming. It will be difficult to convince people in such regions to resist something that benefits them.

Whether climates are warming or cooling, civilizations have arisen by breeding new plant varieties over the millennia. The lesson for possible severe global warming is that breeding new plant varieties can help surmount the resulting challenges. Some of the lessons and future possibilities are as follows.

(1) The "miracle wheat" of Norman Borlaug and the "miracle rice" varieties of M.S. Swaminathan are shorter-stalked higher yielding varieties. Their advances, plus the advances of many others, helped make "The Green Revolution"—improved farm yields so much that the predicted massive famines of the 1970s never happened.[191]

(2) Genetically engineered (GE) or genetically modified organism (GMO) crops entered the market in the 1980s. At that time, there were fears about possible side effects. By 2013, GE crops (mainly corn, cotton, and soybeans) were planted on about half of U.S. crop land in 2013. Their adoption has saved farmers time, reduced insecticide use, and enabled the use of less toxic herbicides. Corn, cotton, and soybean varieties have been bred with either immunity to herbicides or with toxins geared to the pests that attack them.[192] Moreover, there have been no documented medical or environmental problems associated with their use.

(3) Golden rice, another genetically engineered crop was bred to increase beta-carotene (the orange of carrots), a precursor to vitamin A, and thus help protect poor children in rice-based societies from blindness and susceptibility to a number of other diseases. Golden rice might save more than a million lives a year and the sight for a larger number. Unfortunately, there has been a political problem. Leftist and environmentalist groups have promoted conspiracy theories that golden rice is a plan by big business to take over the rice seed markets—actually, the developers planned to give the seed away. Overly complex governmental regulations (many

---

[191] Charles C. Mann, *The Wizard and the Prophet: Two Remarkable Scientists and Their Dueling Visions to Shape Tomorrow's World*, Knopf; 1st edition, January 23, 2018.

[192] Jorge Fernandez-Cornejo, Seth Wechsler, Mike Livingston, and Lorraine Mitchell, *Genetically Engineered Crops in the United States*, Economic Research Report Number 162. United States Department of Agriculture, Feb. 2014. Also available at https://www.ers.usda.gov/webdocs/publications/45179/43668_err162.pdf?v=2442.3

generated in defense against anti-genetic accusations also hurt.[193,194] Meanwhile, the dying continues.[195]

(4) Breed nitrogen-fixing rice and other crops that do not require as much fertilizer as present crops do. Actually, there is an energy cost in plants fixing nitrogen, which is paid in slightly lower productivity; but it has a benefit in less purchased fertilizer and less runoff of fertilizer that over fertilizes downstream bodies of water causing too much algae and resulting fish die-offs. Furthermore, as those concerned about global warming have noted, excess nitrate fertilizer outgasses greenhouse-warming oxides of nitrogen into the atmosphere.

(5) Greater salt-tolerance in crops is becoming increasingly important because the great irrigation schemes of the 1900s are developing the problem of all the irrigation systems of history—salinization or increasing saltiness. Yes, the classical conservationists and historians are correct on that issue. From ancient Mesopotamia to the contemporary Central Valley of California, irrigation systems lacking very careful management eventually lead to fatal salinization and abandonment of the land. And yes, the Nile with its yearly salt-flushing floods was an exception to that rule … until construction of the Aswan Dam. Now, salinization is a growing concern for Egypt also.

(6) As always, drought-tolerant crops can be desperately important when droughts periodically come. If there were to be a significant global warming, as many predict, drought tolerance would become even more important for areas just north of the present desert bands (such as the American Southwest (north of the Sonoran Desert). Warming could increase droughts in that area while allowing more rain in the formerly desert areas closer to the Equator. One research effort suggested that plant productivity can be increased 27% while increasing drought resistance.[196] Another research effort suggests that plant photosynthesis could be increased 40% by using alternate photorespiration pathways.[197] There might be some overlap in

---

[193] Ed Regis, "The True Story of the Genetically Modified Superfood That Almost Saved Millions," *Foreign Policy*, Oct. 17, 2019. https://foreignpolicy.com/2019/10/17/golden-rice-genetically-modified-superfood-almost-saved-millions/ (accessed July 4, 2020)

[194] Ed Regis, *Golden Rice: The Imperiled Birth of a GMO Superfood*, Johns Hopkins University Press; Oct, 2019.

[195] Jayson Lusk, *The Food Police: A Well-Fed Manifesto About the Politics of Your Plate,* Chapter 6, Franken-Fears, p. 104, Crown Forum; April 2013.

[196] Carl R. Woese, Institute for Genomic Biology, University of Illinois at Urbana-Champaign. "Photosynthetic hacks can boost crop yield, conserve water," ScienceDaily, 10 August 10, 2020. www.sciencedaily.com/releases/2020/08/200810113213.htm (accessed Aug. 11, 2020)

[197] Paul F. South, Amanda P. Cavanagh, Helen W. Liu, et al., Synthetic Glycolate Metabolism Pathways Stimulate Crop Growth and Productivity in the Field," *Science*, vol. 363, issue 6422, Jan. 4, 2019. https://science.sciencemag.org/content/363/6422/eaat9077 (accessed Aug. 12, 2020)

these two research results so that they would not be additive. Even so, these two plant revolutions alone might be enough to stall global warming for several decades.

(7) Algae and fungi (both further discussed later) have tremendous potential for increases in food production. However, production for human food must be less protein rich and more carbohydrate rich. Both algae and fungi have high protein levels approaching that of meats, and excessively high protein levels would sicken omnivores such as people, dogs, and pigs (versus true carnivores such as cats).

These are just some of the best-known examples ... for now. If global warming and resulting food scarcity becomes a definite existential threat, we shall need to do more.

Regeneration of soil draws down a major share of the global-warming carbon dioxide that is being released into the atmosphere each year, and it could release much more. The deeper soils of regenerative practices also need less nitrate fertilizer, so there is less global-warming nitrous oxide given off. It is one set of agricultural technologies that is getting steadily better for the environment.

Still, in his *Growing a Revolution*, soil scientist David Montgomery wistfully commented that if only we could just have a program like the Moon Race, then the soil-building technologies of regenerative agriculture could become a real revolution. Besides the lack of high-tech glamor, he worried that many of the large companies would not find a possible profit to interest them. The large fertilizer companies would not fund research leading to decreased sales of nitrate and phosphate fertilizer. The pesticide manufacturers (making insecticides, fungicides, and herbicides) would have less of a market.[198]

But there can be new markets for new industries. There can even be both Moon-shot incentives and smaller incentives. Most of all, there are other more powerful farming methods that could draw down several times more carbon dioxide from the atmosphere each year than goes up in the form of emissions.

It could start with perennial plants that live for many years rather than the one year of annuals. Before farming, most plants were perennial. They might go dormant in dry or cold seasons, but they resumed growth when the missing water or warmth returned.

This has immediate advantages for maturing more quickly during limited growing seasons in colder climates and resistance to droughts and other bad weather. It also has the long-term benefit of preserving and building soils. Perennials protect soil by rarely leaving erodible bare ground and by building additional soil with their leaf litter and their deep roots that keep holding the soil even years after they die. Moreover, the perennials help maintain a local environment that provides a habitat for creatures that eat pests.

As noted earlier, regenerative/conservation soil-building practices could draw down carbon dioxide from the atmosphere and compensate for much of the global warming carbon dioxide

---

[198] David A. Montgomery, *Growing a Revolution: Bringing Our Soil Back to Life*, W. W. Norton & Company, 2017.

emissions. Furthermore, these practices could draw down carbon dioxide for decades. The good of the bad is that most agricultural soils throughout the world have been massively depleted in soil organic carbon over centuries and even millennia of conventional plow-based farming. Thus, the soils throughout the world are a vast largely empty vault into which captured carbon from the air can be inserted … while improving soil productivity and thus yearly crop yields. Much of the carbon captured as soil can be stable for decades or even centuries, and it does not have the blowout risk of pumping carbon dioxide into deep aquifers (which has been proposed).

Most importantly, converting carbon dioxide in the atmosphere into useful soil can be a profitable venture rather than just a government-required and/or government-subsidized activity. Fields can accept massive amounts of carbon from the atmosphere while producing crops. At the same time, such regenerative agriculture can build soil for increased food production, reduce erosion, reduce fertilizer use, reduce pesticide use, and reduce toxic over-blooms of algae from fertilizer runoff.[199]

The potential is even greater in that only about a third of present agricultural production consists of perennials such as alfalfa, grasses, sugar cane (another grass), grapes, and various tree crops. In contrast, the majority of planted land is in annuals, particularly grains, that must be planted again each year.

Still, there is no pizzazz, no Moon-shot excitement. I propose two additional plant breeding initiatives: polycultures of perennials and long-lasting deep roots. Together, they can make a Ten-Foot Moon Shot.

Wes Jackson of the Land Institute of Kansas proposed that we should develop a polyculture of crops for the prairie. It would be a variant of the natural prairie that grew in North America before European agriculture arrived. For discussion in this book, we shall refer to this as the neo-prairie, a combination of perennial plants that can yield grain as well as grazing plants, protect the soil from erosion, and renew soil nutrients with little or no use of plowing, fossil-fuel-based fertilizers, or pesticides.

Reductions in plowing, fertilizer use, and pesticide use are a virtue for environmentalists, but for farmers they are a plus on the bottom line … if they work.[200] David Montgomery (who summarized Jackson's perennial proposal) argues for eliminating pesticide use if possible, but he also commends merely reducing pesticide use.

Jackson's proposed neo-prairie agriculture includes a combination of plants for a reason. Monocultures of annuals (such as corn, wheat, or rice) are perfect targets for pests and blights that increase season after season on a particular field. Perennials, by definition, are the same target year after year. The farmer's best defense is plowing and planting a different crop to do a reset on the field; the pests and blights then have to start up anew. Another defense is chemical pesticides

---

[199] Gabe Brown, "Keynote at Farming for the Future 2020," *YouTube*, March 13, 2020. https://www.youtube.com/watch?v=ExXwGkJ1oGI (accessed Dec. 5, 2020)

[200] David R. Montgomery, *Growing a Revolution: Bringing Our Soil Back to Life*, W. W. Norton & Company, 2017. He summarized: Wes Jackson, *New Roots for Agriculture*, University of Nebraska Press, Lincoln, 2002.

targeting the pests. But both plowing and pesticides disrupt the fungal and bacterial ecology that helps reduce fertilizer need by breaking minerals into more usable nutrient forms and providing them directly to the plants so much less leaches out of the ground to cause pollution.

The prairies evolved a natural defense of having not just the two most common, buffalo grass and blue gama (*Bouteloua gracilis*) but maybe a couple hundred plant species in a single acre (0.01-hectare) area. The Land Institute has developed a perennial wheat called kernza®, so kernza and buffalo grass make two of Jackson's two-hundred varieties for pest-resistant diversity.

Some have questioned whether seed companies would ever become interested in the neo-prairie or other polycultures. However, the seed companies (who are also plant breeder companies) could sell starter seed assortments for the initial transition to neo-prairie and booster assortments to push a neo-prairie field toward certain characteristics such as more nitrogen fixing, better drought resistance.

As for development of farm implements for serving a new type of farming … well, they did it for low-till and no-till farming. There is presently no farm machinery that can harvest several different crops in the middle of multiple plant species field. Companies that want to make a profit will find ways to do it. There will be an extremely high level of artificial intelligence to monitor fields and identify things to harvest. Similarly, the robo workers will need a high degree of dexterity for harvesting.

For species diversity to defend against pests, the advanced genetic techniques available such as CRISPR (short for clustered regularly interspaced short palindromic repeats). That is microbiological bafflegab for snipping out small portions of a gene and either deleting it or moving it to elsewhere in the DNA double helix. It can be used to identify the detailed genetic-traits for a particular characteristic, such as suberin production, and build those traits in other species—without any transfer of genetic material between species, which has worried many people. Thus, a polyculture field could be planted with many features.

Soil boosted with a number of perennial nitrogen fixers (such as alfalfa, *Medicago sativa*) might support grazing livestock and an early sunflower (*Helianthus annuus L.*) harvest.[201,202] Then, after the early harvest, cereal-producing crops could catch up for grain production. Finally, a full hay cut might take the field down to near the ground while producing a hay with some of the high-protein alfalfa, but not too rich for livestock health. Finally, several deep-rooted species could provide moisture during late summer in exchange for some of that fixed nitrogen from shallow-rooted nitrogen fixers. And yes, some of processes might require different farm machinery than most farmers have today.

Polyculture with both perennials and annuals has long been used in tropical countries:

---

[201] "Sunflower: A Native Oilseed with Growing Markets," the Jefferson Institute, https://www.agmrc.org/media/cms/sunflower_guide_69AF73CC348B6.pdf (accessed April 1, 2019)

[202] D. H. Putnam, E. S. Oplinger, D. R. Hicks, et al., "Sunflower," *Alternative Crops Manual*, University of Minnesota, 1990. https://hort.purdue.edu/newcrop/afcm/sunflower.html (accessed April 1, 2019)

Among potential advantages of intercropping systems are weed suppression through shading or natural plant toxins, reduction of insect damage by improving the balance of insect pests and associated natural predators, better use of available soil nutrients, water conservation, erosion control, and greater productivity per unit of land.

> *Among potential advantages of intercropping systems are weed suppression through shading or natural plant toxins (allelopathy), reduction of insect damage by improving the balance of insect pests and associated natural enemies, better use of available soil nutrients, water conservation, erosion control, and greater productivity per unit of land.*
>
> "Polyculture Has Advantages"[203]

Applying polyculture in more temperate climates could have benefits of higher net profit via salable products minus decreased costs in fertilizer, pesticides, and feed supplements for livestock.

An even stronger case can be made for increasing polyculture and deep roots in farming fruit trees, grape vines, and trees in general (silviculture). Silver maples (*Acer saachariunum*) have a symbiotic relationship with goldenrod. The deep roots of silver maples access water far below ground level. The maples provide some of that water to goldenrod plants near the surface. In return, the goldenrods protect the soil around each maple from erosion.

Teaching that same trick to other perennial trees and vines would have benefits for those crops as well as improving others. Farmers often interplant beans, snow peas, and other legumes among apple, walnut, or almond trees.[204] Vineyard farmers often intercrop with the familiar benefits of erosion protection, better soil structure, better water-holding capability of the soil, and more pest predators to allow decreased pesticide use.[205] How much better would it be to provide additional water to these secondary crops? How much better yet would it be for some percentage of the secondary crops (annual or perennial) to be nitrogen-fixing plants so that they could exchange nutrients for water?

This is particularly important because the vineyards and olive groves of the ancient Mediterranean were a major source of soil erosion in antiquity.[206] Moreover, similar erosion

---

[203] Stephen R. Gliessman and Miguel A. Altieri, "Polyculture Has Advantages," *California Agriculture*, pp. 14–16, July 1982.
http://calag.ucanr.edu/archive/?type=pdf&article=ca.v036n07p14 (accessed April 1, 2019)

[204] Sagar Maitrai, "Potential of Intercropping System in Sustaining Crop Productivity," *International Journal of Agriculture Environment and Biotechnology*, vol. 12, no. 1, pp. 39-45, March 2019.
https://www.researchgate.net/publication/332852522_Potential_of_Intercropping_System_in_Sustaining_Crop_Productivity (accessed June 30, 2020)

[205] Urska, "Cover Cropping: Alternative to Herbicide Use in the Vineyard," eVinyeard, Nov. 23, 2016. http://www.evineyardapp.com/blog/2016/11/23/cover-cropping-alternative-to-herbicide-use-in-the-vineyard/ (accessed April 1, 2019)

[206] David R. Montgomery, Chapter 4: "Graveyard of Empires," *Dirt: The Erosion of Civilizations*, University of California Press, Berkeley, California, 2007.

continues to this day. A *Science News* article blared, "Prosecco production takes a toll on northeast Italy's environment,"[207] although the issue was bare ground between the grape plants on steep slopes rather than a particular type of grape.[208]

The early Romans did not have that problem because they manually worked what today's farmers would call polyculture or intercropped farms. Early Roman farms were intensively worked operations where diversified fields were carefully hoed, weeded, and manured—all by hand. The earliest Roman farmers planted a multistory canopy of olives, grapes, cereals, and fodder crops referred to as *cultura promiscua*. Interplanting of understory plants that would be shaded and overstory crops above them smothered weeds, saved labor, and prevented erosion by shielding the ground all year. Roots of each crop reached to different depths and did not compete with each other.

Poly cultures, such as the neo-prairie will increase food production, decrease fertilizer and pesticide costs, and store more carbon in the soil. However, it will not be enough for a Green New Deal. Storing carbon in the soil has been considered, but rejected as insufficient, by other researchers.

Physicist and science fiction writer Gregory Benford, along with agricultural scientist Stuart R. Brand (not to be confused with environmental activist and producer of *The First Whole Earth Catalog*, Stewart Brand), analyzed ways to store captured carbon to decrease carbon dioxide emissions into the atmosphere.[209] They considered increased fixing of carbon in the soil, increased fixing by ocean fertilization, and pumping carbon dioxide into saline deposits. All three carbon-fixing methods had problems:

a) Much of the carbon fixed in soil regeneration oxidizes back up to the atmosphere within a few years as dead roots and turned over grasses compost. Soil can hold carbon for decades, but if left open to the air (say by a return to plow-based agriculture), much of the hard-won carbon can be lost back to the atmosphere.

b) Much of the carbon fixed in the ocean by tiny drifting algae (phytoplankton) stays in the upper plankton-filled waters where the plankton are eaten by predators, and

---

[207] Cassie Martin, "Prosecco Production Takes a Toll on Northeast Italy's Environment," *Science News*, Jan. 18, 2019. https://www.sciencenews.org/article/prosecco-production-toll-northeast-italy-soil-environment (accessed June 7, 2020)

[208] S. E. Pappalardo, L. Gislimberti, V. Ferrarese, et al., "Drinking Earth for Wine. Estimation of Soil Erosion in the Prosecco DOCG Area (NE Italy), toward a Soil footprint of bottled Sparkling Wine Production," *bioRixiv* preprint, Jan. 10, 2019. https://www.biorxiv.org/content/biorxiv/early/2019/01/10/516245.full.pdf (accessed April 2, 2019)

[209] Stuart E. Brand and Gregory Benford, "Ocean Sequestration of Crop Residue Carbon: Recycling Fossil Fuel Carbon Back to Deep Sediments," *Environmental Science and Technology*, vol. 43, no. 4, pp. 1000–1007, Jan. 12, 2009.
Note: Stewart Brand's *The Whole Earth Catalog* was an American counterculture magazine and product catalog published several times a year between 1968 and 1972, and occasionally thereafter, until 1998. It had some great articles, but it primarily focused on product reviews keyed to self-sufficiency, ecology, and alternative education.

the predators respire carbon dioxide back into those surface waters where much of it transfers back to the atmosphere.

c) High-pressure carbon dioxide in saline deposits could easily leak back to the surface and fizz away like the bubbles from a shaken soda pop bottle except that a massive leak could be dangerous (as demonstrated by the toxic carbon dioxide leak from the volcanic Lake Nyos that killed 1,700 people).[210]

Because all the best-known options considered had problems, Benford and Brand devised their own proposal of gathering crop residues autumn leaves, and so forth.; packaging those materials; transporting the packages to seaports, shipping the packages to locations over areas of deep ocean, and then ballasting the packages to sink into the deep waters. Their estimated cost of carbon stored in 2009 was $304/ton ($334 per metric ton).

Calculating that 1 ton of carbon as about a 4 tons of carbon dioxide, the Crop Residue Oceanic Permanent Sequestration (or CROPS) would cost more than a trillion dollars each year to remove those 4 billion tons of carbon dioxide, which is about a tenth of the yearly carbon dioxide emitted by human activities, so the Benford-and-Brand proposal might require several trillion dollars per year to balance carbon dioxide emissions. That trillion-dollar price would yield absolutely no return on the investment.

Moreover, removing all crop residues off the ground after harvest would leave the land bare—prone to oxidation and erosion of the soil. The result might be a net increase in carbon dioxide emissions from the ground, thus replacing some or all of the organic material sunk (at great expense!) into the ocean depths.

That leads directly to the Ten-Foot Moon Shot. As noted earlier in this subsection, plant-breeding technologies have radically advanced to more quickly and efficiently change plant characteristics. Researchers can now map genetic material for desired plant characteristics in one plant and, by use of genetic engineering, insert those characteristics in other plants rather than breeding characteristics and checking them over a period of years as in conventional plant breeding.[211] Even more advanced technologies are being developed.[212]

Discussion in this subsection has already had the phrase "deep roots." Now, these roots can be bred to be more useful in several ways. Researchers at the Salk Institute in La Jolla, California are using both conventional breeding and genetic engineering to develop plants with deeper bushier roots containing high amounts of suberin, a material found in specialized plant cell walls, wherever insulation or protection from the surroundings is needed.

---

[210] David Bressan, "The Deadly Cloud at Lake Nyos," *Forbes*, Aug. 21, 2019. https://www.forbes.com/sites/davidbressan/2019/08/21/the-deadly-cloud-at-lake-nyos/#2e248c6f5dbf (accessed Aug. 18, 2020)

[211] Kai P. Voss-Fels, Rod J. Snowdon, and Lee T. Hickey, "Designer Roots for Future Crops," *Trends in Plant Science*, vol. 23, issue 11, pp. 957–960, Nov. 2018.

[212] Kai Peter Voss-Fels, Mark Cooper, and Ben John Hayes, "Accelerating crop genetic gains with genomic selection, *Theoretical and Applied Genetics*, Dec. 19, 2018.

Suberin is a major component of cork.[213,214] Suberin-heavy roots could resist decomposition for not just decades, but centuries, so they could hold carbon for that time even if the plants that grew them had died.[215]

If long-term retention of carbon were the only benefit of suberin, it would probably benefit planetary society overall but not yield any profit to the farmers. However, these deep long-lasting roots could maintain considerably more water and nutrients than more shallow-root species.[216] The suberin-rich deep roots could store years' worth of water for use in a seriously global-warmed year. This would be a key benefit for crops in the central United States, an area that has been predicted to become hotter and more prone to droughts in a global-warming world.

Another benefit would be in erosion resistance. The rhizosphere, the area of interaction between roots and soil, would be much deeper, allowing access to deeper ground nutrients and extending a healthier soil structure that could take and hold more water. Living or dead, the long-lasting roots would better hold the soil against wind and water erosion.[217]

Therefore, I propose the 10-foot (3-meter) Moon Shot of breeding plants with much deeper long-lasting roots. Most annual crops send down roots only about 1 meter, roughly 3 feet. Developing perennial crops extending their roots 3 meters, or roughly 10 feet, could conceivably compensate for most or all of the carbon dioxide emissions from burning fossil fuels for a proportionately longer time.[218] To make this planetary service pay, such crops would also better resist drought and heat from climate warming.

Furthermore, deep-rooted perennial agriculture, combined with more robotic equipment (to be described later), could allow more hilly or dry land to be converted from grazing land to crop land. Besides the greater monetary return for the farmers and rancher who could convert to cropping, this would reduce land in grazing. That might keep grazing profitable to those staying

---

[213] José Graça, "Suberin: the Biopolyester at the Frontier of Plants," *Frontiers in Chemistry*, vol. 3, p. 62, Oct. 30, 2015. https://www.ncbi.nlm.nih.gov/pmc/articles/PMC4626755/pdf/fchem-03-00062.pdf (accessed March 6, 2015)

[214] Brad Jones, "Scientists Are Engineering Carbon-Storing Plants to Fight Climate Change," *Futurism.com*, Dec. 21, 2017. https://futurism.com/scientists-engineering-carbon-storing-plants-fight-climate-change (accessed Jan. 19, 2019)

[215] Hillary Rosner, "Plants Are Great at Storing CO2, These Scientists Aim to Make Them Even Better," ensia.com, April 9, 2018. https://ensia.com/articles/plants-co2/ (accessed Jan. 19, 2019)

[216] Douglas B. Kell, "Breeding Crop Plants with Deep Roots: Their Role in Sustainable Carbon, Nutrient and Water Sequestration," *Annals of Botany*, vol. 108, issue 3, pp. 407–418, Sept. 1, 2011.

[217] David R. Montgomery, "Soil erosion and agricultural sustainability," *Proceedings of the National Academy of Sciences* of the United States, vol. 104, pp. 13268–13272, 2007. https://www.pnas.org/content/104/33/13268 (accessed March 14, 2019)

[218] Josh Falzone, "Could This Biotechnology Innovation Help Combat Climate Change?" *BIOtechNOW*, Feb. 6, 2019. http://www.biotech-now.org/environmental-industrial/2019/02/could-this-biotechnology-innovation-help-combat-climate-change (accessed Feb. 22, 2019)

in it because society might become more vegetarian, as we shall see in the next subsection (7.4 Eat Your Veggies).

Major Things for Governments and Businesses to Do

- Governmentally fund a "modest Moon shot" of a quarter billion dollars per year to advance plant breeding capabilities for deep roots
- Governmentally fund a quarter billion dollars per year to advance plant breeding capabilities for deep roots and new-prairie agriculture.
- Governmentally mandate protection of crop ancestor species (including prairie grasses), and fund preserves for such plants.
- Business: Develop and market polyculture growing kits for both gardening and for full-scale agriculture.

What You Can Do

- Study plant breeding technologies and speak up against the agents of fear, uncertainty, and doubt (FUD) who attempt to scare people away from advanced plant breeding.[219] Genetically engineered crops have never caused any health or environmental problems, and they are feeding the world now. Abolishing them would cause famines within one season.
- If you have a garden, grow a little area of some ancestor crop species from your area. (Warning: Be sure any ancestor crop comes from your area—local authorities can get quite irate about any plants they consider invading species.)

7.4 Eat Your Veggies

For millennia, practitioners of some religions, health-fanatics, and vegetarians have argued that meat is bad. Socrates of ancient Athens argued from a geopolitical view that increased land requirements for meat inevitably led to a drive for territorial increases and war with the neighbors. More recently (the past two centuries), vegetarians of various stripes have argued that little or no meat is better for one's health. In the early 1800s various utopians argued that four times as many people could live on vegetarian diets as on meat diets. William Vogt, a pioneer of Twentieth Century environmentalism, referred to it as getting more food by going lower on the food chain. Thomas Robert Malthus, who thought that population would always be controlled by famine,

---

[219] Julie Gunlock, *From Cupcakes to Chemicals: How the Culture of Alarmism makes Us Afraid of Everything and How to Fight Back*, WF Press, Oct. 31, 2013. The author explains how scaring people with FUD (fear, uncertainty, and doubt) makes money and power for some people and causes their victims to spend money on things they don't need.

agreed about the utility of a vegetarian diet; however, he did not believe that such a vegetarian life was worth living.

Aside from the environmental movement, futurist Arthur C. Clarke noted in his 1962 *Profiles of the Future* that beef and pork required roughly eight times the weight in grains to yield one weight of meat (chicken was better but still bad at five times the weight in grain).[220] Eventually, environmentalists focused on similar statistics as well as statistics about energy intensity and water intensity, and they condemned meat eating.

Since the 1980s, there has been a global-warming argument about meat. Production of meat, particularly feedlot production (versus range-fed), generates a vast amount of carbon dioxide per unit of (tasty) meat produced.

Sad to say, the vegetarians and environmentalists who also signed up for those arguments were correct to a considerable degree. The health argument, in particular is a strong one. Beyond a certain amount, meat is too rich for healthy living; it shortens life with heart attacks, diabetes, high blood pressure, and the almost unlimited number of symptoms that arise from the latter two conditions.

However, the issue is deep-seated and probably even genetically hard wired. For many millennia, untold ancestors looked across the grassland at a lions' kill and thought, "If somebody (other than me because I want to live to see tomorrow) can just distract those lions for a couple seconds, I'll grab a gazelle haunch, and our clan will be in fat city tonight!"

Whenever, major prosperity arrives—for example, when some measure of economic activity returned to Europe after World War II—the amount of meat in the diet increases. Likewise, rising prosperity in poor countries correlates with increased meat in the diet. Traditional-cultures popularizer Gregg Braden noted that 21st Century people in Tibet often go to MacDonald's when they have a "big yak attack." MacDonald's restaurants in the high Andes service analogous hankerings with llama burgers.[221] The recurring theme of Braden's two vignettes, meat.

Only education and generations of prosperity can convince people that limited (but not zero without careful diet planning) amounts of meat in the diet are healthier for the people. Of course, such diets also put less stress on climate and the environment in general.

Eventually, people will better appreciate vegetables, nuts, and legumes that can provide proteins and/or meaty tastes. For decades, there have been articles such as, "10 Vegetables That Can Substitute for Meat,"[222] that extol the virtues of soybean products, nuts (yes, not a vegetable),

---

[220] Arthur C. Clarke, *Profiles of the Future: An Inquiry into the Limits of the Possible*, Arthur C. Clarke, Harper & Rowe, February 1963.

[221] Gregg Braden., *The Turning Point: Creating Resilience in a Time of Extremes*, Hay House, 2014.

[222] Rhea Parsons, "10 Vegetables That Can Substitute for Meat," *One Green Planet*, 2016. https://www.onegreenplanet.org/vegan-food/vegetables-that-can-substitute-for-meat/ (accessed Dec. 14, 2018)

mushrooms, eggplant, jackfruit, lentils, and other kinds of legumes, and studies calling for "… a save-the-world diet.[223,224]

In the meantime, religious and environmental zealots, along with health diet aficionados (who might also be termed religious zealots in their own way) have been working for their own reasons to develop substitutes for meat. Entrepreneurs have produced some meat analogs over the years. Whatever their reasons, they have succeeded many times with their (that word again!) innovation.

In the late 1970s, this author personally enjoyed "Wham and Cheese" and "Vegetarian Chili" that tasted exactly like the real meat products at one of the stores in the John Henry Weidner health-food chain in Southern California. (Weidner himself was a Dutch Seventh Day hero of World War II who rescued downed Allied airmen and Jews fleeing the Nazis before settling quietly into the health-food trade.)[225]

The 2000s have seen a new surge of entrepreneurs seeking to market vegetable matter products that taste and possess the texture of meats.[226] Impossible Foods make their product from wheat protein, coconut oil, potato protein, natural flavors [what?], leghemoglobin (an iron-containing component of soybean roots also known as heme), soy protein isolate, and yeast extract. Heme is abundant in animal muscle, and it provides the meat flavor. Impossible Burgers use fermentation to produce heme from other vegetable matter — "…similar to the method that's been used to make Belgian beer for nearly a thousand years, in the words of the Impossible Burgers website.[227] They are even attempting the holy grail of meatless meat, a world-class meatless steak.[228]

Beyond Meat sells patties called Beyond Burger that contain pea protein (which is often used in Vegan meat substitutes) and coconut oil.[229] A reporter from *The Inspired Home* declared

---

[223] Lentils with a Side of Rice: The Save-the-World Diet?" *Yahoo News*, Jan. 16, 2019. https://news.yahoo.com/lentils-side-rice-save-world-diet-235519205.html (accessed Jan. 16, 2019)

[224] Walter Willett, Johan Rockström, Brent Loken, et al., "Food in the Anthropocene: the EAT–Lancet Commission on Healthy Diets from Sustainable Food Systems," *The Lancet*, Jan. 16, 2019. DOI:https://doi.org/10.1016/S0140-6736(18)31788-4 (accessed Jan. 16, 2019)

[225] Kurt Ganter, "A Heart Open to the Suffering of Others," *The Journal of Adventist Education*, pp. 28–32. http://circle.adventist.org/files/jae/en/jae201375052805.pdf (accessed June 11, 2020)

[226] "Eat What You Love," Beyond Meat website. https://www.beyondmeat.com/ (accessed Dec. 12, 2018)

[227] *The Impossible Is Here and over There*, Impossible Foods website. https://impossiblefoods.com/ (accessed Dec. 13, 2018)

[228] Allison Shoemaker, "Impossible Foods Tackles the Most Impossible Meatless Meat: The Steak," *Takeout*, Jan. 9, 2019. https://thetakeout.com/impossible-foods-steak-meatless-alt-meat-vegetarian-1831616304 (accessed Jan. 15, 2019)

[229] "Eat What You Love," Beyond Meat website. https://www.beyondmeat.com/ (accessed Dec. 12, 2018)

that the patties were indistinguishable from regular ground hamburger,[230] but another source reported that a package [probably a half pound, about 230 grams] cost $5.99. That is nearly $12 per pound or a little more than $26 per kilogram—more expensive than real hamburger meat but getting closer in price. Beyond Meat products have appeared in markets nationwide, and they have a growing list of competitors.

There is also a commercial effort to market a vegetarian substitute for eggs in baking and scrambled eggs with comparable protein and other properties needed for baking, although they won't look the same sunny side up.

Major Things for Governments and Businesses to Do

- This is a tough one because the United States Department of Agriculture has always been slow to research things that might decrease the market for American farm products. An appeal to India and other largely vegetarian countries: As a return favor for the initial development of miracle wheat and miracle rice, have national development programs for better vegetable production and competitive prizes for the best recipes from each province that might substitute more vegetable protein for meat in India … and be carried as a missionary work to the MacDonald's heathen.

- For the United States. Have a NASA project for a Space Station farm module that could replace much of the food lofted (expensively!) into orbit, and it would be fresher food for low-g degraded taste buds.

What You Can Do

- Research vegetarian diets and dishes.
- Yes, of course, try more vegetarian dishes. Many are superb.
- Treat yourself to one veggie-protein main course each week.

7.5 Rise of the Robo Farm Machines

For thousands of years after the first hunter gatherers actually planted a seed they had gathered, famers worked with only human labor. Then, they hitched cattle and horses to their digging sticks to help prepare the ground for planting. They improved the sticks until they became plows and then better plows. The harnesses improved so the animals could pull more efficiently. Carts and wagons appeared to carry tools, seed, and harvested crops. This process fed more people on the same area of land using fewer laborers, but bigger revolutions were coming, blossoming out of the Industrial Revolution.

---

[230] L. Brook, *The Inspired Home*, Feb. 14, 2018.
https://sso.godaddy.com/?realm=pass&app=o365 (accessed Jan. 13, 2018)

In the 1840s, horse-drawn reapers appeared to replace hand cutting with scythes. The mechanization then extended to winnowing (separating the grain from the stems), and dozens of other farming operations, such as planting and weeding. Then steam power and internal-combustion engines started another farming revolution. Almost every farming operation could be, and was, mechanized for greater productivity. That productivity multiplied again with power from fossil-fueled engines.

The thrust of 20th and 21st Century agricultural machinery has been to replace the equivalent of an ever-larger gang of people and a larger team of draft animals with steadily larger engine-powered vehicles each operated by one person. One machine with one driver now does the work of hundreds. That is why the cost of food compared to hours worked is the lowest it has ever been. This is why the farm percentage of the population has dropped from more than one in two to one in twenty in the most advanced countries.

However, there are costs. The giant machines are fabulously expensive, and as they get larger, there is a point of diminishing returns for increases in productivity.

Greater size also comes with the cost of not being able to easily intercrop where there is typically one tall crop—say corn—with melons or pumpkins growing at a lower level. Another example is shade-loving coffee beans growing in the shade of almond trees. With intercropping, there are fewer "weeds" competing with the crop; there is just another crop. Furthermore, two or more crops make a field much less attractive to pests. Also, intercropping is more flexible for dealing with more or less rain—one crop or the other does better under different weather conditions. Gigantic farm equipment cannot maneuver in small spaces to realize these benefits.

Finally, massive machinery compacts the soil. Compaction reduces soil productivity, reduces water soaking into the ground (which increases erosion), makes it harder for roots to grow, and limits oxygen below ground.[231]

This first became an issue when Ford tractors started competing with mules in the early decades of the 1900s. Then it got worse as weights for tractors and other implements increased dramatically since the 1950s.[232]

Robotics allows an entirely new mechanical approach of relatively small farm implements, say in the hundreds of pounds, tens of pounds, or even fractions of a pound rather than multiple tons. If such implements can run autonomously, then a single operator needs only plan, start the robo workers, and monitor their operation. Instead of having a single 120-foot-wide sprayer, maybe in the future there might be 120 1-foot sprayers that are traveling up and down the field in a swarm.[233]

---

[231] Sjoerd W. Duiker, *The Effects of Soil Compaction*, CAT USC188, Pennsylvania State University, University Park, PA. https://www.agriculture.com/the-truth-about-compaction (accessed Oct. 6, 2018)

[232] B. D. Soane and C. Van Ouwerkerk, "Soil Compaction: A Global Threat to Sustainable Land Use," *Advances in GeoEcology*, vol. 31, pp. 517–525, 1998.

[233] Robert Saik, *The Agriculture Manifesto, Ten Key Drivers That Will Shape Agriculture in the Next Decade*, CreateSpace Independent Publishing, ISBN 9781499382709, 2014.

But there is more. A robotic herbicide applicator can reduce weed killer use by as much as ninety percent by only attacking individual weeds—good for the environment and good for the farmer's bottom line.[234]

And, there is still more. In the times before herbicides and pesticides, people dealt manually with pestiferous competitors for human food. The hoe, particularly the infamously back-destroying short hoe (with a human worker!), removes weeds competing with crops without the cost of tractors and the inevitable lost plants when a tractor operator goes a little off course. Likewise, a gardener can manually collect and kill slow-moving pests. Worms and insects are smashed or dropped into a can of kerosene. For weeding, a mini-robot can work a short hoe all day long without any back problems—no back.[235]

A collie-sized (or smaller) machine could lightly step between plants, distinguish crops, from weeds, and cut or cultivate as appropriate. But there's more. Advanced agricultural robot vacuums could collect pests such as aphids, potato bugs, and cut worms—and provide them as feed for chickens or other livestock. At the same time, sensors could inventory levels of moisture and plant nutrients for each small area of ground, and this information could then be used for a database for deciding the need (or lack of need) for fertilizer and water. A farmer with a trailer full of robo farm workers would put them out at the edge of a field, input the instructions, monitor operations, take in breakdowns for maintenance, and collect for storage when the job was done.

Precision agriculture is a growing multibillion-dollar industry that uses global positioning system (GPS) accuracy for monitoring many small areas of a field and varying application of fertilizers, pesticides, and irrigation water where needed and when needed.[236] It reduces costs, and it reduces the environmental effects of runoff to other areas. Precision agriculture today generally works with distance variations of multiple feet/meters. However, evolving technology for computation and data storage will make it affordable to be exquisitely more precise with navigation in areas of a few inches or less.

Robo farm workers can be light enough to avoid being mired in mud during early-season wet weather. Thus, they may get an additional week or more of growing time compared to heavy implements. That's a bonus after the benefit of avoiding compaction damage to the soil.

---

[234] Ludwig Berger and Tom Polansek, "Robots Fight Weeds in Challenge to Agrochemical Giants," Reuters, May 20, 2018. https://www.reuters.com/article/us-farming-tech-chemicals-insight/robots-fight-weeds-in-challenge-to-agrochemical-giants-idUSKCN1IN0IK (accessed June 29, 2020)

[235] Anmar Frangoul, "An Australian Start-up Is Using Robots to Pull Weeds and Herd Cattle," CNBC, April 11, 2019. https://www.cnbc.com/2019/04/10/an-australian-start-up-is-using-robots-to-pull-weeds-and-herd-cattle.html?utm_campaign=the_download.unpaid.engagement&utm_source=hs_email&utm_medium=email&utm_content=71660486&_hsenc=p2ANqtz--sDS0rah-PBIngRrCqHU4NvHSPWHTz7IRSjp_59qCcRJqxJVxOwGBER8IK9IJw5Zib_rGb-CAa-Z5oEpllnZS8EQgWbg&_hsmi=71660486 (accessed April 12, 2019)

[236] Remi Schmaltz, "What is Precision Agriculture?" Agfunder News, April 24, 2017. https://agfundernews.com/what-is-precision-agriculture.html (accessed July 22, 2019)

A final benefit is that small robo workers can be low to the ground. Hence, they can work on many slopes that would be too steep for a tractor to work safely. This too would make more grazing land available for growing crops.

The agri-food industry—farming, to processing, to retail stocking shelves—is predicted to be a major area of robotic innovation in the near future because it is highly labor intensive now.[237]

- Entrepreneurial start-ups are running experimental farms with experimental robotic implements. There are automated picking machines for leafy greens, peppers, strawberries, cotton, and tomatoes.[238,239,240]

- A company called the Small Robot Company has a stated goal of less chemical pesticide use, but used with greater precision, for high-density permaculture.[241]

- Entrepreneurial start-ups are running experimental farms with experimental robotic implements. One prediction is that the world market for agricultural robots will grow to $75 billion by 2025.[242]

Robo farm workers have the crucial potential of multiplying the practicality (read profitability) of several families of crops and technologies that are discussed in the following pages. Multiple food and energy revolutions will spring from robo farm workers.

---

[237] *Agricultural Robotics: The Future of Robotic Agriculture*, UK-RAS Network, 2018. https://www.ukras.org/wp-content/uploads/2018/10/UK_RAS_wp_Agri_web-res_single.pdf (accessed July 5, 2020)

[238] Erin Winick, "New Autonomous Farm Wants to Produce Food without Human Workers," *MIT Technology Review*, Oct. 3, 2018. https://www.technologyreview.com/s/612230/new-autonomous-farm-wants-to-produce-food-without-human-workers/?utm_campaign=the_download.unpaid.engagement&utm_source=hs_email&utm_medium=email&utm_content=71660486&_hsenc=p2ANqtz--sDS0rah-PBIngRrCqHU4NvHSPWHTz7IRSjp_59qCcRJqxJVxOwGBER8IK9IJw5Zib_rGb-CAa-Z5oEplInZS8EQgWbg&_hsmi=71660486 (accessed April 12, 2019)

[239] Magdalena Petrova, "This Robot Picks a Pepper in 24 seconds Using a Tiny Saw, and Could Help Combat a Shortage of Farm Labor," CNBC, Dec. 20, 2018. https://www.cnbc.com/2018/12/20/watch-this-robot-pick-a-peck-of-peppers-with-a-tiny-saw.html (accessed Dec. 22, 2018)

[240] "How Do Big Farmers Hope to Pick the Next Crop? Carefully -- But with Robots," *The Washington Post*, Feb. 17, 2019. https://www.nola.com/homegarden/2019/02/how-do-big-farmers-hope-to-pick-the-next-crop-carefully-but-with-robots.html (accessed Feb. 19, 2019)

[241] John Harris, "We'll Have Space Bots with Lasers, Killing Plants: The Rise of the Robot Farmer," *The Guardian*, Oct. 20, 2018. https://www.theguardian.com/environment/2018/oct/20/space-robots-lasers-rise-robot-farmer (accessed May 22, 2019)

[242] "Agricultural Robots Market to Touch US$ 75.00 Billion by 2025," *MarketWatch*, Oct. 2, 2018. https://www.marketwatch.com/press-release/agricultural-robots-market-to-touch-us-7500-billion-by-2025-2018-10-02 (accessed Dec. 2, 2018)

Do robo farm workers have failings of complexity, limited application, and breakdowns? Yes! But, remember that for many years the drivers of malfunctioning tractors, farm trucks, and other machinery earned the sneering catcall, "Buy a horse!"

Major Things for Governments and Businesses to Do

- Government: Make robot shop part of the core curriculum in high schools.

- Government: Levy a modest (but noticeable) tax on pesticide use to favor robo weed and bug control.

- Government: Much like the X-Prize for the first two launches to space within 48 hours, offer a series of prizes for the best and most affordable robo equipment for managing a particular farming task.

- Business: Develop those robotic systems for planting, monitoring, tending, and harvesting.

What You Can Do

- Make life more relaxed by purchasing your robot vacuum cleaner, lawn mower, and other commercial robotic appliances that come on the market.

- If you are technically oriented, buy and use your own 3D printer.

7.6 Controlled-Atmosphere Agriculture: Oases, Hanging Gardens, Greenhouses, Vertical Farms, and the Shoebox Paradox

Babylon... the hanging gardens are called one of the Seven Wonders of the World.

—Geographies, Book 16, Strabo, 7 B.C.

There is an unusual sea in the south of Spain that is not a sea. *El Mar de Plástico*, or the Sea of Plastic, is a 300-square-kilometer (120-square-mile) complex of greenhouses that supplies Western Europe with 3 million tons (2.7 metric tons) of fresh vegetables per year. There is also a byproduct waste stream that could support culturing of mushrooms, feeding minilivestock, and biorefining (see 7.10Biochar, Remaining Materials, Extracted Juices, and Leaf-Protein Concentrate in Biorefineries). Besides the efficient food production, the white plastic roofs also reflect enough sunlight to cause a noticeable cooling effect in the region.

For workers (who are mostly immigrants), the temperatures are high, and the pay is low. Conditions are like something from a Charles Dickens novel. Yet, the Sea of Plastic is a harbinger

7 Food, Fear, and Climate Fixes from the Land

of the things to come. Plastics, servo controls, and robo farm workers are starting a great new revolution in production from controlled-atmosphere agricultural environments.[243]

It started with irrigation in places such as the Nile delta of Egypt where a little digging during the spring floods on the Nile River could multiply the well-watered crop land and resulting harvest. Irrigation in Mesopotamia, the area of present-day Iraq and Syria, had a more difficult time than the dependable Nile. The Tigris and Euphrates Rivers were low in drought years or roaring floods in over-wet years. Still their irrigation systems fed great empires, and possibly led to one of the Seven Wonders of the ancient world—the hanging gardens of Babylon with flowers and trees surrounding and above a great courtyard. Visitors who had passed through deserts, found the gardens an awe-inspiring site especially because they were high above the courtyard.

For the Romans, it would have still been impressive but well within their capabilities. By the mid-1600s, the French could have built the famous gardens in Versailles on multiple levels at some significant (but not unreasonable) expense. The pumps used for the fountains would only have needed upgrading. Today, high-rise buildings often sport rooftop gardens.

But there is another aspect to pumping water inside a structure; the structure can protect against cold, dryness, weeds, insects, and plant diseases.

Ever since there has been oiled paper, glass, and (more recently) clear plastics, there have been greenhouses or hot houses to produce longer or even unlimited growing seasons. Indeed, a major feature of the Versailles gardens is the *Orangiere*, a king-sized greenhouse to which potted plants (particularly sweet oranges) are moved when fall frosts are approaching. (This is an early example of a greenhouse-controlled environment agriculture.)

However, greenhouses have generally been limited to more expensive and hard-to-ship items such as tomatoes, greens, flowers, and berries. These crops are at a premium in cities, but so too is the cost of urban land. Meanwhile, there is always the competition of nearby farms—the 20th Century term "truck gardens" refers to the fact that internal combustion trucks and refrigeration extended "nearby" to tens, hundreds, or even thousands of miles away.

Three ways to get more produce out of the expensive urban land are artificial light, special tender care, and higher levels of … yes, carbon dioxide. However, as the environmentalists say, there is no such thing as a free lunch. The light costs expensive electricity, and the tender care costs more labor (a high-cost item), and the high levels of carbon dioxide are not free and present safety issues in controlled greenhouse spaces.

Advances in technology are dealing with these costs to a point where there will be a revolution comparable to the first ancient Egyptian farmers who cut channels in the banks of the Nile River to begin irrigation. First, the basic 20th-Century in servo systems reduced the workforce needed—timers to switch lights on and off, thermostats to control temperatures, water sensors to determine when to open and shut the faucets.

---

[243] "Reimagining Almeria's Plastic Sea of Greenhouses: An Industry in the Spotlight," Association of Vertical Farming website, April 2, 2018. https://vertical-farming.net/blog/2018/04/02/reimagining-almerias-agriculture/ (accessed June 14, 2020)

Now, the 21st-Century advances in robo farm workers mean that a small work crew of human technicians can maintain the robots to get that tender loving care yielding the most production per unit surface. Meanwhile, light-emitting diodes (LEDs) have become more efficient and cheaper than fluorescents. Moreover, their color output can be adjusted to match the best color bands for plant growth, even varying for different plant species being grown. Furthermore, solid-state radio-frequency transmitters have allowed plasma lights to provide strong competition to the LEDs.[244] These lighting improvements makes the production per unit of land area much greater at lower electrical cost.

Most importantly, the use of artificial lights means one is not limited to one level of cultivated area. The one area can be multiplied many times when stacked vertically. That thought has led to a number of entrepreneurial ventures in stacked greenhouses several stories high using water flow (hydroponics) or mist flow (aeroponics) to avoid the weight of a full soil structure. They, along with the aforementioned robotics and LEDs allow multi-story farms, also known as vertical farms. The best summary of this comment is still the quirky book, *The Vertical Farm: Feeding the World in the 21st Century* by Dickson Despommier.[245]

A lesson in innovative irony is that Despommier was not an agronomist but a lecturer teaching obscure pre-med classes to aspiring doctors. As an extra-credit exercise to peak his students' interest, he asked them to find a way for New York City to feed itself. The first idea of rooftop gardens was examined and then rejected—not enough roof area. The assignment was more difficult than Despommier had imagined. Increasingly desperate intellectual exercises eventually led to the concept of vertical high rises with production using flowing nutrient-rich water (hydroponics) or a mist version of that water (aeroponics).

The new concept would trade dependence on the whims of rain, drought, heat, and cold for the steady crush of electric bills, costs for robotic systems, and high labor intensity. Those cost factors drove the initial utopian dreams toward higher priced crops with shorter growing times (such as salad greens instead of tomatoes) with the hope of working down to lower priced items and up to the higher production costs of more complex crops and flowers.

The wild ideas coalesced into a 2010 book that was all theoretical, but the idea sparked a craze. By the time Despommier published a revised edition in 2011, vertical farms had been built in England, Holland, and Japan, while others were in the planning stage.[246]

---

[244] Jacob A. Nelson and Bruce Bugbee, "Economic Analysis of Greenhouse Lighting: Light Emitting Diodes vs. High Intensity Discharge Fixtures," *PLOS One*, June 6, 2014. https://journals.plos.org/plosone/article?id=10.1371/journal.pone.0099010 (accessed Dec. 3, 2020)

[245] Dickson Despommier, *The Vertical Farm: Feeding the World in the 21st Century*, Picador, 2011.

[246] Ian Fraser, "High-Rise Greens: Growing Crops in the City without Soil or Natural Light, *The New Yorker*, pp. 52–59, Jan. 9, 2017.

7 Food, Fear, and Climate Fixes from the Land

The concept has become a media darling with stories about various startups,[247] and production of green leafy vegetables seems to be earning enough for operations to be expanding in 2021.[248] An article entitled "Is Vertical Farming Really the Future of Agriculture?" recounted that an indoor farming company had raised $200 million in funding and hoped to extend operations from greens to tomatoes and strawberries.

However, the article went on to raise the questions of whether indoor farms can increase efficiency of lighting, automated equipment, and labor enough to be more than boutique suppliers for well healed socially conscious customers.[249] Furthermore, the concern about energy costs limiting vertical farms to high-cost boutique foods has continued even as many vertical farms have become established ventures. Widespread use of vertical farms may only work with low electricity rates—much like water desalination.[250]

One approach to achieving that efficiency is developing a mutually beneficial (symbiotic) use with some high-rise structure with their own revenue streams but some less used (hence, more affordable) areas.

That leads to another source of less expensive land area—well no; call it "volume and surface." Ever since affordable steel structural skeletons and Otis' elevators made high-rise structures practical in the late 1800s, many architects (including Frank Lloyd Wright and Buckminster Fuller) have proposed much larger buildings, megastructures. Fuller proposed several variants of a pyramid structure. He proposed essentially an outer surface of apartments or condominiums so that the residents would have a view and a small green yard and/or patio. Fuller's designs tried to fill the inner no-view low-value voids with shopping malls, theaters, and schools.[251]

That was a classic architect/futurist arm-waving response to a problem with, well, people. People don't want to live buried in caves; they get claustrophobic. They want a room with a view. Sad to say, most developers just give up and make bizarrely thin, narrow skyscraper structures with the maximum surface area. The central issue is really a surface issue. Hence, large structures

---

[247] Meagan Flynn, "The Promise of Indoor, Hurricane-Proof 'Vertical' Farms, *The Atlantic*, Feb. 12, 2018. https://www.theatlantic.com/business/archive/2018/02/vertical-farming-houston/552665/ (accessed June 11, 2020)

[248] Dave Borlace, "Vertical Farming: Growing fast! -But Are Green Leafy Vegetables All They Can Do?" *Just Have a Think* (YouTube channel), no. 135, Jan. 3, 2021. https://www.youtube.com/watch?v=zCoC5o6x_Sw&feature=emb_logo (accessed Jan. 5, 2021)

[249] Steve Holt, "Is Vertical Farming Really the Future of Agriculture?" *Eater*, July 3, 2018. https://www.eater.com/2018/7/3/17531192/vertical-farming-agriculture-hydroponic-greens (accessed Jan. 7, 2019)

[250] Jess Shankleman, "Vertical Farming's Success Depends on the Cheapest Lightbulb," Bloomberg, Jan. 15, 2021. https://finance.yahoo.com/news/vertical-farming-success-depends-cheapest-050017419.html (accessed Jan. 17, 2021)

[251] R.B. Fuller, "Triton Floating City," undated (1960s). https://www.behance.net/gallery/2971307/Richard-Buckminster-Fullers-Triton-City-project (accessed Aug. 23, 2020)

have the high-value surface units and the considerably less valuable interior spaces that are often empty. Developers have referred to this phenomenon as the "shoebox effect" or the "shoebox paradox." Another use, demonstrated by a number of Embassy Suites hotels (and on a much larger scale by the Westin Bonaventure in Los Angeles, California), is to have a large atrium of trees and fish ponds for an interior vista to be seen from guest rooms, restaurants, and lounges.

Controlled-atmosphere agriculture (including vertical farms) is one way to fill that shoebox void in tall structures with a major potential revenue source. It can also help save the world with a number of side benefits. Vertical farms can

- Provide fresh high value produce such as tomatoes, greens, sweet peas, and flowers.
- Use off-peak electricity to smooth the demand curve.
- Provide part-time employment for young and semi-skilled workers.
- Lend themselves to associated sprout culture food production (see 7.7 Eat Your Little Veggies—Sprouts)
- Lend themselves to associated fungi food production (see 7.8 Magic Shrooms—Bacteria and Fungi Alone Could Feed the World!)
- Lend themselves to minilivestock (see 7.9 Minilivestock: Chickens Down to Bugs, McBugs, McCrawleys, & McWigglies—Grazers of Small Prairies and Small Ponds) to use the scraps of stems, leaves, and damaged produce that can be an expensive waste stream … or another feed stream for the minilivestock.

Naturally, such vertical greenhouse ventures would be more practical (read competitive) if global warming becomes severe. However, they already have a marketing niche in hot areas with high money (Persian Gulf states and parts of Australia), low temperature (Arctic regions), or total lack of atmosphere (outer space). These ventures have found a niche and will doubtless find more.

The advantages of controlled-atmosphere agriculture are that they can produce much more food per unit land area while using less water, fuel, fertilizer, and pesticides. These advantages can translate into less global warming and/or continued food supplies even if the world were to experience seriously hotter temperatures.

There are only two things slowing adoption of vertical greenhouses and the many other exotic technologies—competition from the always-improving mainstream agriculture and competition from cheaper faster transportation. Which technologies will win and when? Time will tell.

Major Things for Governments and Businesses to Do

- Government: Contract for mixed use structures of food production, commercial venues, and living units in remote-locations (such as North Slope Alaska and Antarctica) living spaces and internal vertical farming facilities.

- Business: Controlled-environment agriculture is definitely the place that needs robotic monitoring and work to make a profit. Develop, market, and/or use those roboes.
- Business: Build some modest variants of Motel 6 pyramids with power, indoor beach, and vertical farms below. These would be for citing in areas of particularly cold climate or hot-and-dry climate.

What You Can Do
- For gardeners, build your own greenhouse. It can be hydroponic, aeroponic, or conventional with ordinary soil.

7.7 Eat Your Little Veggies—Sprouts

Wanted! A vegetable that will grow in any climate, will rival meat in nutritive value, will mature in 3 to 5 days, may be planted any day of the year, will require neither soil nor sunshine, will rival tomatoes in Vitamin C, will be free of waste in preparation and can be cooked with little fuel ....

Dr. Clive M. McKay, arguing for greater use of sprouts during World War II[252]

Creatures that eat meat are carnivores, those that eat plants are herbivores, and those that eat both (for example, people, pigs, and dogs) are omnivores. In recent years, environmentalists have proposed that people eat locally produced food; hence, such people would be "locavores."

Skeptics of such a policy have noted that fresh vegetables are only grown in certain seasons of the year. At other times, people must get shipped-in produce or do without. Moreover, urban areas do not have sufficient land area at any time to support the produce needs of their people.

However, there is already an active market in a class of vegetable crops that can be grown within days in small non-soil areas—sprouts. The initial sprouts from various seeds (often requiring little or no light) have a wide variety of flavors and a wide variety of uses—from salads, to cooking by themselves, to casseroles. They can serve every vegetable function from taste to health.[253]

Indeed, during the Indian famine of 1938 to 1941, low food supply led to malnutrition conditions such as scurvy. Unintentional experiments were run on the efficacy of sprouts. Because of the approaching and then occurring war, colonial authorities could not arrange a massive grain aid effort to India like that of the 1960s, which prevented a famine during that later time. However,

---

[252] Quoted in "Sprout History" in ISGA Internationals Sprout Growers Association website. https://isga-sprouts.org/about-sprouts/sprout-history/ (accessed Dec. 18, 2018)
[253] Steve Meyerowitz, *Sprouts: The Miracle Food: The Complete Guide to Sprouting*, 6th Edition, July 1998.

they allowed educational programs to encourage growing and eating of sprouts. These educational programs greatly reduced symptoms of malnutrition. When the programs slackened or were dropped in any particular province, malnutrition returned. Conclusion: sprouts in a human diet have a major supporting effect even with a restricted caloric intake ... and they taste um-yum good!

The lesson of history is that people have tended to return to more conventional food when it becomes more available. Sprouts have been more labor intensive, and the green varieties have needed expensive light. Again, better robotics and more efficient lighting (especially light-emitting diodes, LEDs) may make sprouts more affordable. For instance, it has been suggested that bamboo sprouts could be multiplied by culturing cells—micropropagation.[254]

The sprout markets are marginal today. However, they will probably secure wider adoption under the same more constrained environments such as those of severe global warming, (conversely) global cooling, massively urban areas, or ... outer space.

Major Things for Governments and Businesses to Do

- Government: Provide a prize to the company that provides a functional sprouts-growing appliance for use at sea, in cold climates, and in space.
- Business: market a sprouting appliance that could be set to periodically water, shake, check, and report sprouts are ready to eat or any problem.

What You Can Do

- Try some recipes with sprouts at home and in restaurants.
- Buy a sprouting kit and a sprouting manual to produce your own gourmet sprouts for salads, soups, or casseroles.

7.8 Magic Shrooms—Bacteria and Fungi Alone Could Feed the World!

Fungi (or funguses as you please) are neither plants nor animals but they look like plants to us, including mushrooms (which is French for fungi).

---

[254] P. Nongdam and Leimapokpam Tikendra, "The Nutritional Facts of Bamboo Shoots and Their Usage as Important Traditional Foods of Northeast India," *International Scholarly Research Notices*, 2014: 679073, July 20, 2014 https://www.ncbi.nlm.nih.gov/pmc/articles/PMC4897250/ (accessed June 8, 2019)

7 Food, Fear, and Climate Fixes from the Land

Back in the dark days of 1975, when the 1972 Club of Rome simulations[255] and many agricultural experts predicted eminent worldwide famines, a scientist from Uppsala, Sweden proclaimed that bacteria and fungi could feed the world twice or three-times over. The inputs to produce high-protein and often highly prized foods are often waste streams such as wood chips, grass cuttings, or straw (dried grass cuttings), yet there are many more. Fungi are an ancient, powerful, and anything but simple form of life.

Farmers in Europe, Asia, and the Americas have cultured mushrooms since at least Roman times. However, the best production came from limited areas such as caves and unused mines that could easily maintain cool damp conditions favored by mushrooms. Even in these places, the mushrooms could be attacked by other fungi and insect pests. Thus, mushrooms were an expensive food.

Only in the mid-1900s did mushroom farmers began to use controlled-atmosphere structures that could be built anywhere. Such mushroom-culturing structures do not have the transparent walls and ceilings of greenhouses to get maximum light. Instead, they are the opposite. They are closed against light, insulated, tightly sealed and maintained at thermostatically controlled temperature and humidity to provide the steady cool climate comparable to caves.

As with greenhouses, labor costs are high. However, the U.S. Department of Agriculture is encouraging producers to start mushroom operations. That is especially so because mushroom operations can be run on small areas of land, mushrooms can grow on waste products such as straw or old logs, and startup costs are low. Referring back to the vegetarian trend, mushrooms are a high-protein non-meat food.[256] (Subversive Note: mushrooms also go well with meats.)

In the wider arena of fungal foods, there are hundreds of ways around the world to make food more digestible, tastier, and/or more nutritious. Such fungi-processed foods include milk cheese, yogurt, soy cheese, soy sauce, bulgur (fermented grain), sauerkraut (fermented cabbage), and kimchi (fermented cabbage, and other vegetables).

Another major potential fungus food source for humans is yeast. Yeast takes any source of sugar, from fruit juice and sugar cane to grains that have been cracked and cooked (malted) and makes alcohol—as has been done from the wines and beers of ancient Egypt and Mesopotamia to today. The leftover yeast is high in protein and has been used for cattle feed for decades.

Not as commonly, people have eaten protein produced from bacteria and yeasts feeding on petroleum, methanol, sulfite liquor, or … dung. During World War II in the 1940s, the Germans used waste sulfite liquor from wood pulping to make artificial (or ersatz) protein, which was referred to as "beefsteak yeast." Its use did not continue after the war, but the U.S. Forest Service

---

[255] Donella H. Meadows, Dennis L. Meadows, Jørgen Randers, and William W. Behrens III, *The Limits to Growth; A Report for the Club of Rome's Project on the Predicament of Mankind*, New York: Universe Books. ISBN 0876631650. (accessed Nov. 26, 2017.

[256] A. C. Shilton, "Get Ready for Mushrooms to Take Over the World," *Outside*, Jan. 18, 2018. https://www.outsideonline.com/2273891/how-mushrooms-can-create-better-future (accessed June 15, 2020)

released a document suggesting that lumber mills use the process instead of dumping the waste sulfite liquor into nearby rivers and streams[257]—a practice apparently common at the time.

However, there were fears that either farm production could not grow fast enough or that a major new environmental crisis (possibly a nuclear war) might cause shortages. Hence, there were a number of development programs. In the 1960s, researchers at British Petroleum developed a "proteins-from-oil process" whereby paraffin (an oil-refinery byproduct still used in candles) was fed to yeasts to produce single-cell proteins, and this high-protein yeast was used as a feed supplement for livestock.

The old Marxist Soviet Union (which collapsed in 1992) could never produce enough grain and meat. As agricultural expert and sometimes future-food worrier himself, Lester Brown, once quipped, "Marx was a city boy."[258] (Brown was gently echoing W. W. Rostow's prescient and strident essay from 1955 indicting socialist agriculture long before Mao's "Great Leap Forward" famine.)[259]

Still, the Soviets were innovators, particularly with industrial processes. They built factories to process petroleum byproducts into artificial feed supplements for livestock (um, yum?).[260]

Unfortunately for the consumers, the Soviet Union was something like a company town, only bigger, a company country. Because the government so wanted edible yeast to succeed, they ignored reports about excessively rich components of yeast (certain nucleic acids) that can accumulate in meat and trigger medical problems associated with rich diets).

If the project had continued, they might have found a solution to the excess-nucleics problem, but further development of the edible yeast ended with the collapse of the Soviet Union in 1992. The less socialistic economy resulted in increased grain and meat production, which underpriced the yeasty revolution—the capitalist counter revolutionaries struck again destroying

---

[257] *Food-Yeast Production from Wood-Processing by Products*, U.S. Forest Service Research Note FPL-065, U.S. Department of Agriculture, Forest Service, Forest Products Laboratory, Madison, WS, Nov. 1964. https://www.fpl.fs.fed.us/documnts/fplrn/fplrn065.pdf (accessed Dec. 25, 2018)

[258] Lester R. Brown, "Marx Was a City Boy," *Science*, vol. 209, number 4462, p. 1187, Sept. 12, 1980. https://science.sciencemag.org/content/sci/209/4462/1187.full.pdf (accessed Sept. 18, 2020)

[259] W. W. (Walt Whitman) Rostow, "Marx Was a City Boy: Or Why Communism May Fail," *Harper's Magazine*, pp. 25–30, Feb. 1955. https://harpers.org/archive/1955/02/marx-was-a-city-boy/ (accessed Sept. 18, 2020)

[260] *The Soviet Hydrocarbon-Based Single Cell Protein Program*, United States Central Intelligence Agency, April 1977. https://www.cia.gov/library/readingroom/docs/DOC_0000498552.pdf (accessed Dec. 12, 2020)

another utopian dream! (By 2020, several companies in Russia were developing improved "bioprotein" processes for feed production.) [261]

For the future, the technology problems that Soviet problems could be remedied by developing better processing or breeding of bacteria more suitable for human food, … if a society had the need and the will to do so. As with the all-vegetarian diets, convincing people to make yeast a major component in their diets would require some work.

Capitalism in the United Kingdom (UK) did succeed with factory food. Since 1985, Quorn (spelled with a "qu" instead of a "c" has been sold as a meat substitute in Europe. Quorn is sold as both a cooking ingredient and as the meat substitute used in a range of prepackaged meals. [262]

Quorn foods contain mostly mycoprotein, which is derived from the *Fusarium venenatum* fungus (also referred to as a soil mold) and is grown by fermentation. The fungus is grown in continually oxygenated water in large, fermentation tanks. Glucose sugar and nitrogen compounds are added to feed the fungus, as well as vitamins and minerals to improve the food value. The resulting mycoprotein is then extracted and heat-treated to remove the excess levels of nucleic acids that doomed the ill-fated Soviet food products. Without the heat treatment, purines, found in nucleic acids, are metabolized by humans to produce uric acid, which can lead to gout.

In most Quorn products, the fungus culture is dried and mixed with egg albumen, which acts as a binder, and then is adjusted in texture and pressed into various forms. An alternate vegan formulation replaces the egg albumin with potato protein.

Another approach to factory food has come from methane-eating bacteria (methanotrophs). The bacteria produce single-celled protein that is dried into food pellets. The product has been approved for use as animal feed to replace soy and fish meal. United States-based Calysta opened a plant producing the protein in Teesside in the United Kingdom in 2016, and plans to open a United States facility that can produce as much as 200,000 metric tons (240,000 U.S. tons) of animal feed a year. A Danish competitor, Unibio, is also developing the technology. [263]

Another potential task for yeast is duplicating the power of termites and ruminants like cattle, breaking resistant plant molecules into usable food and other resources. Plants, particularly woody plants, are structurally strong for a reason. They have very complex sugars (particularly cellulose, lignocellulose, and hemicellulose) that can serve as structural materials and are complex enough to resist attack by most herbivores.

---

[261] Vladislav Vorotnikov, "Bioprotein Production Is Perking up in Russia," *All About Feed*, Sept. 16, 2020. https://www.allaboutfeed.net/New-Proteins/Articles/2020/9/Bioprotein-production-is-perking-up-in-Russia-641362E/ (accessed Dec. 12, 2020)

[262] The company explanation of the name an English village with that name. However, there is an obvious marketing advantage of the similarity to "corn," the European term for grain. The company website is at https://www.quorn.co.uk/. (accessed Dec. 16, 2018)

[263] Maija Palmer, "Methane-Based Animal Feed Is More Than Just Hot Air," *Financial Times*, Jan. 31, 2017. https://www.ft.com/content/f520cebc-dbe5-11e6-9d7c-be108f1c1dce
Also see the Calysta and Unibio websites: http://calysta.com/company/ and https://www.unibio.dk/ (all websites accessed Dec. 25, 2018)

Fungi can break all four of these molecules into simpler sugars to make food, burnable fuels, or materials comparable to plastics. Under different culturing conditions, fungi can grow materials with vastly different properties from plastic car parts to synthetic bone implants.[264]

Koji refers to both an ancient fermentation process used in East Asia and Southeast Asia and the *Aspergillus oryzae* fungi that releases fermentable sugar from rice, millet, or soy. The fungi convert the raw sugar and carbohydrates into high-protein materials, such as miso, and the sugar is often used for making wines.

Terramino Foods hopes to use koji to produce a substitute for salmon. The fungus itself is nearly tasteless, but adding some algae supplies the oceanic fishlike taste.[265]

In contrast to fungi food production, food preservation (slowing the actions of fungi and other organisms) could save massive quantities of produce on shelves and perhaps greater quantities of crops in the field. The Irish Potato Famine of the 1840s resulted from a fungus; hence, methods to control or just slow the actions of fungi could have a major impact on agricultural production.

Throughout history, there has been an ongoing battle of farmers against another fungus, wheat rust. (The ancient Romans even made sacrifices to placate the minor but very malignant wheat-rust god.) Again, this is an area of breeding not the fungi, but breeding plants with resistance. This is a function of maintaining the old native species that have resistance mechanisms for almost anything and combining those mechanisms with new high-performing species. This resistance to the rust fungus blight was one of the main thrusts of the Green Revolution that started with a development program in Mexico to reduce heavy losses from rust blight and ended with dwarf wheat and rice that multiplied crop yields.[266]

Another service provided by fungi comes from enhanced soil activity that pulls down atmospheric carbon dioxide and fixes it in the soil. As mentioned in subsection 7.1 (Beyond the Plow: Regenerative Agriculture), this is the network of mycelium tendrils that allows plants and other soil organisms to trade nutrients for a more productive ecology.

Back at the controlled-atmosphere exotic future, extra stems and leaves could provide growing media for mushrooms and other fungi. Moreover, the fungi would not need any light; they would only need to be cooler and sealed off from possible blights.

Now, in the early 2000s, entrepreneurs are developing an extended array of marketable nonfood products from mushroom crops.

---

[264] Richard Webb, "Magic Mushrooms: Far from Being a Load of Old Rot, Fungi Could Save the World," *New Scientist*, vol. 220, no. 2946, pp. 39–41, Dec. 7, 2013.

[265] Adele Peters, "This Salmon Burger Tastes Like the Real Thing (But It's Actually Fungi and Algae)," *Fast Company*, March 17, 2018. https://www.fastcompany.com/40559474/this-salmon-burger-tastes-like-the-real-thing-but-its-actually-fungi-and-algae (accessed Dec. 12, 2018)

[266] Charles C. Mann, *The Wizard and the Prophet: Two Remarkable Scientists and Their Dueling Visions to Shape Tomorrow's World*, Knopf; 1st edition, January 23, 2018.

- Mushroom "leather." production rather than leather from livestock makes more land available for food crops, has a smaller carbon footprint, and requires less chemical processing.[267,268]

- Likewise, it has been suggested that mushroom-based materials could replace many petroleum or natural-gas-based plastics.[269],[270]

- Fungi are also a key input for new construction materials. The two most common organic compounds on Earth are cellulose (the structural material of plants) and chitin (the structural material of crab and shrimp shells. It has been proposed that 3D manufacturing could be used to reassemble small particles of cellulose into a synthetic wood. Furthermore, incorporating chitin into the process can make an artificial product that is considerably stronger and more rot resistant than natural wood.[271,272,273,274] Conceivably, such engineered wood could be produced with a smaller environmental footprint of carbon dioxide emissions and mineral usage than the competing steel and concrete.

- Another construction approach uses the *Trametes versicolor* fungus, which is baked with waste rice hulls and small glass waste particles (fines) to make lightweight,

---

[267] Eillie Anzilotti, "This Very Realistic Fake Leather Is Made from Mushrooms, Not Cows," *Fast Company*, April 24, 2018. https://www.fastcompany.com/40562633/this-leather-is-made-from-mushrooms-not-cows (accessed June 16, 2020)

[268] Mitchell Jones, Antoni Gandia, Sabu John, and Alexander Bismarck, "Leather-Like Material Biofabrication Using Fungi," *Nature Sustainability*, Sept. 7, 2020. https://www.nature.com/articles/s41893-020-00606-1 (accessed Sept. 12, 2020)

[269] Erin Demuth Judd, "Can Fungi Replace Plastics," *Phys.org*, March 12, 2013. https://phys.org/news/2013-03-fungi-plastics.html (accessed May 25, 2019)

[270] Jennifer Nalewicki, "Is Fungus the Material of the Future? *Smithsonian*, April 5, 2017. https://www.smithsonianmag.com/innovation/fungus-material-future-180962791/ (accessed May 25, 2019)

[271] Naresh D. Sanandiya, Vijay Yadunund, Marina Dimopoulou, Stylianos Dritsas and Javier G. Fernandez, "Large-Scale Additive Manufacturing with Bioinspired Cellulosic Materials, " https://www.nature.com/articles/s41598-018-26985-2 (accessed June 15, 2020_

[272] Sid Perkins, "Stronger Than Steel, Able to Stop a Speeding Bullet—It's Super Wood!" *Scientific American*, Feb. 7, 2018.

[273] Jianwei Song, Chaoji Chen, Shuze Zhu, et al., "Processing Bulk Natural Wood into a High-Performance Structural Material," *Nature*,| vol. 554, Feb. 8, 2018. https://www.nature.com/articles/nature25476.epdf?referrer_access_token=5mj_CNBELPticT aKG2veLtRgN0jAjWel9jnR3ZoTv0PCy5i9Q4zwaBEuUCd3sS81sFOkMPY6JV8-cMPMgshGlpqeb8NlAzagnSvNqLykgZSPrOaaiWr4tXCA--_vAP74G4181DeCGB_erHVKkzrMzJO29a0-Ypv1xJxdfurzdlRJ-jdQmm00mGrW3s7BH2_moyJ7Dt5l6-2f9I9XQahxA5eqsJtfOYbSzCk_whj8lEsN8gYnMlWKlrXHP0Ktp3d2&tracking_referrer=blogs.discovermagazine.com (accessed Oct. 2, 2018)

[274] Jeff Spross, "How to Build a Skyscraper out of Wood," *The Week*, Jan. 14, 2019. https://theweek.com/articles/816653/how-build-skyscraper-wood (accessed Jan. 15, 2019)

strong and fire-resistant strong bricks. The bricks are cheaper than synthetic plastics or engineered wood, and their fabrication reduces the amount of waste that goes to landfill. Besides transforming waste materials into salable product, the process prevents the carbon-dioxide emission from burning of rice husks.[275] Australian workers also incorporated glass fines into such fungal bricks to discourage termites and avoid dump tipping fees for waste glass.[276]

- Another service comes from enhanced soil fungi that pull carbon from atmospheric carbon dioxide and fix it in the soil.[277]

Major Things for Governments and Businesses to Do

- Government: Initiate a series of small research grants and prizes for more affordable and larger engineered wood products such as park benches, outdoor decking, telephone poles, and railroad ties.

- Government: Run a Small Business Administration program to subsidize fungi growing operations in both rural and urban areas.

- Business: Start fungi (not just mushroom) operations, growers associations, and a movement for making mushrooms and other fungi a larger part of people's diets.

What You Can Do

- Enjoy more mushrooms in your cooking.

- Sample a few more exotic things such as sauerkraut, cheeses that you've never eaten before such as bulger, yogurt, and (if you are feeling prosperous) truffles.

- And yes, there are mushroom-growing kits available for purchase.

---

[275] Richard Gray, "The Unexpected Magic of Mushrooms," *BBC Future*, March 14, 2019. https://www.bbc.com/future/article/20190314-the-unexpected-magic-of-mushrooms (accessed June 15, 2020)

[276] Tien Huynh And Mitchell Jones, "Scientists Create New Building Material out of Fungus, Rice and Glass," *Phys.Org*, June 20, 2018. https://phys.org/news/2018-06-scientists-material-fungus-rice-glass.html (accessed June 15, 2020)

[277] T. T. Mukasa Mugerwa and P. A. McGee, "Potential Effect of Melanised Endophytic Fungi on Levels of Organic Carbon within an Alfisol," *Soil Research*, vol. 55, no. 3, 2017. https://www.publish.csiro.au/sr/issue/8502 (accessed Jan. 3, 2019)

7.9 Minilivestock: Chickens Down to Bugs, McBugs, McCrawleys, & McWigglies—Grazers of Small Prairies and Small Ponds

Minilivestock encompasses small indigenous vertebrates and invertebrates which can be produced on a sustainable basis for food, animal feed and as a source of income. It includes bush rodents, guinea-pigs, frogs, giant snails, manure worms, insects and many other small species. Minilivestock production is suitable for backyard family production and can contribute to increased food security.

J. Hardouin, É. Thys, V. Joiris, and D. Fielding[278]

People have cultured minilivestock (also called micro-livestock or unconventional livestock) for centuries. Stone age peoples may have cultured snails. Chickens, rabbits, and other small rodents have been raised for thousands of years.

Just as commonly over the millennia, people have eaten various arthropods (including insects). In the *Bible* story of John the Baptist (Mathew 3:4), it says that, "… John had his raiment of camel's hair, and a leathern girdle about his loins; and his meat was locusts and wild honey."

Deniers like to say that John actually ate beans from the locust tree, not locust insects. However, locust beans are a substitute for chocolate while the insect locusts are roughly 60% protein. They were considered a delicacy by the ancient Greeks and Romans, and to this day some Arab tribes in the region cook and eat locusts.

In fact, 80% of the world's population eat arthropods (including insects) of various types.[279] That includes crickets, grasshoppers, termites, and larval worms. Some cynical folks have referred to the restaurant delicacy of lobsters as cockroaches of the sea (expensive roaches I might add), and the even more expensive king crabs look like giant spiders. Insects often blend well into recipes where the cook would otherwise use meat.

The idea of a society-wide movement of adding insects to our diets is not new. In 1885, Vincent M. Holt published a book arguing *Why Not Eat Insects?*[280] Holt argued that people would be better off trading their beef, chicken, pork, and fish—for a diet of beetles, wasps, and caterpillars.

---

[278] J. Hardouin, É. Thys, V. Joiris, and D. Fielding, "Mini-livestock breeding with indigenous species in the tropics, *Livestock Research for Rural Development*, Vol. 15, Article #30, 2003. http://www.lrrd.org/lrrd15/4/hard154.htm (accessed Dec. 18, 2018)

[279] For biologists, insects are one type of arthropod, but the public often blurs the definition to many things being called insects.

[280] Vincent M. Holt, *Why Not Eat Insects?* 1885. (available at http://bugsandbeasts.com/whynoteatinsects/?page=Why-Not (accessed Dec. 18, 2018) Holt's book was cited and commented upon by David George Gordon, "Three Recipes That Could Help End World Hunger," *National Geographic*, July 1, 2014. https://www.nationalgeographic.com/people-and-culture/food/the-plate/2014/07/01/3-recipes-that-could-help-end-world-hunger/ (accessed Dec. 18, 2018)

In 2013, the United Nations released a report suggesting development of an insect-based agriculture.[281] In that report they suggest that farming insects is more efficient, and consequently better than conventional meat sources, such as chicken, pigs or cows because as cold-blooded creatures they use less water and feed. Even better, many of them feed on any available plant material. Yes, many are ruminants—cellulose digesters—much like bigger ruminants such as cattle and sheep.

Entomophagy or insectivory is the human consumption of insects as food. The eggs, larvae, pupae, and adults of certain insect species have been eaten by humans since prehistoric times and continue to be an item of the human nutrition in contemporary times, and that also includes eating grasshoppers, roaches, ants, termites, spiders, and many more.

Being small, insects were hard to corral, so they were never raised on a large scale. However, that is beginning to change.[282]

Um, … uh … yum?

Human insect-eating is common to cultures in most parts of the world, including North, Central, and South America; and Africa, Asia, Australia, and New Zealand. More than a thousand species of insects are known to be eaten in 80% of the world's nations. The total number of ethnic groups recorded to practice entomophagy is around three thousand. Although insect eating is rare in the developed world, insects remain a popular food in many developing regions of Latin America, Africa, Asia, and Polynesia.

In addition, the UN report lumped in worms, such as angleworms, with insects even though they are an entirely different category of life. In fact, they are even more efficient than insects because they are almost entirely digestible rather than having exo-skeleton shells made of chitin.

For all of them, they are several times smaller and faster in growth than conventional livestock so their production areas could be compressed into a much smaller volume than production areas such as a cattle feed lot; hence, the term minilivestock was invented, and one group of Nigerian researchers argued that minilivestock is already a practical food and income

---

[281] *Edible insects: Future Prospects for Food and Feed Security*, United Nations, Rome, 2013. http://www.fao.org/docrep/018/i3253e/i3253e00.htm (accessed Dec. 13, 2015)

[282] Aaron T. Dossey, Juan A. Morales-Ramos, and M. Guadalupe Rojas, *Insects as Sustainable Food Ingredients: Production, Processing and Food Applications*, editors, Elsevier, Inc., 2016.

supplement for households as well as businesses.[283] Entrepreneurs are pioneering food production in such facilities with names such as Aspire, Tiny Farms, and Bitty Foods.[284,285]

Yes, there is a "yuck" factor especially for Europeans and North Americans. However, consider comparable species eaten by some uncouth northern barbarians: lobsters, shrimp, crabs, crayfish, snails, and others.

Um, a little more melted butter, please!

For the less adventurous northern palates, the insect minilivestock can be processed into a high protein flour.[286]

Referring back to the oceanic farming that will be discussed in Chapter 9, the cultured minilivestock of shrimp farms have largely supplanted fishing for wild shrimp. However, a serious environmental complaint about the relatively cheap shrimp-farm production of the past several decades is that the cultured shrimp production is being attained by destroying semi-tropical and tropical mangrove seaside swamps. Those areas of mangrove swamp are one of the most important sinks for absorbing carbon dioxide, a major buffer against surges of too much organic carbon reaching the ocean at one time, and a barrier against waves from hurricanes and tsunamis. The most powerful way to save the mangroves is a cheaper means of shrimp production in other locations—ponds on land can be both freshwater or saltwater.

Speaking of the land, a compressed McBug feedlot would work well as another complement to the controlled atmosphere farms of Section 3.2.10 *El Mar de Plástico* because that area has a major disposal problem with spoiled or blemished fruits and vegetables, as well as the unused leaves and stems—just the sort of rations McBugs would enjoy.

One company that could serve this market is Agritech. Their process crushes organic material into a paste. That paste feeds soldier-fly larvae. The larvae can then replace fishmeal for chicken and seafood farming. In 2018, they secured a hundred-million-dollar capitalization for expanding their operations.[287]

The business model of Agritech can apply for feeding many types of both carnivorous and the omnivorous minilivestock and even conventional livestock. The unused stems and leaves from

---

[283] Tabassum-Abbasi, Tasneem Abbasi, and s. a. Abbasi, "Reducing the Global Environmental Impact of Livestock Production: the Minilivestock Option," *Journal of Cleaner Production*, vol. 112, part 2, pp. 1754–1766, Jan. 20, 2016.

[284] Carl Engelking, "A New Animal Farm: Billions of People Eat Bugs. Entrepreneurs Hope You Will, Too," *Discover*, p. 11, July/August 2018.

[285] Guiomar Melgar-Lalanne, Alan-Javier Hernández-Álvarez, and Alejandro Salinas-Castro, "Edible Insects Processing: Traditional and Innovative Technologies," *Comprehensive Reviews in Food Science and Food Safety*, June 30, 2019. https://onlinelibrary.wiley.com/doi/full/10.1111/1541-4337.12463 (accessed July 2, 2019)

[286] Entomo Farms website, http://entomofarms.com (accessed June 15, 2020)

[287] Louisa Burwood-Taylor, "AgriProtein Raises $105m for Insect Farms," *Ag Funder Network News*, June 4, 2018. https://agfundernews.com/breaking-agriprotein-raises-105m-insect-farms.html (accessed April 7, 2020)

the greenhouse rooms would be not just disposed of but used for grazing by the minilivestock herbivores such as chickens, guinea pigs, snails, crickets, and tilapia, the tropical herbivore fish that has become a major market item. Shrimp and crawfish are omnivores, so they could eat some of both.

Major Things for Governments and Businesses to Do

- Government: Conduct studies of the various minilivestock so that intelligent decisions can be made. Note: In the early 1960s, an entrepreneur started a business of providing predatory bugs to protect grain in storage from bugs that would eat the grain (and there are always such pests); the Food and Drug Administration wanted to arrest him for adulterating the grain.

- Government: Develop a set of best practices for minilivestock regulation in urban areas.

- Business: Invest in the worms, larvae, and full-grown arthropods that can be processed into feed for other livestock.

- Business: Robotics developers, tiny livestock is another area that can only work with robotics.

What You Can Do

- Enjoy some escargot.

- If you are brave, sample some other minilivestock dishes.

7.10 Biochar, Remaining Materials, Extracted Juices, and Leaf-Protein Concentrate in Biorefineries

Another method of storing carbon, increasing production, and soil regeneration is called biochar. Before the Europeans arrived in the Amazon Basin, there were millions of people there. They put biochar and organic matter onto the soil to make and maintain dark rich soil. Those people are mostly gone—probably due to European diseases—but their *terra preta de Indio* (Portuguese for Indian black earth) remains.[288]

This lost Amazonian civilization baked wood and agricultural waste without air (pyrolyzed it) to make a rough charcoal that is now called biochar. They mixed the biochar with manure and kept adding those inputs to the soil. The resulting dark soils have held their nutrients in spite of the tropical heat and rains that often leach tropical soils to rusty barren clay called laterite. Nearly

---

[288] Albert Bates, *The Biochar Solution: Carbon Farming and Climate Change*, New Society Publishers; 2nd edition, 1994.

five hundred years after the Amazonian civilization disappeared, these areas still yield lush crops … and a side benefit is that they continue to draw down more carbon dioxide from the atmosphere.

Modern use of biochar is another option for reducing carbon dioxide emissions to the atmosphere while rebuilding soil, and increasing crop yields.[289]

The applied biochar does not provide nutrients itself, but its vast area of passages provides habitat for beneficial microscopic plants and animals. Also, it helps prevent soils from getting acidic, which reduces crop yields. Finally, for those concerned with warming from higher levels of carbon in the air, biochar remains in the soil for centuries or even millennia. For today's issues of soil building and combatting global warming, biochar could be useful in the following ways.

- It can be combined with waste streams from feed lots for composting into soil fertilizer rather than a flow of toxic waste.

- Does not provide nutrients itself, but its vast area of passages provides habitat for beneficial microscopic plants and animals.

- Holds fertilizer nutrients in its network of pores so that nutrients leach away only gradually, which is better for crops.

- Helps prevent soils from getting acidic, thus helping maintain crop yields.

- Remains in the soil for centuries or even millennia, thus mitigating the increase in atmospheric global-warming carbon dioxide.

- If fashioned into smooth charcoal briquettes, it allows local peoples in non-industrialized countries to have cleaner cooking fires and allows them to leave dung and twigs in the soil for soil building, which also helps mitigate the increase in carbon dioxide.

- Can serve as aggregate to improve the properties of concrete and road black top.

Biochar by itself does not provide highly profitable immediate benefits. The limitation is that biochar systems have been too expensive in energy for transportation and process heat to be profitable except for disposing of waste biomass.[290] Fortunately, there are existing and experimental processes that demonstrate how a number of activities can be ganged to increase profits and climate mitigation. In all cases, some of the carbon, in biochar form, should be returned to the soil for maintaining soil structure and mitigating warming.

---

[289] Paul Taylor, Hugh McLaughlin, and Tim Flannery, *The Biochar Revolution: Transforming Agriculture & Environment*, Global Publishing Group, Dec. 2010.

[290] Kelli G. Roberts, Brent A. Gloy, Stephen Joseph, Norman R. Scott, and Johannes Lehmann, "Life Cycle Assessment of Biochar Systems: Estimating the Energetic, Economic, and Climate Change Potential," *Environmental Science and Technology*, vol. 44, no. 2, pp 827–833, 2010.

- When sugar cane is pressed to yield sugar-rich juice that is refined into sugar, two tons of fibrous bagasse remains for every ton of sugar produced. The remaining pulpy residue (bagasse) is often processed into paper. It can also be burned as fuel to run the sugar mill or mixed with molasses and fermented to make cattle feed. Any unusable material can be pyrolyzed and returned as soil-regenerating biochar to improve the fields.[291]

- Note, pressing juices out before or instead of pyrolyzing increases the number of potential products and reduces processing costs. These are some of the concerns for maximizing biorefinery value chains.

- Sawdust and pressed-out materials can be processed into plastics to replace petroleum inputs. These are better than carbon-neutral plastics because their carbon was drawn down from the sky instead of being burned off as agricultural wastes often are.

- Elephant grass is a tropical forage crop, but it can produce more biofuel per unit of land area than sugar cane. Extracted liquids from elephant grass can be processed for diesel fuel, and the remaining carbon can be processed into briquettes for cooking with much greater productivity than the competing eucalyptus.[292]

- Alfalfa is a high-protein plant that is raised for forage and as a cover crop that fixes nitrogen. Processing plants dry and compress cut alfalfa for use as a high-protein livestock feed supplement, but it is only for ruminants like cattle and sheep because it has a high cellulose fiber content. One way around this limitation is to squeeze out the high-protein juice, coagulate it, and dry it. The resulting green powder is nutritious and digestible for people, pigs, and chickens. It was estimated in the 1970s that an area the size of Texas could feed the world.[293] Also, once again, the high-fiber remainder can be processed for fiber or chemicals, or it can be pyrolyzed and returned to the fields that produced the alfalfa for continued soil regeneration.

- Processes are being commercialized for both mixing plant materials (including sawdust) with plastics to make structural composites and for synthesizing

---

[291] Institut de Recherche Pour le Développement, "Developing Uses for Sugar-Cane Bagasse: Biotechnology Applied to the Paper Industry," Eurekalert! Public Release, Nov. 13, 2006. https://www.eurekalert.org/pub_releases/2006-11/idrp-duf111306.php.= (accessed July 20, 2019)

[292] Mario Osava, "Energy-Brazil: Elephant Grass for Biomass," Interpress Service, Oct. 2007. http://www.ipsnews.net/2007/10/energy-brazil-elephant-grass-for-biomass/ (accessed July 21, 2019)

[293] Barbara Ford, Chapter 6, "It Ain't (Just) Hay," *Future Food: Alternate Protein for the Year 2000*, Barbara Ford, William Morrow & Co., New York, 1978.

biopolymers directly from plant material.[294,295] This is important because roughly 5% of all petroleum used is for synthesizing plastics. Because the plants generate their material by drawing carbon dioxide out of the atmosphere, producing those plastics would be carbon negative.

- Previously described farm operations of mushroom growing and minilivestock all still leave some unusable materials. Anything remaining can be pyrolyzed to biochar for soil building in the local area.

All these techniques can be combined in various combinations to increase productivity for the most food while still mitigating potential global warming.

The disastrous 2020 fires in the U.S. and Canadian West and Northwest forests have demonstrated that the forests are already in crisis. The crisis is only partly from the slight warming and drying that has been observed in the early 2000s. That warming and drying is less than the climate shift that caused the collapse of the Anasazi settlements, and they are themselves less than the probable warming and drying that global warming may cause in coming decades.

We already have a perfect storm caused by decades of few or no controlled burns (as opposed to the first nations peoples in North America who burned regularly[296] and Australian Aborigines who burned regularly[297]), people building in high-fire-risk areas, lax enforcement of brush clearance around those buildings, often Byzantine environmental regulations for brush clearance, and (in the United States) federal government regulations that largely destroyed the forest thinning previously done by the lumber industry.[298] In addition to all that, the still-milder drought than what is probably coming, natural periodic bark-beetle infestation. Only, the natural cycle of beetle infestations has combines with all the other factors of drought and bad management. The failure to clear old growth, bad logging practices where it was still allowed, and those bad

---

[294] Dennis Jones and Christian Brischke (editors), *Performance of Bio-based Building Materials*, Woodhead Publishing, 2017.

[295] Aine Quinn, "Sawdust Might Be One Answer to the World's Plastic Problem," Blomberg, July 21, 2019. https://finance.yahoo.com/news/sawdust-might-one-answer-world-040004916.html (accessed July 26, 2019)

[296] Lauren Sommer, "To Manage Wildfire, California Looks to What Tribes Have Known All Along," *NPR*, Aug. 24, 2020. https://www.npr.org/2020/08/24/899422710/to-manage-wildfire-california-looks-to-what-tribes-have-known-all-along (accessed Dec. 7, 2020)

[297] Vic Jurskis, Roger Underwood, and Neil Burrows, How Australian Aborigines Shaped and Maintained Fire Regimes and the Biota," *Ecology and Evolutionary Biology*, vol. 5, number 4, pp. 164-172, 2020.
http://www.sciencepublishinggroup.com/journal/paperinfo?journalid=231&doi=10.11648/j.eeb.20200504.17 (accessed Dec. 7, 2020)

[298] Katy Grimes and Megan Barth, "California Burns: The 'New Normal' Thanks to Obama Era Environmental Regulations," Canada Free Press, Aug. 6, 2018.
https://canadafreepress.com/article/california-burns-the-new-normal-thanks-to-obama-era-environmental-regulatio (accessed Sept. 24, 2020)

environmental policies made a super infestation that killed millions of trees, making the forests ripe for mega-fires.[299,300]

The latest crescendo of fires has led to many suggestions,[301] and the massive burned-out areas presents an opportunity to implement some of them. For the forests, a Better Green New Deal might have some additions as follows.

- Simplify and ease the restrictions on clearing of brush around structures.

- Enforce requirements for clearing of brush around structures.

- Ease restrictions on lumbering to thin trees and controlled burns rather than waiting for the next apocalyptic infernos.

- Enforce better lumbering practices such as not leaving slash material in the forest and not leaving trees felled trees with bark still on stacked in the woods before winter. (Such felled timber with bark caused at least one major bark-beetle outbreak in the past.)

- Because there are now vast burned-out areas, begin replanting these areas with trees or other plants that are better adapted to dryer and warmer climates than the present species.

- Process unusable slash into biochar for soil building in the forest and for use of conventional agriculture in the fire breaks.

- Plan and maintain wide strips of fire break in the forests with less combustible agriculture, tourist stops, and planned combustion-resistant housing. Wide fire breaks will reduce the risks of in conducting controlled burns. These would all be revenue-generating operations for a government heavily into deficit spending. These areas would also help finance building and maintaining the fire roads.

- The wide firebreaks would also be a good place to develop the potential benefits of co-located agriculture and solar photovoltaic infrastructure (dubbed "agrivoltaics") on food production, irrigation water requirements, and energy production. Photovoltaic panels are expensive, and panels surrounded by the typical gravel surrounding such panels produce a heat-island effect that reduces their efficiency and are susceptible to wind erosion. Co-locating crops among the

---

[299] Andrew Nikiforuk, *Empire of the Beetle: How Human Folly and a Tiny Bug Are Killing North America's Great Forests*, Greystone Books, Vancouver, Canada, 2011.

[300] Chuck Devore, "Wildfires Caused by Bad Environmental Policy Are Causing California Forests to Be Net CO2 Emitters," *Forbes*, Feb. 26, 2019. https://www.forbes.com/sites/chuckdevore/2019/02/25/wildfires-caused-by-bad-environmental-policy-are-causing-california-forests-to-be-net-co2-emitters/#466af2aa5e30 (accessed Sept. 25, 2020)

[301] Edward Ring, "Long-Term Solutions for California Wildfire Prevention," *California Globe*, Sept. 24, 2020. https://californiaglobe.com/section-2/long-term-solutions-for-california-wildfire-prevention/ (accessed Sept. 24, 2020)

panels decreases the heat-island effect and the erosion around the panels. At the same time, shade from the panels reduces water needs and heat stress on the plants. Beneath the photovoltaic panels, experimental systems have used chiltepin pepper, jalapeño, and cherry tomato to demonstrate agrivoltaics in three different dryland environments, beneath the panels.[302,303] Clover, alfalfa, and other nitrogen-fixing cover crops could also be grown near the end of the growing year—again, with mini robo workers to trim around the panel fixtures and deliver green materials for pressing out high-protein liquid and biochar production.

Major Things for Governments and Businesses to Do

- Government: Rebuild the lumber industry with regular thinning by harvests rather than simply cutting fire breaks. Then process waste material into biochar for soil regeneration in the forests.

- Government: Develop and maintain wide fire breaks in the forests with strips maintained for less combustible agriculture, tourist stops, photovoltaic panels and planned combustion-resistant housing. These would all be revenue-generating operations for a government heavily into deficit spending. These areas would also help finance building and maintaining the fire roads.

- Business: Plan and do the initial building for the fire-break communities.

What You Can Do

- If you are a gardener, check with your nursery about fertilizers containing biochar.

- If you live in fire country, cut back the brush around your property per the guidelines from your local fire department. Then cut some more.

7.11 The Unnoticed Frontier: Lawns and Other Turf-Grass Areas

Another set of revolutions is one I call Farming New Kentucky. I submit for your consideration, and possible amusement, a 1988 statistic—probably a larger number now. The area of America's single largest crop—about 25 million acres or 63,000 square miles (100,000 square kilometers) is nearly the size of Kentucky. This new Kentucky was the land area producing … turf grass. Yes, grassy lawns in residential yards, golf courses, parks, and cemeteries embody the

---

[302] Greg A. Barron-Gafford, Mitchell A. Pavao-Zuckerman, Rebecca L. Minor, et al., "Agrivoltaics Provide Mutual Benefits Across the Food–Energy–Water Nexus in Drylands," *Nature Sustainability*, vol. 2, pp. 848–855 (2019). https://doi.org/10.1038/s41893-019-0364-5 (accessed Sept. 29, 2020)

[303] "Benefits of Agrivoltaics Across the Food-Energy-Water Nexus, NREL website, National renewable Energy Laboratory, Boulder, Colorado, Sept. 11, 2019. https://www.nrel.gov/news/program/2019/benefits-of-agrivoltaics-across-the-food-energy-water-nexus.html (accessed Sept. 29, 2020)

largest single crop in America—covering a much greater land area than corn or wheat. Considering that, the authors of *Redesigning the American Lawn* outlined the history and made suggestions.[304]

Landscaping with grass came from northwestern Europe where cool rainy weather makes grassland one of the native sets of plants. Aristocrats in France and England had large areas of land around their great houses ringed with trimmed grass in lawns for showing off the great houses and for socializing. Lawn games developed for play on these areas of turf grass include croquet from France and golf from Scotland.

Meanwhile, towns had their common areas where farmers could graze animals in common and the townsfolk could hold fairs and festivals. Many of these areas eventually became parks, such as the Boston Common in Massachusetts in North America, with their mixed areas of grass, flowers, statures, and picnic areas.

Speaking of North America, settlers from these turf-grass areas of Europe brought these traditions to North America where the areas of turf grass grew even bigger in damp areas that also favored grass. Less logically, many people who settled dry areas often made large impractical investments in adding patches of English lawn in desert areas. In *Redesigning the American Lawn*, the authors argued for a number of modestly practical proposals.

- In desert areas, replace lawns and gardens with desert plants that require little or no water (xeriscaping) and rock gardens because the financial and ecological costs are highest there.

- In areas with more rainfall, replace the "industrial lawn" with the "freedom lawn." By that, they meant reducing the intensity of watering, fertilizer, and pesticides. The intended result would be generally adequate lawns looking not quite as good with a few bare spots.

Many of the proposed changes have come to desert yardscapes. Furthermore, any significant increase in North American temperatures could significantly increase the desert areas in the lower 48 states.

Another proposed improvement is to replace the lawn with a victory garden.[305] Retired veterinary pathologist Kevin Morgan has proposed replacing the lawn grass with garden plants. (Good luck with the homeowner's association and/or the zoning authorities.) He provides a good deal of information on minimal-chemical to organic gardening. Also, he notes the health benefits of actually working the soil and growing some of your own food. Still, unless one is exceptionally hard working and has a really big yard, the economic and environmental effects will be minimal.

---

[304] F. Herbert Bormann, Diana Balmori, and Gordon T. Geballe, Chapter 3, "The Economic Juggernaut," *Redesigning the American Lawn: A Search for Environmental Harmony*, Yale University Press, 1993, citing E. C. Roberts and B. C. Roberts, *Lawn and Turf Sports Benefits*, Lawn Institute, Pleasant Hill, Tennessee, 1988.

[305] Kevin Thomas Morgan, *We Can't Eat Grass: For Food Security Trade Your Lawn for an Ecologically Sustainable Victory Garden*, independently published, May 4, 2020.

Most non-desert areas seem to be continuing their love affair with turf grass. Many home owners will only stop maintaining lush green lawns when their weed whackers are pried out of their cold dead hands. Considering that, might turf grass and other yard plants be used as crops?

Wherever it is located, turf grass is a significant investment in water, fertilizer, pesticides, and grounds-maintenance energy. Even while leaving some clippings on the ground to maintain soil structure, these investments could still yield a huge amount of plant production. Indeed, yard care with composting and leaving grass trimmings on yards without any removal eventually causes the turf to rise over sidewalks and eventually cause drainage problems and ponding on the sidewalks after rains.

Furthermore, a 2012 study estimated that turf grass used three times more irrigation water than the next largest water-using crop. However, the study did note the good result that the turf grass might be a significant carbon sink.[306] A 2015 report increased the estimated turf area in the United States to about more than 63,000 square miles (about 163,000 square kilometers)—nearly the size of Wisconsin[307] (a significant growth in land area since 1988).

Might there be some combination of urban grass and farming? Yes! As noted earlier in this text, researchers in the 1970s estimated that alfalfa grown in an area the size of Texas could provide protein concentrate sufficient to feed the world at that time. Farming another Kentucky would be getting close to a quarter of that, so it's a start.

If America were hungry enough, ambitious enough, and … already had an urban farm infrastructure of fungi raising, minilivestock grazers, and/or the process of breaking cellulose into sugar (saccharification), another vast increase in food could be provided. Thus, some have proclaimed, "Lawn Farming, the Next Big Thing."[308]

So, how might we farm this new Pennsylvania? As in all good conventional land farming, a fraction goes to harvest, but some stays on the land to maintain the soil. Furthermore, the harvest can extend to care and use of trimmings from other yard plants such as flowers, trees, bushes, and

---

[306] Milesi Cristina, Christopher D. Elvidge, J. B. Dietz et al., "A strategy for Mapping and Modeling the Ecological Effects of US Lawns," *The International Society for Photogrammetry and Remote Sensing*, May 7, 2012. file:///C:/Users/VALUED~1/AppData/Local/Temp/A_strategy_for_Mapping_and_Modeling_the_Ecological.pdf (accessed July 5, 2020)

[307] Rob Wile, "The American Lawn Is Now the Largest Single 'Crop' In The U.S," *The Huffington Post*, Aug. 17, 2015. https://www.huffpost.com/entry/lawn-largest-crop-america_n_55d0dc06e4b07addcb43435d?guccounter=1&guce_referrer=aHR0cHM6Ly93d3cuZ29vZ2xlLmNvbS8&guce_referrer_sig=AQAAAC-wtQni9TUtZ-e4BEyW9W50x4bu_s6Uolx6wIIFALPKcHV2W1kpYTqhR5e2tMqpH_MxtViwCyb_hUq2kEDz3TbmBaqk5dJ4falmj-MBkEgLskUQIzyi9JNS5B_1wVKS_04MqcsXEHWGLAa57b91dGtCXx2dnpJ_JmW6txl7_m8Z (accessed July 20, 2019)

[308] Gene Logsdon, "Lawn Farming, the Next Big Thing," *Resilience*, May 12, 2016. https://www.resilience.org/stories/2016-05-12/lawn-farming-the-next-big-thing/ (accessed Dec. 25, 2018)

hedges. Some of the strategies for doing using both farm and ornamental yard care might include the following:

- As noted earlier, researchers in the 1970s concluded that high-protein concentrate from alfalfa grown in an area the size of Texas could feed the world. Gathered turf grass cuttings would have a lower nutrient level, but would still yield an immense harvest.

- Turf grass clippings could provide feed for minilivestock such as snails, tilapia (fish), and various worms and bugs that can be fed to chickens, shrimp, and crawfish.

- Turf grass clippings and trimmed wood could provide growing media for mushrooms.

- Any remaining turf grass clippings and trimmed wood cooked into biochar could regenerate better soil while doing long-term carbon storage.

- More ambitious urban landscaping/agriculture could plant fruit trees and vegetables in otherwise unused back yards. The recent arrival of ubiquitous video cameras and computer monitoring would protect against neighborhood thieves.

This revolution would work using utility-scale landscaping services or landscapers working with a yard-waste-using utility. Such arrangements would have profit centers to reduce costs or even provide payment for properties serviced. At the very least, they would decrease the cost of yard wastes, which are now about 20% of all material going into municipal landfills.

Such urban agrobusiness would be impractical for individual home owners, but it could be implemented by commercial or government enterprises on a utility scale. Such utilities would

- Maintain lawns, bushes, and trees for the property owners as per negotiated agreements with the owners.

- Use the cuttings gathered for some combination of squeezing out juice for leaf protein juice to be concentrated, mushroom culture, minilivestock culture, and biochar for a final product (some of which would be returned to the ground for soil building).

- Grow bushes and trees that could also provide products such as chestnuts, walnuts, maple sap, and wood when removed/replaced. (As a practical matter, property owners allowing some portion of their yard to be used as crop land would get some return from the maintenance company.)

- Use global positioning system monitoring tied to a geographic-information system to tend the crops and generate bills and/or payments (as applicable) for the property owners.

- Negotiate with property owners for small areas of conventional crops in under-used areas such as back yards and unused areas of commercial sites.

Benefits of farming New Kentucky would be as follows:

- Increased food production in general
- Increased locally produced fresh produce and the jobs to go with that production
- Soil building and placement of biochar in the soil to drawdown the carbon dioxide level in the atmosphere
- An incentive for an increasingly verdant park-like setting in urban areas
- For property owners who opt into the system, elimination of the costs for lawn mowers, tree trimming, grass seeding, and assorted yard-maintenance chemicals
- For municipalities, reduction in landfill fees (yard waste and lumber are about a fifth of landfill mass)
- For the environment in general, maintenance of urban landscapes with less chemical use and less polluting groundskeeping equipment. (A two-stroke lawn mower or leaf blower that runs on a mixture of gasoline and oil produces as much pollution as a squadron of cars.[309] Moreover, yard farming for food production and increased storage of carbon, as well as looks, requires a disciplined restrained use of chemical fertilizers and pesticides often lacking from conventional landscaping services and (much worse) the individual home owners who are usually amateurs.

Major Things for Governments and Businesses to Do

- Government: Develop standard regulations of how business and/or government might run a combined landscape/farming/yard-waste-disposal service.
- Business: Offer services of integrated landscaping and agriculture.

What You Can Do

- If you have over-risen turf above your sidewalks because you have been leaving your clippings on the lawn and composting, call a service selling sod and ask for a quote on selling enough sod to make the lawn lower than the sidewalks (plus get yourself a little cash).

---

[309] Nicholas Wade, Cornelia Dean, and William A. Dicke, editors, "Lawn Mower Wars," *The Environment from Your Backyard to the Ocean Floor*, *The New York Times Book of Science Literacy*, Vol. II, The New York Times Company, pp. 262–265., 1994.

7.12 Waste Not, Want Not

> *... If food waste were a country, it would be the third largest producer of greenhouse gases in the world, after China and the U.S. On a planet of finite resources, with the expectation of at least two billion more residents by 2050, this profligacy, [Tristram] Stuart argues in his book, <u>Waste: Uncovering the Global Food Scandal</u>, is obscene.*
>
> Elizabeth Royte [310] referring to Stuart's book [311]

People want to produce and save as much food as possible for themselves and to trade with others. They waste not because of evil intent, but only if they have not yet acquired better methods. We can both increase food supply by reducing waste, and, as a byproduct, we can reduce global-warming emissions.

A classic example of early food waste is in an area now called Havre, Montana. The native people called it *Wahkpa Chu'gn*. They lived among millions of North American bison (which Americans call buffalo), a powerful, larger, and more dangerous relative of domestic cattle. The buffalo could charge in groups at puny humans who were on foot and armed only with spear throwers or bow and arrow. (Horses, which evened the odds for hunters, came later with the European-descended immigrants.)

Being at such a disadvantage, the native hunters devised what might today be called a scandalously inefficient hunting method. A group of hunters in a line would wave blankets and make noises to stampede a herd. On the other side of the herd would be one of the tribe's swiftest runners in a buffalo hide to lure the stampeding buffalo in a certain direction away from the disturbance.[312]

In 1970, I flew in a light plane following close to the ground in the path of the buffalo stampedes. The ground was nearly level with a slight rise going forward. The rise continued until there was a small peak, a gentle slope down, ... and then a cliff.

This was a buffalo jump, a technique used in many places throughout the Great Plains. To produce a photographic reenactment, our plane hugged the level ground, edged up at a slight rise, dove down over a cliff immediately behind the rise, and then swooped back up toward the clouds with the image series complete. Two millennia of buffalo did not swoop back up to the sky. They continued their fall toward the rocky hillside below and died by the tens, hundreds, or even thousands at different times.

The tribe at the buffalo jump would butcher as many animals as they could use and dry and/or smoke the meat to jerky, but if more animals fell than they could use, they left the rest to

---

[310] Elizabeth Royte, "Waste Not," *National Geographic*, vol. 229, no. 3, pp. 31–55, March 2017. referring to *Waste: Uncovering the Global Food Scandal*, W. W. Norton & Company, Oct, 12, 2009.

[311] Tristram Stuart, *Waste: Uncovering the Global Food Scandal*, W. W. Norton & Company, Oct, 12, 2009.

[312] Ojibwa, "Buffalo Hunting on the Northern Plains," Native American Net Roots web page, March 2, 2011. https://nativeamericannetroots.net/diary/889 (accessed Feb. 19, 2019)

rot or be eaten by animals. The tribes had limited capability for storage and only human-carried or human-pulled transportation. Thus, the food waste was massive.

Later, pemmican was invented. Dried meat was mixed with hot grease and various berries to make hard dried loaves that could be sown into leather bags and transported long distances.[313] Pemmican could be transported by canoe and later by horse and then steamboat when these better transports arrived. If this seems strange, think of the old-world variants of pemmican: sausages have been made since ancient Babylonia and Rome—bolognas, frankfurters, wursts, and many others.

The same issues of better preservation methods and better transportation versus wastage and pests have played out endlessly over the ages. The preservation improvements have included smoking, salting, pickling, canning, pasteurization, and refrigeration. Moreover, advancing means of transport created distant markets for preserved items so that they would not be wasted. Transportation improvements grew from the animal-drawn drags, to carts and wagons on finished roads of Roman days, to refrigerated trucks, ships, and aircraft of the present.[314]

John Steinbeck's novel, *East of Eden*[315] shows one advance in food transportation, refrigerated rail cars taking Salinas, California lettuce to New York when lettuce was out of season in the East. Steinbeck's view through a fictional lens shows the potential profit … or loss of such an entrepreneurial revolution. In the novel, Steinbeck's Adam Trask parlays a large fortune into a small one when his good idea fails through no fault of his own.

Despite such real and fictional failures, by the 1920s Salinas had become "The Salad Bowl of the World."[316] Comparable events include mutton from New Zealand, bananas from Central America, and oranges from Florida. Many condemned the "obscene profits of the entrepreneurs who succeeded; few offered to cover the losses of those who failed.

Meanwhile, the equally great revolution of ice boxes and then refrigerators entering homes is a largely unappreciated revolution for the developed world. Home refrigeration allows in-home access to frozen or fresh foods without daily trips to the market and with considerably less danger

---

[313] John E. Foster and Daniel Baird, "Pemmican," *Encyclopedia Britannica*, article May 3, 2016. https://www.britannica.com/topic/pemmican (accessed Feb. 19, 2019)

[314] Sue Shepard, "Introduction," *Pickled, Potted, and Canned: How the Art and Science of Food Preserving Changed the World*, Simon & Schuster, 2001.

[315] John Steinbeck, *East of Eden*, The Viking Press, Sept. 19, 1952.

[316] "East of Eden and the modern miracle of iceberg lettuce," Arable Natural Resources Management website, Nov. 21, 2016. https://medium.com/@ArableLabs/the-modern-miracle-of-iceberg-lettuce-66363282308f (accessed Feb. 26, 2019)

of food poisoning from spoiled food.[317,318] Refrigeration saves major amounts of food from a wasteful rendezvous with the garbage pail.

In fact, the developed world has a margin or cushion of inefficiencies. Roughly 30% to even 40% of food produced is lost to these inefficiencies in the United States.[319] Machine harvesters drop a certain amount of grain; if there were a food shortage, people (or small robotic implements) working along behind could be gleaners and get more of the harvest that was dropped or in awkward corners that the big machines miss.

Of course, the biofuels of bioethanol and biodiesel literally burn food. Soybean oil for diesel engines produces more than five times the fossil energy required to produce it, which is fair. (The process only consists of separating oil from soy meal and some refining of the oil.) However, biodiesel (required by law) is more expensive than petroleum-derived diesel.[320] The biodiesel requirements were enacted because of fears of peak petroleum (now debunked) and fears of excess carbon dioxide causing dangerous levels of global warming—a credible argument unless people get hungry.

Bioethanol (or ethanol produced from sugar cane, sugar beets, or corn, or sorghum) requires much more energy and processing. Besides growing and harvesting, the crop must be cooked to start fermenting, fermented, and then distilled to be concentrated enough to burn. That makes the energy and carbon dioxide balance much poorer. An ethanol plant that produces alcohol fuel and byproduct animal feed produces slightly more than twice the energy in the alcohol fuel than it cost to fertilize the land, grow the corn, transport the grain, cook the grain to malt it for fermentation, and distill out high-purity alcohol.[321]

The study proposed that bioethanol producers might improve the energy return by burning crop residue (the stalks and leaves of the crop left after harvesting). However, stripping the ground bare of this material leaves the soil unprotected from the elements, which causes carbon in the soil to oxidize—putting more of the carbon dioxide into the atmosphere, just what biofuels production was supposed to prevent!

---

[317] Barry Donaldson, Bernard Nagengast, and Gershon Meckle, *Heat and Cold: Mastering the Great Indoors: A Selective History*, American Society of Heating, Refrigerating and Air-Conditioning Engineers, April 1, 1995.

[318] Jonathan Rees, *Refrigeration Nation: A History of Ice, Appliances, and Enterprise in America*, Johns Hopkins University Press, Dec. 15, 2013.

[319] Gabriela Robles, "Waste Not, Want Not," *Food Quality & Safety*, March 3, 2017. https://www.foodqualityandsafety.com/article/waste-not-want-not/ (accessed March 10, 2019)

[320] Dan Charles, "Turning Soybeans into Diesel Fuel Is Costing Us Billions," National Public Radio, Jan. 16, 2018. https://www.npr.org/sections/thesalt/2018/01/16/577649838/turning-soybeans-into-diesel-fuel-is-costing-us-billions (accessed March 10, 2019)

[321] Paul W. Gallagher, Winnie C. Yee, and Harry S. Baumes, *2015 Energy Balance for the Corn-Ethanol Industry, United States Department of Agriculture*, Office of Energy Policy and New Uses, Feb. 2016. https://www.usda.gov/oce/reports/energy/2015EnergyBalanceCornEthanol.pdf (accessed March 10, 2019)

7 Food, Fear, and Climate Fixes from the Land

Biofuel skeptics suggest that the government requirements for use of these fuels is a massive welfare program for farmers. It maintains a massive excess capacity—but good to have in case there were to be a catastrophic drop in food production. Once again, it shows the incredible abundance in the developed world.

Another major food waste in developed countries results from a perceived demand for perfect appearance of fruits and vegetables. Such societies could make a cultural shift to accept more blemishes. Certainly, that would happen if there were an environmental crisis that reduced food supplies.

Meanwhile, there are two immediately available ways to help developed societies while feeding people better and wasting less food in landfills: food banks and minilivestock described earlier.

For food banks, consider that many cultures today have a tradition in etiquette that a certain amount of food should be left on one's plate at meals. This display of "good manners" is a holdover from earlier times when table scraps from the rich were given to the poor. This is not done today because it would spread disease and is demeaning. Fortunately, most developed societies are much richer than in the bad old days. Food banks are a much better policy.

Markets have those slightly blemished items and are often willing to give them away. Food banks gather such cast-off but still good food—called seconds. Variations on food banks have helped people for centuries, and better technologies can make them even more useful. Computer data bases can now coordinate supplies and wants (or lack of wants) much better than before. The emerging technology of robotic deliveries will connect those two much more affordably.[322]

In passing, socialists often attack the idea of food banks. Their main argument is that the charity of food banks (and food banks are often run by charitable organizations) is a distraction from solving the underlying causes of food insecurity—poverty. Poverty should be solved first, and then food security would naturally follow.[323] By that reasoning, the "worse is better" of eliminating food banks would be the logical policy. "Worse is better," was the phrase of the Bolshevik Marxist in Aleksandr Solzhenitsyn's *August 1914*.[324] He wanted German invaders to defeat his Russian countrymen because defeat would bring on the revolution and an eventual better world.

The fallacy of that thinking is that there may be stumbles on the way to the socialist workers' paradise. Indeed, Stalin's Marxist socialist state had many famines. If any comparable

---

[322] Bernard Marr, "5 Major Robotics Trends to Watch For in 2019," *Forbes*, March 8, 2019. https://www.forbes.com/sites/bernardmarr/2019/03/08/5-major-robotics-trends-to-watch-for-in-2019/#4c91a2b65650 (accessed June 18, 2020)

[323] Lynn Anderson, "Sending Surplus Food to Charity Is Not the Way to Reduce Greenhouse Gas Emissions," *The Conversation*, May 23, 2019. https://theconversation.com/sending-surplus-food-to-charity-is-not-the-way-to-reduce-greenhouse-gas-emissions-115685 (accessed May 23, 2019)

[324] Aleksandr Solzhenitsyn, *August 1914*, The Bodley Head, 1972.

crisis were to happen, not having food banks would make things even worse. Far better it would be to continue improving food banks until (but not before!) poverty is erased.

In a non-crisis world, developed countries have tons of produce that go to landfills. Again, this can be ameliorated by the advances in robotics, raising minilivestock, chemical processing, and ultimately biochar for anything remaining. Unusable produce from Spain's giant greenhouse area called the Sea of Plastic is a case in point. Such agro-industrial areas will grow more common, and developing profitable uses for these so-called waste materials is much better than arguing about what to do with the garbage.

Currently, many developing areas are one crop failure away from hardship and two failures from famine. They are working their way toward having problems like the developing world. They are building infrastructure such as roads, grain storage silos, and refrigeration facilities (both commercial and at home). Growing economies will give them more reserves like the developed world to deal with the potential dooms of disasters real or artificial. Hopefully, they will succeed.

Major Things for Governments and Businesses to Do

- Government: Explore zoning options for areas of minilivestock within urban areas
- Business: Consider offering services of integrated landscaping and agriculture.

What You Can Do

- Support your local food bank.

7.13 The Sweet and Woody Solution: Artificial Meat, Fruits, and Vegetables

We shall escape the absurdity of growing a whole chicken in order to eat the breast or wing, by growing these parts separately under a suitable medium."

Winston Churchill[325]

During the Great Depression in the 1930s, a cash-strapped political outcast named Winston Churchill had an incentive to write bold essays on history and even predictions of the future. One of his predictions, artificial meat, has not yet been marketed, but it seems to be on the cusp of becoming a range of marketing products.

In a caustic dystopian counterpoint to Sir Winston, Frederick Pohl and Frederick Kornbluth in the science fiction novel *The Space Merchants*, described a world in which the protagonist

---

[325] Winston Churchill, "Fifty Years Hence," *Popular Mechanics*, March 1932

worked in an unusual packing plant job. He and his coworkers used chain saws to harvest chicken flesh from a mass of growing chicken culture, which the workers referred to as "Chicken Little."[326]

Such a high protein/meat production method seems eminently possible. The food and growing energy of animals would be some type of sugar flowing in their blood vessels providing the energy fuel. Researchers have experimented with culturing artificial meat. It is still impractically expensive ... but then, so were automobiles, trans-oceanic plane flights, telephone calls, and television until innovations made them commonplace.

The production from a given amount of sugar energy would be theoretically a several-fold improvement of production when the "animal" does not need to have bones, teeth, horns, hide, hoofs, barns, pastures, manure disposal, and trucking to market.

By one estimate, cultured meat production might use 7% to 45% less energy, 99% less land, and 82% to 96% less water than traditional methods. For those who worry about greenhouse warming, it would emit 78% to 96% less carbon dioxide than conventional meat production.[327]

In a comparable manner, plants also run-on sugars to produce vegetable and fruit matter containing a number of sugars. Just as with animal tissue, plant tissue can be cultured without the extraneous roots, stems, and leaves.

Of course, competing head-to-head with field-grown grains, potatoes and turnips, and with pasture-grazed animal meat might be difficult. More expensive products, such as the fruit flesh of oranges, kiwis, strawberries, tomatoes, boysenberries, lingonberries, and grapes could become industrial commodities. Likewise, cloth fabrics, such as cotton could be produced—and in various colors without fear of contaminating neighboring field with pink, blue, or any color whatsoever. Likewise, silk or even more exotic fabrics could be cultured.

As with most innovations, cultured meats and vegetable products will probably start with the most expensive materials and then work down to cheaper products as the technology improves. Handily, there are a number of very expensive products.[328] For instance:

- White truffles $15,400 per pound (7000 Euros per kilogram)
- Bluefin tuna $4000 per pound ($8,800 Euros per kilogram)
- Saffron   thousands of dollars per pound

---

[326] Frederik Pohl and Cyril M. Kornbluth, *The Space Merchants*, (first printed by Ballantine Books, 1953) St. Martin's Griffin; Revised, 21st Century edition, December 6, 2011.

[327] Brian Wang, "Lab Grown Meat Will be Served in Restaurants by the End of 2018 and It Will Be Healthier," *Nextbigfuture.com*, March 3, 2018. https://www.nextbigfuture.com/2018/03/lab-grown-meat-will-be-served-in-restaurants-by-the-end-of-2018-and-it-will-be-healthier.html#more-142759 (accessed July 18, 2019)

[328] Zain Kazi, "20-most-expensive-food-items-in-the-world," *Parlo*, July 7, 2017. /https://www.parhlo.com/20-most-expensive-food-items-in-the-world/ (accessed Nov. 16, 2018)

In 2013, Dutch scientist Mark Post of Mosa Meats became the first person in the world to make a beef burger from cultured cattle cells … for $313,000 per pound.[329]

One business startup is working to "catch" bluefin tuna in the laboratory." The entrepreneurs of Finless Foods in the San Francisco (CA) Bay Area are working to culture bluefin flesh. Their first Bluefin product is expected to be a paste that can be used in sushi rolls. Ultimately, cultured Bluefin tuna would protect wild stocks by pricing oceangoing fishing boats out of the water.[330]

Ultimately, one could culture not just food but entire structures in structural materials such as cellulose and hemicellulose or even more exotic molecules. Why nail a house together when you can grow it?

All this goes back to the energy crises of 1973 (Arab-Israeli war), 1979 (revolution in Iran), and early 2000s (peak oil hysteria) (all instances of rising prices before the resulting increased efficiency and then increased production, such as hydrofracturing, drove the prices of oil and natural gas down). During each of these crises, a proposed solution was processing ethyl alcohol (ethanol) fuel out of plant mass, bioethanol, and they all required sugar feedstocks. Moreover, this bioethanol production would theoretically be achieved by pulling carbon dioxide from the air, so cars could be driven without the fear of more global warming from increased carbon dioxide.

Bioethanol could be attained in several ways. First, some sugar sources (such as honey, grapes, sugar cane, and sugar beets) have enough available sugar to easily ferment into wine or beer. Second, grains can be malted by sprouting them or baking them to crack the husks and increase sugar content and then drying in an oven. Both processes have been well known since antiquity. Wines generally have less than 12% ethanol, and beers are generally about 5%. However, both can be distilled into nearly pure alcohol. (Ethanol can never be 100% pure because it pulls water from the air, and that water in the bioethanol often causes problems with oxidation of the engine.) Also, people can drink the distilled products, or they can run automobiles. Therein starts to show the weakness of this approach—the recreational liquid drug sells for a much higher price than vehicle fuel.

Sugar cane has been nearly the ideal plant available for supplying sugar. It grows like a giant grass without annual planting and weeding. It needs only to be periodically trimmed, which is also harvesting. Meanwhile, the roots left in the ground are a net absorber (sink) of carbon dioxide from the sky for those who worry about such things. However, sugar cane is limited to

---

[329] Erin Brodwin, "The Startup behind the Bill Gates-Backed Veggie Burger That 'Bleeds' Is Part of a Transition to Animal-Free Meat — Here Are the Other Frontrunners," *Business Insider*, July 19, 2018. https://www.businessinsider.com/lab-grown-meat-startups-burgers-chicken-fish-plants-cells-2018-7 (accessed Nov. 15, 2019)

[330] Tim Carman, "A New Way to Fish," *The Washington Post*, Nov. 13, 2018. https://www.washingtonpost.com/graphics/2018/lifestyle/cultured-bluefin-tuna/?utm_campaign=WEX_Examiner%20Today&utm_medium=email&utm_source=Examiner%20Today_11/14/2018&utm_term=.ce424cfd895d (accessed Nov. 15, 2018)

hot climates. Sugar beets are also a net carbon sink, and they can grow in cooler temperate climates; however, they are not as productive as sugar cane.

The next feedstock for ethyl alcohol is considerably less practical than sugar cane. Plant grains such as corn (maize) store sugar in the form of starch. At some energy expense, starches can be broken into sugars to make "high fructose corn syrup" among other high-sugar feed stocks. Ethyl alcohol from corn is the ethanol required by law in American gasoline fuel. It is more expensive and less efficient than gasoline, but it provides an excellent way to sop up extra grain production that had reduced grain prices for many years. That is good until lower harvests elsewhere drive grain prices higher still, Of course, the drastic increase in grain prices may have touched off a number of revolutions in the Arab world starting in early 2011.[331] As for saving the planet from global warming or fuel shortage, several analysts have uncharitably calculated that plowing, planting, weeding, cooking starches into sugar, processing the fermentation, and distilling the resulting grain wine may cost nearly as much energy and carbon dioxide emissions as the energy content of the finished bioethanol—spoiled sports.

Worse, fermenting and distillation of sugar cane, grain, or potatoes destroys food that could be eaten by people.

The long-sought goal for cheap and energy-positive bioethanol has been ethanol production from a much larger percentage of plant matter, cellulose, lignocellulose, and hemicellulose. These highly complex sugars provide the woody structure plants. The logic of the biofuels program has been that eventually breaking those complex sugars into simpler ones would allow production of much greater amounts of ethyl alcohol.

Moreover, the feedstocks could be much cheaper—wood wastes such as sawdust and production of otherwise weed plants such as switch grass grown on marginal lands.

The problem is that breaking cellulose and other complex sugars into simpler more usable sugars is very difficult—it is like trying to digest wood. Nature has developed such processes in termite and cattle stomachs, … but it required a development cycle of millions of years. Until now, this cellulose-breaking process (saccharification) has required large amounts of heat and/or caustic chemicals, both of which cost a large fraction of the energy in the fuel produced.

Chemists and microbiologists are working to develop those more cost-effective and energy-effective methods. There are always new hopes—my cellulose cracking favorite of the week uses an infrared laser tuned to the right frequency.[332]

When, and if, they succeed, the first revolutions will not be in energy. As noted earlier, burning something and/or generating electricity are the least valuable uses for materials, and markets have demonstrated that fact. The biomass-to-sugar production industry might first

---

[331] "The Impact of Rising Food Prices on Arab Unrest," *NPR*, Feb. 18, 2011. https://www.npr.org/2011/02/18/133852810/the-impact-of-rising-food-prices-on-arab-unrest (accessed Oct. 25, 2020)

[332] "Towards a green future: Efficient laser technique can convert cellulose into biofuel," *Science Daily*, June 23, 2020. https://www.sciencedaily.com/releases/2020/06/200623111337.htm (accessed June 28, 2020)

generate revolutions in the higher-value products of food, fiber, chemicals, and even construction. As noted above, such sugar-cultured agriculture would cost much less in terms of land, energy, water, and greenhouse-gas emissions. Just teasing out the various chemicals of woody plants may soon provide a green source competitive with petroleum for chemical feedstocks.[333]

Later, there could be more. Sugar-powered energy could replace fossil fuels. If we want sugar power, why bother with fermentation and distillation. All life runs on the electric charging of converting sugar into carbon dioxide. Studies suggest that sugars (or related starches) could be used for fuel cells with as much as ten times the energy density of lithium batteries.[334,335,336]

Major Things for Governments and Businesses to Do

- Government: Continue researching and offer prizes for entrepreneurial advances in cracking complex plant sugars. The revolution of sugar from otherwise unused plants (such as switch grass) has not come, but when it does…

- Business: Keep working to find profitable ways to break the highly complex plant sugars into more usable small-molecule sugars that are more usable. Remember entrepreneurs, government and workers for large companies will go home at 5 o'clock and give up on the dead ends. You could be the entrepreneurs who find the way and make history along with a lot of money.

What You Can Do

- If you like wine or beer, start making your own. The type of fungus called yeast makes it happen.

---

[333] Yuhe Liao, Steven-Friso Koelewijn, Gil Van den Bossche, et al., "A sustainable Wood Biorefinery for Low-Carbon Footprint Chemicals Production. *Science*, Feb. 13, 2020. DOI: 10.1126/science.aau1567

[334] Urba Ziyauddin Siddiqui and Anand K. Pathrikar, "The Future of Energy Bio Battery," *IJRET: International Journal of Research in Engineering and Technology*, vol. 2, issue: 11, Nov.-2013. https://pdfs.semanticscholar.org/2077/4ddc5d9089955a95ed79245fe8e44fbc88a7.pdf (accessed July 20, 2019)

[335] Zhiguang Zhu, Tsz Kin Tam, Fangfang Sun, Chun You, and Y.-H. Percival Zhang, "A High-Energy-Density Sugar Biobattery Based on a Synthetic Enzymatic Pathway," *Nature Communications*, vol. 5, article number: 3026, Jan. 21, 2014. https://www.nature.com/articles/ncomms4026 (accessed July 20, 2019)

[336] Kun Cheng, Fei Zhang, Fangfang Sun, Hongge Chen, and Y-H Percival Zhang, "Doubling Power Output of Starch Biobattery Treated by the Most Thermostable Isoamylase from an Archaeon Sulfolobus tokodaii," Nature Scientific Reports, vol. 5, article number: 13184, Aug. 20, 2015. https://www.nature.com/articles/srep13184 (accessed July 20, 2019)

7.14 Carbonated Fields

The biggest climate concern of the Green New Deal relates to the carbon cycle. We usually think of that as burning and exhaling putting global-warming carbon dioxide into the atmosphere, and then plants draw the carbon dioxide back down. The worry is that the bulge of extra carbon dioxide in the atmosphere from burning fossil fuels could result in more greenhouse warming than would be good for us.

However, there is also a bigger geological component to the carbon cycle. Heat in magma underground drives off carbon dioxide, which also works its way up into the atmosphere. That leaves silicon oxide (silicate) linked to various metals, particularly iron, calcium, potassium, and magnesium. Then, when the volcanic rocks (such as basalt, olivine, and serpentine) reach the cooler surface, these silicates reabsorb carbon dioxide from the air and return back to being carbonates such as limestone (calcium carbonate, $CaCO_3$) and magnesite (magnesium carbonate, $MgCO_3$). These processes happen much faster when the surface is warm and damp so that bacteria, fungi, and mosses can better catalyze the reactions to speed carbonation.

Besides the drawdown of carbon dioxide, carbonation makes the soil more fertile by providing more soluble minerals with potassium and by increasing the pH to less acidic levels. To get that less-acidic condition, there is already a common farm practice of spreading limestone on fields. Volcanic rocks (and steel-mill slag), which are essentially pre-limestone plus potassium, are the main contributors to the highly productive soils around volcanoes. Consequently, spreading volcanic rock might be at least approaching the status of a cost-effective method for drawing down a half to as much as two billion metric tons (0.55 to 1.1 billion tons) of carbon dioxide per year.[337,338]

The limitations on spreading of volcanic rock on fields are primarily financial.

1. As with transmitting sand, lime, and cement for concrete, transportation is a major cost factor increasing cost with greater distance.

2. Mining the volcanic rock deposits is energy intensive, hence costly.

3. The conversion from silicate to carbonate is a slow process that happens only on the surface of the rocks; hence, the rock must be milled to the size of small pebbles or even sand to carbonate in months or years rather than millennia.

4. The carbonation works quickest with higher temperatures and greater humidity, so that tropical and semi-tropical areas would be the best locations for soil carbonation Brazil, much of India, and the American Southeast.

---

[337] "Farming Crops with Rocks to Reduce CO_2 and Improve Global Food Security," *Phys.org*, Feb. 19, 2018. https://phys.org/news/2018-02-farming-crops-co2-global-food.html (accessed July 10, 2020)

[338] David J. Beerling, Euripides P. Kantzas, Mark R. Lomas, et al., "Potential for Large-Scale CO_2 Removal via Enhanced Rock Weathering with Croplands," *Nature*, vol. 583, pp.242–248, July 8, 2020. https://www.nature.com/articles/s41586-020-2448-9 (accessed July 10, 2020)

5. Existing arable crop land would be the best areas to apply this technique because those lands already have infrastructure of placing the rock particles and because such areas would show the greatest increase in food production to help pay for the rock powder.

Cropland carbonation of volcanic rock is another concept that is still theoretical. It should work, but it would be best to have field-scale demonstrations over several years to test delivery methods and observe results for rock mining, rock milling, crop production, and carbon drawdown. Moreover, these demonstrations would be needed in a number of climate types.

Major Things for Governments and Businesses to Do

- Government: Initiate a program of test fields over several years with several agricultural universities in different climate zones to distribute volcanic powder and measure crop productivity over time.

- Government: Issue modest research grants to seek radically new more efficient ways for breaking, scraping, and milling hard rock. This could lead not only to cheaper delivery of volcanic rock powders but delivery of other ores as well.

- Government: In areas with acid soils, explore regulations and safety issues for grinding basaltic rock, steel-mill slag, and concrete waste; then examine the issues for sharing these materials with farmers. It might be considerably cheaper than landfill disposal

- Business: Prepare marketing plans for selling and delivering such materials to farmers.

7.15 Conclusion: People Can Grow Enough Food and Help Cool the Climate

The cornucopia is a symbol of prosperity from Roman days. It is a large hollowed out ram's horn overflowing with fruits and vegetables.

The cornucopia symbol applies to the techniques described in this chapter. Even with severe climate warming, these techniques could still produce food for many times the world's present human population. The trick is to also use them for reducing the net carbon dioxide and other global-warming agents in the atmosphere.

That is even excluding agricultural advances not covered due to page limits. Even more so, it does not include advances not even thought of today.

Hopefully, many of those advances will not be needed because of those many potential advances. Still, there are no guarantees. We might need them all.

In fact, in developed countries, the situation will probably be just the opposite. For decades, these countries have often had embarrassing and expensive excess production. America has had overflowing granaries. Europe has contended with "butter mountains."[339]

Different parts of the world will probably need some or all of the potential from agricultural advances.

---

[339] Nick Faggie, "EU's 'Butter Mountain' Costs Taxpayers £236m," *Express*, Jan 23, 2009. https://www.express.co.uk/news/uk/81314/EU-s-butter-mountain-costs-taxpayers-236m (accessed July 10, 2020)

8 Fixes from Energy and Material

On August 6, 1945, people in Hiroshima, Japan saw a single American bomber approaching. They were not unduly concerned because it was only a single bomber, and Japan had endured attacks by hundreds of bombers at one time in the great war that had stretched across the Pacific since 1941.

They soon learned that the lone airplane was a special bomber carrying the beginning of the nuclear age. A single bomb powered by splitting atoms flattened a major part of the city. That, and a second atomic bomb over Nagasaki, ended World War II.

Now, the power of fission (splitting atoms) is the best hope for providing humanity enough peaceful energy while reducing the danger of global warming. This is so because of four reasons.

1. Mass production of smaller units makes nuclear fission cheaper,
2. Use of next generation design concepts makes nuclear fission cheaper,
3. The progressive increase in global-warming aerosols resulting from use of the competitive fossil fuels makes the need steadily greater, and
4. The limitations of the other noncarbon alternative energies (solar, wind, and geothermal) ... with the caveat that technology could surprise us at any time.

8.1 Energy Power, Limits, Threats, and Possibilities

For maybe as many as a million years, the fire of burning carbon or carbon-containing materials (coal, oil, natural gas, or biofuels such as wood) energy has been the key to human progress.[340,341] In Greek myth, humans were poor, cold, wretched creatures until one of the rogue gods (Prometheus) brought them fire. Fire warmed cold nights and drove off fierce animals. Later, it allowed firing of pottery and smelting of metals. By modern times, fire enabled powered machines traveled on land, water, and sky—even beyond the sky. Lastly, it has allowed fabrication of multiple materials and electronic devices.

Yet, there has also been a steadily rising potential cost from fire. As noted elsewhere in this book, the byproducts of fire (especially carbon dioxide) have been increasing in the

---

[340] University of Toronto, "Evidence That Human Ancestors Used Fire One Million Years Ago," *ScienceDaily*, April 2, 2012.
https://www.sciencedaily.com/releases/2012/04/120402162548.htm (accessed June 30, 2020)

[341] Ryan McKenna, "Fire and It's Value to Early Man," Swarthmore College, update then listed as Feb. 6, 2007. http://fubini.swarthmore.edu/~ENVS2/S2007/rmckenn1/FirstEssay.htm (accessed Nov. 24, 2018)

atmosphere, and they could lead to catastrophic global warming. Conversely, reducing the amount of carbon burning would be a major way to reduce the dangers of global warming.

Unfortunately, the systems for reducing carbon burning all have their problems and limitations. –Michael H. Fox gave an excellent summary of the problems and limitations.[342] However, Reese Palley, an advocate of small modular reactors, summarized them best.[343]

- **Nuclear Fission:** Splitting large atoms such as uranium, is the power for the atom bomb and for nuclear fission power reactors. Despite concerns about nuclear fission, none of the other noncarbon energy systems in this list seem adequate for preventing catastrophic global warming. The following subsection provides details on how nuclear fission has had problems and how those problems might be surmounted to mitigate or reverse global warming.

- **Nuclear Fusion:** Combining four hydrogen atoms into one helium atom is the power of the Sun. Fusion on Earth would theoretically yield infinite power, but the Sun is a million miles across; its resulting giant gravity field compresses the hot reacting fluid mass to allow fusion. Bottling the Sun in Earth's puny gravity field is very difficult. Because the pressure is less, the reaction temperature must be hotter than the Sun … without melting the surrounding structure. Furthermore, the most plausible types of fusion reactors have an immense flux of neutrons, so many materials in a fusion reactor would break down in a few years or even months. Fusion is the power source of the future, … and has been since 1962.

- **Efficiency:** Cars, furnaces, power plants, and every other piece of fire technology (and other equipment using electricity generated by that fire technology) have all grown tremendously more efficient over the centuries. However, it is not enough. Efficiency has slowed the growth of carbon burning but it has not stopped the growth—even after the Paris Climate Accord of 2017. Moreover, increased efficiency costs money for things such as more insulation, double-paned instead of single-paned windows, heat pumps instead of furnaces, and more efficient cars all add to the costs. Consequently, efficiency improvements can help, but they won't solve the problem by themselves.

- **Cheaper Natural Gas from Unconventional Sources:** Cheaper natural gas (methane and CH_4) production from fracking tightly-held deposits have displaced much of the coal combustion. Other unconventional methods (including deep gas and the frozen mixture of ice and methane called methane clathrate might yield considerably more. However, even though methane burning produces only 55% of the carbon dioxide as burning coal for the same amount of energy, and 73% that of oil Switching to methane only slows the increase in emissions of global warming

---

[342] , Michael H. Fox, *Why We Need Nuclear Power: The Environmental Case*, Oxford University Press, New York, 2014.

[343] Reese Palley, *The Answer: Why Only Inherently Safe, Mini Nuclear Power Plants Can Save Our World*, The Quantuck Lane Press, New York, 2011.

carbon dioxide. Also, fracturing the gas-bearing rock and pumping in the fluids involved has triggered minor earthquakes; bigger fracking ventures might trigger bigger quakes.

- **Solar:** Solar power uses that massive potential of nuclear fusion combining hydrogen into helium by collecting the energy from the solar fusion reactor at a fabulously safe distance of 93 million miles (150 million kilometers). The sunlight energy on Earth is free, but the collecting systems definitely do cost. The collectors need stands to hold the solar panels (or mirrors, depending on the system), and the stands are better if they can mechanically track the Sun in the sky. Building and maintaining tracking systems also adds to costs. Furthermore, because the sunlight is spread out (diffuse) after that safe 93 million miles, the collectors must cover large areas. The land is not free, and enough electricity to power a city means that it must be located in some remote area with cheap land, usually far from the electricity users. Spanning that distance requires high-power transmission lines. Both large land areas and long transmission lines raise both cost and environmental issues. Last, but not least, the Sun only supplies energy for a third of the day; the rest requires massive electrical storage or an entire alternate power source. (Note: power satellites in high Earth orbit would receive twice the energy nearly all the time. However, an economy with enough spare energy to position giant power satellites in orbit probably would not need additional energy in the first place.)

- **Wind:** Wind energy is also diffuse, so maximum efficiency has been harnessed by arrays of skyscraper-sized wind mills spread out over many square miles (although you can farm or graze livestock around the windmills). As with solar, these arrays of megastructures tend to be in remote sites requiring power lines to connect to the transformer station and a high-power transmission line to reach the user area. A crucial difference from solar is that the schedule of wind speeds is only an average; the actual wind for harvesting varies considerably, and it cannot be scheduled. Thus, the storage and/or alternate generating capacity must be even larger than solar. Another issue problem is increasing rural resistance to the noise and bird kills caused by windmills.[344]

- **Geothermal:** There is essentially a slow-burning nuclear-fission reactor inside the Earth as radioactive materials such as uranium and thorium give up heat while decaying into smaller atoms. This provides an immense flux of heat from below to the surface. This heat has been tapped for power in some more volcanically active areas. Recent improvements in drilling technology suggest that geothermal energy could become a major power source. However, geothermal energy has its problems. Water and steam from below may contain toxic and/or corrosive materials. Collecting the heat often requires pumping water down into the hot rock

---

[344] Vince Bielsky, "American Towns Don't Want to Be Big Cities' 'Green Energy' Graveyards," *The Federalist*, July 29, 2020. https://thefederalist.com/2020/07/29/american-towns-dont-want-to-be-big-cities-green-energy-graveyards/ (accessed July 29, 2020)

to get steam back up; this might be more water than what is used in fracking, and it may be sent deeper into the ground. Consequently, geothermal energy might generate significantly more earthquake activity than fracking. Furthermore, the returning steam may also carry more corrosive and/or toxic minerals than from fracking.

- **Hydroelectric Power Dams:** Power dams were a major icon of Twentieth Century progress. Today, they are still a major percentage of the "alternative energy" being used. Unfortunately, there are only so many major rivers to dam. Another overlooked issue is that drowned forests and washed-in plant material behind many dams may be bubbling up enough global-warming methane as they decompose to offset the carbon-dioxide emissions that they prevent.

- **Tidal Power:** There are only a few places in the world where the coastal geography focuses tides enough for practical electrical power generation.

Nuclear fission, was never mentioned in the Green New Deal, but it is the best option so far. How to make it work is in the next subsection.

8.2 Problems, Limitations, and Fixes from Nuclear Fission to Compete

There is no such thing as a free lunch![345]

That adage applies to energy sources and how they affect all energies involved with a potential Green New Deal. In particular, it applies to nuclear fission.

8.2.1 Problems with The Best Energy Hope: Nuclear Fission Reactors

Humanity has embraced and then run away from atom-splitting nuclear fission several times. The draw has been tremendous potential stacked against, tremendous fears and tremendous costs. The potential is defined in Albert Einstein's equation of $E = Mc^2$ that translates into energy equals mass times the speed of light squared. Since the speed of light is a large number, a tiny amount of mass yields a tremendous amount of energy. For a simple example, two tablespoons of mass converted to energy could power a cruise ship from America to Europe and back. There are enough materials susceptible to fission to power all human society for thousands of years, but there have been problems obstructing the use of nuclear fission reactors.

1. Uranium fission reactors have been correctly associated with breeding plutonium for those scary nuclear bombs.

---

[345] That motto has been ascribed to both economist Milton Friedman and science fiction writer Robert Heinlein. They both probably took it from a common saying.

2. For that reason, anti-war activists and many environmentalists deliberately associated nuclear fission power reactors with nuclear explosions.[346,347] (In the 1979 disaster-warning movie *The China Syndrome*, some of the movie posters had a bright mushroom-shaped cloud in the background.)[348]

3. Fission reactors have been associated with real and potential nuclear radiation leaks—again, environmentalists have exaggerated the size and medical effects of the leaks. Even crusading environmental journalist George Monbiot issued a shocked denunciation of false and/or unsubstantiated claims.[349] Surprisingly, a slight increase in radiation levels might increase levels of health.[350]

4. Fission reactors were originally developed from government crash programs for building compact reactors for submarines. Then that compact design was scaled up for giant power plants. Like high-performance race cars, these reactors require highly skilled operators. Operators with lesser skills have caused problems and full-blown disasters.[351]

5. Because nuclear fission evolved from a series of government procurements, it tended to have higher costs. This also combined with a safety mandated policy of safety measures as low as reasonably achievable (ALARA); consequently, as better safety techniques were developed, regulatory requirements ratcheted up with more costs and more certification costs for the newly required design changes.[352]

6. Because of items 3, 4 and 5, expensive safety add-ons were put into the fission reactor designs, increasing complexity and the chances of mechanical failures.

7. The potential for catastrophic failure meant that terrorists could attack a fission reactor and cause such a catastrophic failure.

---

[346] Michael Shellenberger, "If Nuclear Power Is So Safe, Why Are We So Afraid of It? *Forbes*, June 11, 2018. https://www.forbes.com/sites/michelatindera/2020/07/16/at-least-36-billionaires-made-six-figure-contributions-to-committees-supporting-joe-biden-in-the-last-3-months/#72af04e77151 (accessed July 18, 2020)

[347] Michael Shellenberger, "How Fear of Nuclear Ends," *TEDx Cal Poly*, Jan. 5, 2017. https://www.youtube.com/watch?v=ml6lzPCmlW8 (accessed July 16, 2020)

[348] *The China Syndrome*, producer: Michael Douglas, 1979.

[349] George Monbiot, "The Unpalatable Truth Is That the Anti-Nuclear Lobby Has Misled Us All," *The Guardian*, April 5, 2011. https://www.theguardian.com/commentisfree/2011/apr/05/anti-nuclear-lobby-misled-world (accessed July 18, 2020)

[350] Ed Hiserodt, *Underexposed: What If Radiation Is Actually Good for You?* Laissez Faire Books, Oct. 31, 2005.

[351] Bernard Leonard Cohen, *The Nuclear Energy Option: An Alternative for the 90s*, Springer, softcover reprint of the original 1990 edition, 1990.

[352] Jack Devanney, *Why Nuclear Power has been a Flop at Solving the Gordian Knot of Electricity Poverty and Global Warming*, Book Baby, 2020.

8. Nuclear fuel might be stolen by other governments or even terrorist groups wishing to develop their own nuclear bombs.

9. Because of items 7 and 8, expensive high security has been required at all nuclear facilities.

10. In 1977, President Carter ordered that United States stop the reprocessing (recycling) of nuclear fuel to harness the unused uranium-235 and plutonium-239 fuel rods to decrease the chance of these fissionables being stolen for somebody else's bomb making. However, that has resulted in larger amounts of highly radioactive waste material that must be stored for thousands of years, and this is an additional cost and safety concern.[353] (In 1957, a nuclear waste storage area in the Russian Urals erupted in a geyser of hot radioactive spray killing several hundred people. In those Cold War days, the Soviet Union managed to keep the accident secret for many years.)[354]

11. Environmentalists used law suits and protests to delay construction on each reactor.

12. Meanwhile, the competing coal-fired and natural gas-fired power plants got steadily more efficient, and thus more cost competitive.

13. Items 1 through 12 led to increasingly large nuclear reactors in an attempt to be competitive, but this gigantism made reactors so large that committing to build a reactor has become a betting-the-company event and periodic maintenance shut downs heroic gambles.

14. Worse, the steadily growing sizes meant new designs, each of which had to be expensively approved, and each of which might have new design flaws that might be discovered at great pain and expense.

15. Finally, the giant size of such reactors meant that each became a 'bet the company" investment that might grievously damage the company if it were delayed or had any kind of incident that required a shut-down.

All those fears were reinforced by four major well-known accidents at nuclear power reactors.

- **Detroit, Michigan, 1966:** a metal object broke loose inside the reactor vessel and blocked sodium coolant from reaching a portion of the fuel, allowing about 1% of the core to melt. A popular book declared that *We Almost Lost* Detroit.[355]. Detroit Edison released two editions of a short report saying, *We Did Not Almost Lose*

---

[353] Nolan E. Hertel, "Pro & Con: Should U.S. Lift Ban on Reprocessing Nuclear Fuel?" *AJC Atlanta News Now*, Aug. 11, 2012. https://www.ajc.com/news/opinion/pro-con-should-lift-ban-reprocessing-nuclear-fuel/lmdDHNcRaX3TYluKycbsxO/ (accessed Aug. 25, 2020)

[354] Zhores Medvedev, *Nuclear Disaster in the Urals*, W. W. Norton, 1979.

[355] John G. Fuller, *We Almost Lost Detroit*, Reader's Digest Press, New York; First Edition, 1975.

Detroit."[356] More sober analysis concluded that the containment dome prevented a release of radioactivity. That said, we still need to remember that the plant was shut down during 4 years while Detroit Edison made very expensive repairs in a highly radioactive area inside the reactor.[357] Most importantly, during 4 years of no revenues from no electricity produced, Detroit Edison was still paying taxes on the plant, insurance on the plant, and interest on the loans for building the plant. Thus, the profits took a major hit.

- **Three Mile Island, Pennsylvania, 1979:** An incorrect water sensor caused the operators to remove water. This caused a partial meltdown. Despite serious concerns of an explosion that could breach the containment dome, the dome held. However, that unit was never run again.

- **Chernobyl, Ukraine, 1986:** The reactor overheated during a test and breached the vessel wall. The design in use had no containment dome, so the reactor core (which contained large amounts of graphite) caught fire and burned for several days spewing a plume of highly radioactive ash. Thirty-one firefighters and workers died, and a plume of radioactive ash spread around the world. The eventual toll from increased cancers may be as high as 4,000. Fifty thousand people were permanently evacuated from the 2,600 kilometers^2 (1,000 square miles) permanent exclusion zone.

- **Fukushima Daiichi, Japan, 2011:** After a historic Richter magnitude 9.0 earthquake, a tsunami wave damaged the emergency power generators at a complex of four reactors. The reactors were not damaged, and they were successfully shutdown. However, they needed active cooling for several days to complete the cool down. Without emergency power to circulate cooling water, the reactor cores of three reactors melted down. There was some venting of steam contaminated with radioactive material, but the containment domes held. There were no casualties at the reactors, but thousands of people were evacuated for several years until the area could be decontaminated, and the reactors are now radioactive scrap that must be monitored and actively cooled for decades.

Those incidents led to a general perception that nuclear fission is more dangerous than other energy sources, but a number of environmentalists have taken the heretical view that nuclear

---

[356] Earl M. Page, *We Did Not Almost Lose Detroit: A Critique of the John Fuller Book: We Almost Lost Detroit*, Issue 2, Detroit Edison, May 1976'
http://michiganintheworld.history.lsa.umich.edu/environmentalism/items/show/104 (accessed July 27, 2020)

[357] J.C. Reindl, "Did We Really 'Almost Lose Detroit' in Fermi 1 Mishap 50 Years Ago?" *Detroit Free Press*, Oct. 9, 2016.
https://www.freep.com/story/news/local/michigan/2016/10/09/detroit-fermi-accident-nuclear-plant/91434816/ (accessed July 29, 2019)

fission is actually less harmful than the alternatives both in environmental damage and in lives lost. Long-time environmental activist Michael Schellenberger summarized those arguments.[358]

1. Nuclear fission kills far fewer people than the particulates of fossil-fuel combustion—several thousand fission deaths that will result from nuclear accidents to date versus 1.8 million per year from fossil-fuel emissions (and not counting possible global warming effects in the future).

2. Solar and wind cost more carbon dioxide emissions to produce each unit of electricity (because of steel and concrete in the construction) than nuclear, so the climate benefit is much less than claimed.

3. Photovoltaics contain a great deal of toxic materials, and there are no plans for preventing these wastes from reaching the environment when the units wear out.

4. When solar and wind produce a significant fraction of the electricity, extra (unusable) power produced during unneeded times is wasted. Consequently, that unused power is worthless. Operators can store with batteries or other means, but it is very expensive. The cost per kilowatt for the customer gets more expensive.

5. Germany, the country that has done the most to switch to solar and wind, has electrical costs twice that of neighboring nuclear-using France.

6. Developed countries can afford such higher energy costs, but developing countries often cannot.

8.2.2 The Basic Lesson on Nuclear Fission Reactors

We shall now go into more detail about how nuclear fission reactors work (and sometimes don't work). Then, we shall examine some proposals to make them work better.

Certain isotopes (varieties of an element with different numbers of neutrons) are less stable than others in a nuclear sense. Atoms of such isotopes tend to split or fission into two or more lighter elements and give up neutrons and energy. These isotopes that naturally split are fissile; and their energy can be used in fission reactors for heat to make electricity while their neutrons cause more fissioning to continue the reaction and make more heat and more neutrons in a sustaining reaction.

To make things exciting, the neutrons can also "breed" other isotopes of other elements, some even more fissile. Most commonly, the uranium-235 isotope used for reactor fuel can breed the more common uranium-238 into plutonium-239, which is a better reactor fuel and a better bomb-making material. (Of the two bombs that ended World War II, one used uranium-235, and the other used plutonium-239). The bomb makers were not sure either would work, so because they were in a war-time crash program, they expensively developed both pathways to a bomb.)

---

[358] Michael Shellenberger, "Why I Changed My Mind about Nuclear Power," TEDxBerlin, Nov. 17, 2017. https://www.youtube.com/watch?v=ciStnd9Y2ak (accessed July 16, 2017)

Building and running practical power reactors is more difficult than exploding bombs. Reactors require a controlled energy source to safely produce energy for electricity or chemical processing. Enrico Fermi, one of the key scientists in developing fission bombs of World War II, also speculated on a sword beaten into a peaceful plowshare, a power reactor. Fermi (leader of the team that operated the first self-sustaining nuclear pile) said that a fission reactor should be designed in such a manner that if it malfunctioned, the workers could each get a cup of coffee, and they could sit around for some time to discuss and decide on the best approach to fix the problem.

Such a safe design was not implemented. There was a tremendous momentum to continue with the existing equipment and experience that had built bombs (isotopes of uranium and plutonium). There was also a tremendous financial stake in that technological trajectory. The tantalizing prospect was that mere pounds of fissionable fuels could replace trainloads of coal.

However, the existing uranium–plutonium reactor at that time had crucial advantages. Besides the financial advantages of already-made investments, starting a self-sustaining reaction with them only requires pulling control rods out from the reactor area to allow a self-sustaining fission reaction to commence. (Control rods are composed of chemical elements such as boron, silver, indium, and cadmium that can absorb many neutrons without fissioning themselves; hence, they damp the flow of neutrons and the resulting fission reactions.) With a technology that went from proof that nuclear fission could be induced on a laboratory scale to working bombs in 4 years, it was only natural that some proponents expected that fission reactors would soon deliver electricity, "too cheap to meter."

Ironically, the phrase originated from a speech by Atomic Energy Commission Chairman Lewis Strauss, in 1954 in which Strauss was probably referring to fusion energy. Strauss could not say fusion power because it was a secret project at the time. By 1955, when word of the fusion project was released, Strauss was much less optimistic: "… there has been nothing in the nature of breakthroughs that would warrant anyone assuming that this [fusion power] was anything except a very long range—and I would accent the word 'very'—prospect." [359]

Whatever the reason for Strauss' exuberance about the ambiguous nuclear power source, many in the nuclear fission field were unhappy about the quote fearing that it would damage the credibility of fission. They were correct.

Nuclear fission power has struggled in competing with fossil fuels since its introduction. The reasons were partly developmental and partly a factor of bomb-making versus power-making. The history is instructive for both restarting the advance of nuclear technology and for those expecting to take any technology from theoretical to research demonstration to practical innovation. Some major design and considerations follow.

> 1. The fuel was a potential fission-bomb component (although not as concentrated). Somebody could divert a few tens of kilograms of fuel and derive the critical mass for several fission bombs, allowing such bombs in the

---

[359] Thomas Wellock, "'Too Cheap to Meter': A History of the Phrase," United States Nuclear Regulatory Commission, June 3, 2016. https://public-blog.nrc-gateway.gov/2016/06/03/too-cheap-to-meter-a-history-of-the-phrase/ (accessed Nov. 24, 2020)

world to "proliferate." Hence, worries about "proliferation" caused major cost increases for security and for materials accounting.

2. The original fission reactors in World War II had fuel rods in between bricks of graphite to "moderate" the speed of neutrons so that they could more efficiently "breed" the more energetic plutonium and other heavier elements. For power reactors that needed to move the heat out of the reactor to be used, water made an excellent heat carrier and cooling fluid. However, a power reactor requires temperatures at least several hundred degrees above normal boiling to make an efficient steam cycle. Those temperatures would boil the water away unless the designers would add a massive pressure vessel around the reactor core allowing the water to remain liquid until it expanded out through the turbine. Such a structure is an expensive addition to the cost of a power plant.

3. Another issue of the pressurized chamber is that overheating can cause the steam pressure inside to rise too fast to be vented, and the chamber could explode spewing out tons of radioactive materials in the surrounding area, and the chamber would be susceptible to accidental plane crashes, sabotage, or artillery attack. Thus, Western reactors were protected with tremendously strong (and tremendously expensive) steel-rebar-reinforced concrete domes. If there were to be an explosion of the pressure chamber, most of the radioactive fuel would be contained within the dome. The domes were designed so that even a commercial airliner crashing into the dome would not penetrate.

4. Even within a pressure chamber, the steam around the fuel rods could not go above certain temperature limits. Fossil-fuel power plants did not have this limitation, and the flame could heat a heat exchanger to a steam cycle at higher temperature. Since power plant efficiency is a function of the temperature difference available, the hotter fossil-fuel combustion plants could get a higher efficiency. Even worse for the fission reactor vendors, fossil fuel plants could use even hotter combustion to "top" the plant with a gas turbine that would get a percentage of the available energy followed by the turbine exhaust heating water to steam for the steam cycle. Thus, fossil-fuel electricity tended to produce electricity more cheaply than fission power.

5. Nuclear plant designers sought to outgrow their fossil-fuel competitors by getting the economies of scale from tremendously large power plants, say on the scale of 1 billion watts (1 Gigawatt or 1 million kilowatts). However, these larger plants caused many of the parts to be special order, and each design was a new design requiring more design cost than standardized power plants. Worse, each one-off design had to get approval and had its own new problems to be worked through. In contrast, the French government used one standard design that was improved over several iterations and that provided three-quarters of French electricity by 1990.

6. All of these issues came to a head in one reactor in the complex of three units in Three Mile Island, Pennsylvania in March 1979. A faulty sensor and other system failures caused operators to reduce the level of cooling water in the reactor. The reactor overheated, and there was a partial meltdown of the reactor core. There were grave concerns that melted fissionable materials would pool at the bottom of the chamber in sufficient mass to cause a runaway reaction and enough heat to breach the containment dome or even cause an explosion. Although the containment dome held with minimal release of contaminated steam, that reactor never operated again, and the fission industry in the United States was set back severely.

7. Another approach to get past the temperature limitation of water was to use molten sodium as the coolant/energy carrier. However, sodium has its own problems. Hot sodium spontaneously ignites in air, and it explodes when coming in contact with water or even just the small amount of water in the structural concrete). Thus, repairs are much more complicated than with a water coolant. The Fermi 1 reactor in Detroit, Michigan was an ambitious reactor design that used sodium coolant and was also a breeder for making additional fuel. In 1966, a baffle (added as a safety feature!) came lose and lodged in coolant circulation system. Repairs required 4 years.

8. A third approach is to use nonreactive gases such as helium or carbon dioxide. These gases can operate at much higher temperatures, say with graphite cladded spheres that would produce gas so hot it could run a gas turbine. However, if the reactor area was ever breeched, the graphite cladding would be hot enough to burn like charcoal.

9. The most painful event was the 1986 explosion at the Soviet (now Ukraine) reactor in Chernobyl. That reactor type had three design flaws: (a) It had no containment dome in case of accident because, of course, there were no accidents in a planned socialist economy, (b) it was unstable at low operating levels, and (c) it had major amounts of graphite moderating neutron flow. When a test at a low operating level went bad, the overheated steam exploded through the reactor chamber and the roof—remember, no containment dome. With the hot reactor exposed to air, the graphite burned white hot for several days. There were 56 deaths (mostly among the firefighters), a plume of radioactive ash that went over a large area of Europe, the evacuation of a third of a million people, and the permanent evacuation of a whole city of 50,000 people. The good news was that the accident was not the end of the world. The bad news was that it was not a small event either. Comparisons with deaths caused by soot and release of radioactive materials from coal show that fission is much safer, but that argument is hard to make versus pictures of the abandoned city near Chernobyl.

10. The most expensive long-term issue is storage of the radioactive waste materials left over that are not usable for fuel. In 1971 Nuclear pioneer Alvin Weinberg said that the nuclear people had made a Faustian bargain with society.[360] On the one hand, society got an inexhaustible source of energy. But the price was guarding the nuclear waste for thousands of years.

After refueling or decommissioning of a fission reactor, the vast majority of the uranium and plutonium fuel has not been "burned," in conventional reactors; it is just no longer quite active enough to run in a reactor. The logical alternative would be to reprocess this spent fuel to remove depleted less radioactive material to make the remainder potent enough to be fuel a in reactor again. Multiple refueling would eventually destroy most of the radioactive material, so storage requirements would be minimized.

Unfortunately, nuclear waste removed after use in a reactor includes a mixture of other radioactive isotopes formed by neutron bombardment while the reactor was running. Those other radioactive isotopes make separating them out difficult, read expensive. Furthermore, reprocessing would create another material stream that could be diverted by those seeking to build bombs. Consequently, the United States banned reprocessing of nuclear waste in 1977. Thus, for the United States, much more waste has accumulated than has been reprocessed. (France and several other countries do reprocess nuclear waste.)

Instead, it was optimistically expected that nuclear wastes would be entombed somewhere for ten-thousand years, swept under the rug.

This under-the-ruggism is dangerously impractical. Who guards the thousand-year-old tombs of the Mayans today? Police are often far behind the tomb robbers of ancient tombs in Central America. Likewise, the Etruscan tombs from three thousand years ago have been looted in Italy since antiquity. And the tombs of the pharaohs—with their false walls, booby traps, and curses upon any who would defile them—have almost all failed to protect their treasures. Looters missed the tomb of Tutankhamen because he was a minor pharaoh who died young. Other nuclear waste disposal schemes are even less practical than entombed storage, as noted here:

- Spreading waste into the ocean for safety by dilution is impractical because some plants and animals can significantly concentrate or biomagnify heavy metals.[361] This is particularly so for seafood production in materials concentration, in which there are several levels of plant plankton (phytoplankton), eaten by plankton animals (zooplankton), eaten by forage fish (such as sardines), which are eaten by larger fish such as cod and tuna. Biomagnification of the heavy metal mercury in

---

[360] Weinberg, Alvin, *The First Nuclear Era: The Life and Times of a Technological Fixer*, New York: AIP Press, 1994. ISBN 1-56396-358-2.

[361] Nehreen Majed, Md. Isreq H. Real, Marufa Akter, and Hossain M. Azam, "Food Adulteration and Bio-Magnification of Environmental Contaminants: A Comprehensive Risk Framework for Bangladesh," *Frontiers in Environmental Science*, May 18, 2016. https://www.frontiersin.org/articles/10.3389/fenvs.2016.00034/full (accessed Nov. 24, 2020)

oceanic seafood has been extensively studied,[362,363] and radioactive wastes would probably increase in concentration in a similar manner.

- Permanent entombment in Antarctic glaciers is not permanent because glaciers are really slow-moving rivers of ice. Waste containers would be carried and possibly ground to powder while flowing with rock and ice to eventually be deposited in the surrounding ocean by melting icebergs.

- Rocketing nuclear waste away from Earth and into the Sun is impractical. It takes more energy to shoot material into the Sun than it does to leave the Solar System. A society with enough available energy to do that would not need nuclear fission. Getting free of Earth's gravity is only a small part of the propulsion task. Like Earth, tens of thousands of radioactive bricks escaping Earth orbit would tend to stay in the vicinity Without tremendous amounts of thrust energy, many of those bricks would eventually drift back into Earth orbit and rain down at several miles per second.

- Pumping material deep underground could be secure for decades or even centuries, and a more recent concept would send retrievable canisters of high-level wastes down a mile (1.6 kilometers) or more and then sideways into shale formations.[364,365] However, Steel pipes and cannisters eventually corrode, especially when next to hot heavy-metal oxides. Within decades, the cannisters would no longer be retrievable. The required storage times of millennia would be even more problematic. Once the canisters corroded to failure, any water penetrating the borehole could be heated to steam. There might be sufficient heat to expel radioactive steam back up at the surface in an artificial geyser. (The Soviet radioactive release in the Urals was such a radioactive geyser in a liquid-waste storage area near the surface.)[366] Such liquid pools could provide nasty surprises to wildcat drillers thousands of years hence. More importantly, emplacement of

---

[362] Raphael A. Lavoie, Timothy D. Jardine, Matthew M. Chumchal, et al., "Biomagnification of Mercury in Aquatic Food Webs: A Worldwide Meta-Analysis," *Environmental Science & Technology*, vol. 47, number 23, pp. 13385–13394, October 23, 2013. https://pubs.acs.org/doi/10.1021/es403103t (accessed Nov. 24, 2020)

[363] Rufus K. Guthrie, Ernst M. Davis, Donald S. Cherry, and H. Edward Murray, "Biomagnification of Heavy Metals by Organisms in a Marine Microcosm," *Bulletin of Environmental Contamination and Toxicology*, vol. 21, pp. 53–61, 1979. https://link.springer.com/article/10.1007/BF01685385 (accessed Nov. 24, 2020)

[364] James Conca, "Deep Borehole Nuclear Waste Disposal Just Got a Whole Lot More Likely," *Forbes*, June 24, 2019. https://www.forbes.com/sites/jamesconca/2019/06/24/deep-borehole-nuclear-waste-disposal-just-got-a-whole-lot-more-likely/?sh=1a817e2c67c8 (accessed Nov. 25, 2020)

[365] "Our Story," Deep Isolation, Inc. web page, Berkeley, California, undated. https://www.deepisolation.com/our-story/ (accessed Nov. 25, 2020)

[366] Zhores Medvedev, *Nuclear Disaster in the Urals*, W. W. Norton, 1979.

nuclear "waste" deep underground would render potential future fuel unobtainable short of a major mining operation.

Planning to sweep tens of thousands of tons of highly radioactive material under the rug is imprudent to the point of insanity. If nuclear fission were to become the majority energy source, the bump under the rug would simply grow too large ... and dangerous. Furthermore, it would be a waste of potential fuel. The fissionables must be reprocessed, fed back into the reactors, and reacted away so that they will not be a menace thousands of years in the future.

Moreover, reprocessing and re-use would increase potential fission energy more than fifty-fold. This is crucial for long-term sustainability. Nuclear skeptics have suggested that a world energy economy based on once-through uranium fission might expend all of Earth's readily available uranium by 2050. Even though this is probably much too pessimistic, a shrinking resource base could increase costs and concerns. With reprocessing and re-use, fission could supply energy for millennia.

8.3 The Return to Atom Splitting Power with Better Next-Generation Nuclear Fission Reactors

Nuclear will make the difference between the world missing crucial climate targets or achieving them. We are hopeful in the knowledge that, together with renewables, nuclear can help bridge the 'emissions gap' that bedevils the Paris climate negotiations. The future of our planet and our descendants depends on basing decisions on facts, and letting go of long-held biases when it comes to nuclear power.

Excerpt from 2015 open letter from James Hansen and three other well-known climate scientists arguing for use of nuclear fission to reduce global warming[367]

For all the reasons given above, humanity slowed development of fission reactors, especially after each of the four well-known accidents. Those abrupt stops and market uncertainty disrupted supply chains and increased financing costs. This all made continued development of nuclear fission power even more questionable.

Yet, as noted above, there are no viable alternatives to fission. The clock keeps ticking on increasing amounts of greenhouse agents in the atmosphere. Many of those who warned about global warming declared that nuclear fission must be improved and applied to prevent catastrophic climate change.

According to a 2018 study by the Massachusetts Institute of Technology (MIT), achieving a low-carbon-emission future at a reasonable cost and minimal social impact requires a mix of

---

[367] James Hansen, Kerry Emanuel, Ken Caldeira and Tom Wigley, "Nuclear Power Paves the Only Viable Path Forward on Climate Change," *The* Guardian, Dec. 3, 2015. https://www.theguardian.com/environment/2015/dec/03/nuclear-power-paves-the-only-viable-path-forward-on-climate-change (accessed Feb. 24, 2019)

power sources, with nuclear power as a major component. The Future of Nuclear Energy in a Carbon-Constrained World by the MIT Energy Initiative (MITEI) says that trying to produce a radically low-carbon economy without nuclear fission reactors would cost two to four times as much as one with.[368]

Thus, as Green activist and now heretic nuclear supporter Mark Lynas has said,

"... *A Green Future Needs Nuclear Power.*[369]

8.3.1 The Background Primer on Next-Generation Nuclear Reactors

As mentioned earlier, the first generations of nuclear fission reactors had (and still have) a number of weaknesses that increase costs and decrease safety. The next several generations are inherently better. A quick summary on nuclear engineering should help here.

- Most of the weight of an element (more correctly mass) is from positively charged protons and neutrally charged neutrons in the center or nucleus of each atom, while the negatively charged electrons have almost no mass at all.

- Nuclear engineering, and the naming conventions for elements and isotopes deal with the neutrons and protons at the center of each atom (the nucleus).

- The chemical properties of an element are determined by its number of protons (which are the same as the number of electrons). The number of protons in an element is called the element's "atomic number."

- However, an element with the same number of protons can have isotopes with varying numbers of neutrons, and the element is referred to using the sum of the protons and neutrons (such as uranium-235 and uranium-238). The isotopes often have vastly different nuclear properties. For instance, the natural uranium-235 isotope (which occurs in less than 1% of uranium mined) is a fissile isotope, so the neutrons it gives off can start and maintain a nuclear chain reaction. The uranium-238 isotope that occurs at around 99% of the uranium in natural ores is only fertile (not fissile)—it needs to be bombarded with neutrons to breed into plutonium-239, which is another fissile isotope.

- The naturally occurring isotope of thorium (thorium-232) is another fertile (not fissile) isotope. If it is bombarded with neutrons, it goes through a number of

---

[368] Jacopo Buongiorno, John Parsons, Jacopo Buongiorno, et al., *The Future of Nuclear Energy in a Carbon-Constrained World*, Massachusetts Institute of Technology, Cambridge, Massachusetts, 2018. http://energy.mit.edu/wp-content/uploads/2018/09/The-Future-of-Nuclear-Energy-in-a-Carbon-Constrained-World.pdf (accessed Feb. 3, 2019)

[369] Mark Lynas, *Nuclear 2.0: Why a Green Future Needs Nuclear Power*, UT Cambridge Limited, Cambridge, England, 2013.

intermediate isotopes and produces uranium-233, which is an even more fissile fuel than uranium-235 or plutonium-239. (Handily, it is highly radioactive and decomposes faster than those two other isotopes, so anyone trying to divert it for bomb making would have difficulties in bomb making, and any bombs made would quickly break down.)

- A neutron moderator is a medium that reduces the speed of ordinarily "fast" neutrons, thereby turning them into slower or "thermal" neutrons capable of sustaining certain nuclear chain reaction involving uranium-235, uranium-233, or other fissile isotopes. Water and the carbon of graphite are common moderator materials. (An added complexity is that some isotopes can work better with fast neutrons and breed more fuel—hence, the fast breeder reactor.)

- A nuclear poison is a substance that has a large capacity for absorbing neutrons, but which does not fission and produce more neutrons. Hence, it decreases the nuclear activity. This can be deliberate with control rods (made of materials such as boron or cadmium) that throttle back or stop nuclear activity. It can be an unwanted side effect as when the fission product xenon gas builds up inside a reactor's fuel rods, reducing energy output and eventually stopping the fission reaction unless the reactor is refueled.

8.3.2 Ways to Develop NexGen Reactors

In *How Innovation Works*, Matt Ridley commented that there have been many proposals for better fission reactors. The problem has been a great inertia born out of the high cost of developing new directions in a risky technology.[370] Nevertheless, the need has been rising and supporting technologies have been evolving.[371]

How might some of the innovative proposals and prototypes be transformed into major innovations? Six key principles follow.

1. Standardize into just a few designs used many times.

2. Get small (no gigantism) for mass factory production, maintenance by replacement, and factory disassembly—much easier than on-site disassembly.

3. Use more advanced systems that produce power more efficiently and that incorporate safety features into the design rather than increased control measures and expectation of infallible operator skills.

4. Use thorium as a fuel.

---

[370] Matt Ridley, *How Innovation Works: And Why It Flourishes in Freedom*, Harper, May 19, 2020.

[371] "Advanced Nuclear 101," web page, Third Way, Washington D.C., Dec. 1, 2015. https://www.thirdway.org/report/advanced-nuclear-101 (accessed June 27, 2020)

5. Use transmutation and reprocessing of wastes to minimize material for long-term storage—no more under-the-ruggism.

6. Design the reactors to be inherently safe.

8.3.2.1 Standardize to Just a Few Designs

After World War II, France began a very earnest program to develop nuclear fission for both weapons and electrical power. French President Charles de Gaulle described the reason for the French nuclear development program as the four no's: no coal; no oil; no gas; no choice.

Possibly because they were hungrier for energy, the French nuclear power program has been more successful than the United States, and nuclear reactors supply three quarters of the French electricity.[372] The French managed to largely avoid the long certification delays and ballooning costs that plagued construction of American fission power reactors.

How did the French succeed? They settled on a few standard reactor designs and built the same design repeatedly, often putting several reactors on a single site. That allowed them to standardize their processes and increase efficiencies. Canada, Japan, and India kept costs relatively stable with similar tactics.[373]

Most nuclear reactor installations in the United States to date have been built as one-offs, or at best several reactor units in one complex. Most of these units or complexes have been different enough from any of the others that the developers had to get the new design certified by federal regulators. That approval process often requires years.

The French model has many advantages. Use a limited number of proven designs to reduce costs.

---

[372] "Nuclear Power in France," web page, World Nuclear Association, updated March 2020. https://www.world-nuclear.org/information-library/country-profiles/countries-a-f/france.aspx (accessed June 19, 2020)

[373] Jessica R. Lovering, Arthur Yip, and Ted Nordhaus, "Historical Construction Costs of Global Nuclear Power Reactors," *Energy Policy*, vol. 91, pp. 371–382, April 2016. https://www.sciencedirect.com/science/article/pii/S0301421516300106 (accessed June 19, 2020)

8.3.2.2 Getting Small (No Gigantism) for Mass Factory Production, Maintenance by Replacement, and Factory Disassembly

The economy of scale is offset by the economy of mass production.

Robert Sadlowe, Oak Ridge Laboratory[374]

In his polemic for small modular reactors, Reese Palley recounted a story of Charles Rolls and Henry Royce who went into the car-making business in 1904. Royce noted that many of the car breakdowns at that time resulted from failures around the few large bolts that held a car together. As Royce was quoted as saying, "…many little bolts are better than one big one." Following that dictum, Rolls-Royce used a number of small bolts so that a failure of a single bolt would not cause catastrophic failure. Consequently, Rolls-Royce cars had a much lower failure rate than their competitors.[375]

A comparable situation exists in the fission reactor industry for slightly different reasons. Reactor makers built steadily larger fission reactors to compete with the low costs of large coal-fired power plants. Nuclear pioneer Alvin Weinberg detested those very large reactor plants and complained about what he called gigantism. Such gigantic plants often provide a gigawatt of power (1000 megawatts or a million kilowatts), which decreases many costs because many costs remain the same even if the plant is many times bigger.

Unfortunately, the gigantic sizes also mean that these reactors also require many specially fabricated large parts available from very few vendors and with waiting times of months or even years. Many highly skilled teams arrive at the often-remote construction site at different times to do various tasks that are usually different from the tasks at other reactor sites. The result is that, as with the early 1900s cars, the costs of building a new nuclear reactor become fabulously expensive. Worse than the cost, because the reactors are often built at remote sites some components are often built by local construction crews, so the levels of experience going into those components may be substandard.

Furthermore, once a giant reactor is completed, it may be such a large fraction of the utility's capacity that the possibility of an accident or just a maintenance shut-down requires a source of back-up power.

Next, gigantic power reactors might fail in gigantic ways. Weinberg commented that a 60-megawatt reactor chamber could contain an overheating situation, but he had grave reservations about a reactor many times larger. Current materials may allow that size limitation to be greater, but the principle is the same.

---

[374] Quoted in Chapter 7, "The Answer," in Reese Palley, *The Answer: Why Only Inherently Safe, Mini Nuclear Power Plants Can Save Our World*, The Quantuck Lane Press, New York, 2011.

[375] Also, Chapter 7, "The Answer," in Reese Palley, *The Answer: Why Only Inherently Safe, Mini Nuclear Power Plants Can Save Our World*, The Quantuck Lane Press, New York, 2011.

Finally, gigantic reactors must eventually be disassembled at that same site. Only, for disassembly, many areas of the reactor have become so radioactive that it is dangerous to work with them. In the United States, reactor builders post a large deposit for end-of-life disassembly so disassembly is already paid for. Still, it is an expensive process with highly trained crews going out to (again) often remote sites to perform the breakup and pack-up of material for transportation to the final disposal or storage site.

There is another automobile innovation story that applies. In the early 1900s, there were American shops building cars. However, few cars were sold. Only rich people could afford them because a few highly skilled craftsmen built them one at a time. Then, pioneering automaker Henry Ford toured packing plants. He saw a process where animal carcasses went down the line with items being disassembled into steaks, chops, ribs, etc. in discreet steps that could be repeated quickly. Ford also toured grain warehouses that moved items using conveyor belts. Ford reversed the disassembly process to make it an assembly line and powered the movement with conveyor belts. Standardized parts were added in discrete easily repeatable steps as the body traveled down the line. As a result, hundreds, and then thousands, of cars were built daily, and the car prices dropped so low that even assembly line workers could buy the cars they made.[376]

Once again, the analogous solutions from other technologies suggest many of the solutions for a fresh start in a much better direction.

- Use a number of small reactors that can use commonly available parts (rather than special-builds) that are available on shorter turnaround times, greatly reducing cost and schedule risks.

- Use shipyard, or even smaller, factory construction to provide controlled-construction and volume advantages of lower cost and higher quality compared to building on site.

- Build reactors small enough—perhaps several hundred megawatts (several hundred million watts)—that would be small enough for transport by semi-trailer truck. This would multiply the number of locations that could be served.

- Compete with large reactors or other large power generating sources by ganging many small reactors in an area to provide the power of one large source. This would have an additional advantage that a single unit down for maintenance or repair would not cause a large grid-destabilizing loss of power production.

- Conversely, compete with large reactors or other large power generating sources by spreading small units throughout a region. This would reduce the number (and resulting costs) of high-power transmission lines. Also, it would make the electrical grid more stable.

---

[376] Jennifer L. Goss, "Henry Ford and the Auto Assembly Line," *ThoughtCo.*, Jan. 23, 2020. *https://www.thoughtco.com/henry-ford-and-the-assembly-line-1779201 (accessed June 20, 2020)*

- Seek additional market share for the smaller reactors for smaller process-heat applications such as petroleum refining, baking limestone into cement, and preliminary heating of ores for smelting to reduce use of hydrocarbon fuels.

These features are most commonly referred to as a small modular reactor (SMR), and there are several ambitious development efforts seeking to make SMRs work.[377]

There is already a surprisingly large market for barge-mounted or ship-mounted fossil-fuel power plants. Developing countries are rapidly electrifying, but the areas that most need expansion of their electrical grid often have the least infrastructure and the smallest pool of trained personnel, especially in frontier areas and island countries or provinces. More than sixty power ships with a total power of 4 gigawatts (4,000 megawatts were providing power for areas around the world in 2010.[378] Such power ships are constructed in shipyards and reach their customer by sailing under their own power or being towed to their area to be serviced. These power ships are generally oil fired. Their fuel is expensive, the fuel can spill with many deliveries, and the power ships' operations contribute to global-warming emissions of carbon dioxide.

If affordable and safe, nuclear-fission power ships become available, they would provide electricity with fewer chances of oil spills, less smoke pollution, and less emission of global-warming carbon dioxide.

The Russian Rosatom company towed a shipyard-constructed 70-megawatt fission reactor to a site on the Arctic coast of Siberia where it has replaced a coal-fired power plant.[379] A Chinese firm has also announced plans for shipyard construction of barge-mounted reactors especially for powering remote islands and remote coastal locations.[380]

There was one earlier floating fission power plant and a considerable number of studies in the 1960s and 70s. The U.S.S Sturgis, a converted liberty ship, was probably the first floating nuclear power plant. It supplied 10 megawatts of power from 1968 until 1976 to the Panama Canal

---

[377] Matt Ferrell, "Small Modular Reactors Explained - Nuclear Power's Future?" *Undecided* (YouTube channel), Dec. 8, 2020. https://www.youtube.com/watch?v=cbrT3m89Y3M&feature=emb_rel_pause (accessed Jan. 5, 2021)

[378] Sonal Patel, "Of Floating Power Barges and Ships," *Power*, Feb. 1, 2010. https://www.powermag.com/of-floating-power-barges-and-ships/ (accessed Oct. 27, 2020)

[379] "ROSATOM: World's Only Floating Nuclear Power Plant Enters Full Commercial Exploitation," Rosatom website, May 22, 2020. https://www.rosatom.ru/en/press-centre/news/rosatom-world-s-only-floating-nuclear-power-plant-enters-full-commercial-exploitation/ (accessed Oct. 27, 2020)

[380] "Floating nuclear power plant to light up China," *RT.Com*, March 24, 2019. https://www.rt.com/business/454605-china-nuclear-power-plant/ (accessed June 20, 2020)

Zone while a new onshore power plant was being constructed. Meanwhile, there were serious design studies for near-shore fission reactors in the United States.[381,382,383] and in Europe.[384]

The design company ThorCon, in partnership with the Indonesian government, has one of the most ambitious proposed design for shipyard-constructed molten-salt reactors and life cycle operation.[385] In their proposed ThorCon Isle, each barge-mounted power plant consists of a number of modules for all the functions of a power reactor complex and (very importantly) a large crane. There are two or four power modules, depending on the power required, and each contains two replaceable reactors, each in a sealed "can." At any one time, just one of the cans of each module is producing heat.

Each can produce about 560 megawatts of heat that is transmitted through a secondary loop to a supercritical steam turbine and generator. The generator delivers about 45 percent-efficient electricity or the electrical power of 250 megawatts. Thus, the two-cans version can produce about 500 megawatts of electricity, which is comparable to a typical coal-fired plant. The four-can version could produce about 1000 megawatts electric.)

But, then the small modular reactor advantages kick in. One of the two cans is operating, and the other can is in cooldown mode. Every 4 years, the can that has been cooling is removed and replaced with a new can. Used cans are returned to the shipyard for refurbishment and refueling or decommissioning. Virtually all the modules of the power plant can be removed and replaced so that no complex maintenance is required on site. Conceivably, a regularly refurbished and replenished ThorCon Isle could operate for decades.

An important consideration for powering areas with minimal infrastructure: The ThorCon Isle concept is designed for "walk away safety." Even if there were to be a major malfunction, there would be no catastrophic failure because the molten salt reactor of each reactor fails gracefully with any overheated fuel going into a drain pan, which spreads the fissionable material and has no neutron moderator. Hence, the fission reaction stops.

Two additional features that should be added both for the welfare of people near the power plant and for world ecology. They are:

---

[381] Caroline Delbert, "Floating Nuclear Power Plants Sounded Screwy in 1969. Today? Not So Much," *Popular Mechanics*, Jan. 31, 2020. https://www.popularmechanics.com/science/energy/a30731226/floating-nuclear-power-plants/ (accessed June 24, 2020)

[382] Nick Touran, "That Time We Almost Built 8 Gigawatt-Class Floating Nuclear Power Plants," Whatisnuclear.com website, January 26, 2020. https://whatisnuclear.com/blog/2020-01-26-offshore-power-systems.html (accessed June 20, 2020)

[383] *A Survey of Unique Technical Features of the Floating Nuclear Power Plant Concept*, U.S. Atomic Energy Commission, Directorate of Licensing, March 1974.

[384] Binnie & Partners London, *Islands for Offshore Nuclear Power Station*, published by Graham & Trotman for the Commission of the European Communities, 1982.

[385] "ThorCon: Powering up Our World," website, 2020. http://thorconpower.com/ (accessed June 18, 2020)

- The reactor barge has concrete between double hulls; and air-filled glass beads in the concrete make the concrete buoyant just as volcano-produced pumice (with its air pockets) floats indefinitely. Thus, it is truly unsinkable rather than unsinkable if the hatches all hold (remember, the Titanic was called unsinkable). When installed onsite, the reactor barge is anchored behind a breakwater to prevent it drifting away. Furthermore, at that point, it is ballasted with water, sand, and/or possibly iron oxide to secure the hull in the event of a tsunami. Breakwaters protect against waves. Likewise, the maintenance supply ship must also be double hulled with floating concrete so that it cannot sink. Several fuel cannisters at the bottom of the Pacific slowly leaking fissionable material would rightfully worry many people, and it would slow adoption of next-generation fission power.

- Waste heat from thermal power plants can damage local marine life, particularly in tropical waters that are already warm. This could cause resentment among the local fishermen and might even cause regional ecological damage. The power-plant berth lagoon needs a pipe connection (a penstock) to bring in colder deep water to prevent the thermal bloom. In the ThorCon design, the penstock is already 3 meters (nearly 10 feet) in diameter. The installation contractors must be sure to extend the penstock to reach sufficiently cold water for their inlet to the penstock. This could be a significant initial expense, but the increased temperature difference from the cold would produce more power, and the deeper waters could fertilize the nearby area rather than creating an overheated barren zone.

The same principles of inherent safety apply even more strongly to reactors small enough to fit on a semi-trailer. NuScale Power is pursuing U.S. Nuclear Regulatory Commission certification for its design of a small modular reactor that will be only 50 megawatts.[386,387] However, 12 of the units will be ganged to equal 600 megawatts. NuScale's design has an advantage in that the small size and large surface area-to-volume ratio of each reactor core sitting below ground in a super seismic-resistant heat sink, allows natural processes to cool it indefinitely even without power.

A second NuScale safety advantage is that the control rods for each reactor hang above the reactor chamber, and they are secured electromagnets. In ordinary operation, the operators could move the control rods down or back up to decrease or increase reactor energy production, respectively. However, if there were to be a loss of power to the reactor, the electromagnets would

---

[386] James Conca, NuScale's Small Modular Nuclear Reactor Passes Biggest Hurdle Yet," *Forbes*, May 15, 2018. https://www.forbes.com/sites/jamesconca/2018/05/15/nuscales-small-modular-nuclear-reactor-passes-biggest-hurdle-yet/#4fd3695d5bb5 (accessed June 20, 2020)

[387] Mark Anderson, "Slow, Steady Progress for Two U.S. Nuclear Power Projects," *IEEE Spectrum*, May 20, 2020) https://spectrum.ieee.org/energy/nuclear/slow-steady-progress-for-two-us-nuclear-power-projects (

lose power and release the control rods—shutting down the reactor.[388] (A similar technique was used with the deep-diving submersible, Trieste, that had a steel plate for ballast underneath secured by an electromagnet—loss of power would release the plate, allowing the submersible to bob to the surface.)

Meanwhile, Canadian National Laboratories is working to be a global hub of small modular reactors starting with a design competition.[389] The competitors range from a 4-megawatt (4 million watts) electric high temperature gas-cooled micro-reactor to a 190-megawatt molten salt reactor.

8.3.2.3 Advanced Systems

When power reactors were first designed in the late 1940s and early 1950s, there were too many options to investigate, and some of them might have been better. Since then, there have been more proposed systems. Several of the concepts seem to have advantages over presently used systems for fission power reactors.

One such advanced concept is the integral fast reactor (IFR), and an in-plant reprocessing system for depleted fuel, that were developed at Argonne National Laboratory and the Idaho National Laboratory from 1984 to 1994 when the program and its working demonstration reactor were cancelled because of proliferation concerns.[390] Now, with the nuclear weapons programs and sharing of technology among Pakistan, North Korea, and Iran, it may be too late to stop proliferation, so potential diversion of potential bomb-making materials from American reactors might be a moot point. Also, the reprocessed IFR fuel would not be good for bomb making because it contains a mixture of highly radioactive elements.

The IFR program had a design that would have been a large reactor. Hitachi and GE led a consortium proposing the much less ambitiously sized power reactive innovative-small module or PRISM with sizes varying from 300 to 700 megawatts.) PRISM has several updated design

---

[388] J. Jerrald Hayes, "Mini-Nuclear Reactors Are Coming, and They Could Reinvent the Energy Industry," July 21, 2019. https://www.youtube.com/watch?v=Nh5Tx1QLKBI (accessed Jan. 2, 2021)

[389] Canadian Small Modular Reactor Roadmap Steering Committee, *A Call to Action: A Canadian Roadmap for Small Modular Reactors*, Ottawa, Ontario, Canada, Nov. 2018. https://smrroadmap.ca/wp-content/uploads/2018/11/SMRroadmap_EN_nov6_Web-1.pdf contained in *Canadian Small Modular Reactor: SMR Roadmap* website, undated. https://smrroadmap.ca/ (materials accessed June 20, 2020)

[390] Charles E. Till and Yoon Il Chang, *Plentiful Energy: The Story of the Integral Fast Reactor*, CreateSpace, 2011.

features for greater efficiency and greater safety[391],[392]. A full-up IFR reactor has the following features.

- It is a "fast" reactor in that it does not have a moderator to slow neutrons down. This allows use of the more common uranium-238 rather than the more commonly used fuel of uranium-235. That increases potential available power by sixty times.

- An in-plant separation process concentrates fissionables so that they can be re-used onsite as new fuel rods in the reactor.

- Its coolant is liquid sodium, which can carry heat away at much higher temperatures than steam, and it does not require a pressure vessel. That increases power generated per unit of heat. The sodium allows most neutrons through, and most of the radioactive isotopes generated in the sodium decay away within a few hours. Of course, better materials and staying at a smaller size would be particularly important with sodium because of its combustible nature if it touches water or even just air.

- Its fuel rods are constructed in such a way that they expand and decrease their fission rate if heated beyond a certain temperature. This supplies an inherent safety feature in addition to the conventional control rods that slow the fissioning by absorbing neutrons and releasing none.

- It can take plutonium-239 as part of its fuel. Plutonium-239 was stockpiled for bomb-making for wars (that thankfully never occurred) and as a byproduct of uranium reactors. Now, it must either be part of that hundred-thousand years of storage, ... or it can supply electrical power for decades and be consumed in the process so that it would not be a storage expense and danger for millennia into the future.

A second advanced system is the high temperature gas reactor (HTGR). Fuel pellets the size of ping-pong balls to softballs (depending on the system or proposed system) are encased in graphite or silicon carbide. Nonreactive gases, such as helium or carbon dioxide carry the heat away for use in generating power ... or for providing heat for industrial processes. There have been several demonstration HTGRs and operating HTGRs.

Improved materials capable of withstanding higher temperatures have led to proposals for a very high temperature reactor (VHTRs) with a core outlet temperature greater than 900°Celsius and a goal of 1000°Celsius (1,650° and 1,800° Fahrenheit). That would be heat sufficient to

---

[391] "PRISM: A Promising Near-Term Reactor Option, *Power*, July 31, 2011. https://www.powermag.com/prism-a-promising-near-term-reactor-option/ (accessed June 19, 2020)

[392] Brian S. Triplett, Eric P. Loewen, and Brett J. Dooies, PRISM: *A Competitive Small Modular Sodium-Cooled Reactor*, GE Hitachi Nuclear Energy, Wilmington, North Carolina, July 12, 2010. https://nuclear.gepower.com/content/dam/gepower-nuclear/global/en_US/documents/PRISM%20Technical%20Paper.pdf (accessed June 20, 2020)

support high-temperature processes such as thermochemical breakdown of water to get hydrogen, retorting limestone into cement, and iron-ore processing into steel.[393]

The older pellets can be drained out the bottom for transport to a reprocessing center while replacement pellets are fed in at the top. That eliminates the several-days to several-weeks shut down period normally required for refueling.

If there is a mishap, the reactor structure can survive the additional heat. Meanwhile, smaller pellets of boron (with the usual graphite or silicon coating) can be dropped down with no pumping required to poison the nuclear reaction.

A third possibility is my personal favorite, the molten salt reactor. Alvin Weinberg first invented the light water reactor, which is the most commonly used reactor type in the world today. Later he invented the molten salt reactor, which he considered to be a much better next-generation reactor technology. He even ran a successful demonstration project with a uranium-fueled molten salt reactor from 1964 to 1969.[394,395]

The use of fuel rods in nuclear reactors has hardly been questioned even though they are a major cost item and a major area of possible failure. The fuel "rods" are actually pipes holding pellets of fissionable material while a heat exchanger fluid (usually water or sodium) carries the heat away. As fission reactions occur, they produce some byproduct gases, including xenon-135, which suppresses (poisons) the fission reaction, and radon, which is a highly radioactive but stable gas. Both xenon and radon cause the rods/pipes to expand. The poisoning decreases energy output of the reactor, and too much rod/pipe expansion could cause it to became jammed in the highly radioactive reactor core—a major repair problem. Both of these issues shorten the usable lifetime of the fuel rods. After that estimated usable lifetime, the rods must be removed, replaced, and stored for thousands of years … or recycled (more on that later). Fabricating the rods is expensive. Removing and replacing the rods requires days to weeks, during which the reactor must be shut down.

A molten fuel reactor sidesteps all these problems. It operates at regular atmospheric pressure, so the xenon nuclear poison gas and other gases bubble out of the liquid and can be removed without a shutdown for changing of fuel rods. If the percentage of fissionable fuel drops too low, more concentrated fuel can be added. There is no expense for fabrication of new fuel rods, or removal, and storage of spent fuel rods. Recycling fuel at the end of reactor life can be much easier; hence, it can be much cheaper.

---

[393] I. Pioro and R. Duffey, *Managing Global Warming: An Interface of Technology and Human Issues*, Trevor Letcher (editor), Academic Press, Nov. 8, 2018.

[394] Dave Mosher, "A *Forgotten War Technology Could Safely Power Earth for Millions of Years. Here's Why We Aren't Using It,*" Business Insider, Feb. 25, 2017. https://www.businessinsider.com/thorium-molten-salt-reactors-sorensen-lftr-2017-2 (accessed June 20, 2020)

[395] John Houtari, "A Great Technical Achievement, Molten Salt Reactor Could Be Entombed," *Oak Ridge Today*, Nov. 26, 2017. https://oakridgetoday.com/2017/11/26/great-technical-achievement-molten-salt-reactor-entombed/ (accessed June 20, 2020)

Eliminating the fuel rods also makes higher temperatures possible. Besides the swelling pressure of generated gases on the fuel rods, zirconium cladding on the outside of the fuel rods begins oxidizing at a certain temperature, pulling oxygen out of the water in the typical light water reactor. That oxidation generates hydrogen, which builds up pressure in the reactor chamber and could lead to explosion (possibly the cause of the failures at Three Mile Island, Chernobyl, and Fukushima).

The fuel of molten salt reactors is also the cooling fluid, and it can flow by convection up to a radiator for transfer to a secondary heat-transfer loop, to a steam turbine … or even a gas turbine if the materials can safely withstand the heat to run a gas turbine. The liquid fuel is an inherent safety feature. Even more so than the expanding rods of the PRISM reactors mentioned earlier, the molten salt fuel expands, and so it is less effective as a fuel when heated beyond a certain temperature, and produces fewer neutrons and less heat at that point.

If there were still to be overheating and all systems failed, a freeze plug at the bottom of the reactor would no longer have power flowing to a refrigerating fan keeping the plug cool. Consequently, the plug would melt and allow the fuel to pour into a drain pan, where (without any graphite moderator to slow the neutrons down to the slow speed that works with the thorium) fissioning would stop, and the reactor would safely cool without the need for pumping cooling water as was needed at Fukushima.

Finally, if the salt were to reach the outside world, it would not react with anything; it would simply freeze into a solid. Because the salt is made with highly oxidizing fluorine and/or chlorine and those are combined with highly metallic elements, the salts are nonreactive with other materials.

8.3.2.4 Use Thorium, the Super Fuel, Along with Those Other Technologies

Fritz Zwicky, a great pioneer of both astrophysics and aerospace, once compared research and development to rain falling on a hillside. At first, it flows as broad sheets and only erodes shallowly. Then sheets divide into many tiny rivulets, which flow together into streams and then mighty rivers. The rivers cut deep like the Grand Canyon, but in concentrating in those deep canyons, the water flow leaves other broad areas almost untouched save by the broadest and shallowest of streams.

It is the same with the development of nuclear energy. Great potential technologies are hidden in those ignored highlands. Much of the original research on radiation and nuclear fission was done with a presently little-mentioned element in nuclear technology—thorium. The element thorium would have been the better choice for fission power production, but uranium and plutonium were much better suited for the bombs that the Allies were desperately building during World War II lest Nazi Germany build such bombs first. Then, the momentum of the uranium and plutonium development kept most of the research and development focused on technology approaches favoring those two elements.

Yet, over time, the requirements for practical (read profitable) reactors, and several generations of reactor development also led to new concepts including use of that other radioactive element, thorium, because it has significant advantages compared to the conventional uranium/plutonium conventional nuclear system.[396,397,398,399,400,401]

- Thorium reserves are at least three times those of uranium, and thorium is found in a concentrations 500 times greater than fissile uranium-235.[402] The actual thorium availability is probably considerably more because reserves are based on economically minable materials (ores) calculated from geological surveys. Thorium has been little used, so there has only been limited thorium prospecting. If the price of thorium were to rise significantly, the amount of reserves would probably increase immensely.

- In contrast to conventional uranium/plutonium "burning" of nuclear fuel, the thorium breakdown uses a greater percentage than the uranium does.[403] transmutes to uranium-233, and that product decays down into much smaller elements with mostly heat and alpha and beta particles, neither of which do damage to the reactor structure as happens with uranium-235 and plutonium-239 high-neutron processes.

- However, the uranium-233 does release two neutrons when it decays, so it breeds more fissionable uranium-233 from the thorium to continue the reaction.

- The breakdown paths for thorium use a greater percentage of the fuel, so the usable energy from three times the mass of available uranium can be multiplied again—maybe 9 or 10 times that of uranium ... unless the uranium fuel is reprocessed to recover the unused uranium (more on that soon).

---

[396] Richard Martin, *Super Fuel: Thorium, the Green Energy Source for the Future*, St. Martin's Press LLC, New York, NY, 2012.

[397] Robert Hargraves, *Thorium: Energy Cheaper Than Coal*, CreateSpace Independent Publishing Platform, July 25, 2012.

[398] Kirk Sorensen, "Thorium can give humanity clean, pollution free energy," *TEDx Colorado Springs*, Jan. 8, 2015. https://www.youtube.com/watch?v=kybenSq0KPo (accessed July 18, 2020)

[399] Marvin Baker Schafer, "Abundant Thorium as an Alternative Nuclear Fuel: Important Waste Disposal and Weapon Proliferation Advantages," *Energy Policy*, vol. 60, pp. 4–12, Sept. 2013.

[400] Srikumar Banerjee, "Thorium to light up the world," *TEDxCERN*, Oct. 24, 2014. https://www.youtube.com/watch?v=yGhEdcwXxdE (accessed July 19, 2020)

[401] Uguru Edwin Humphrey and Mayeen Uddin Khandaker, "Viability of Thorium-Based Nuclear Fuel Cycle for the Next Generation Nuclear Reactor: Issues and Prospects," *Renewable and Sustainable Energy Reviews*, vol. 97, pp. 259–275, Dec. 2018.

[402] Sameer Surampalli, "Is Thorium the Fuel of the Future to Revitalize Nuclear?" *Power Engineering*, Aug. 13, 20019. https://www.power-eng.com/2019/08/13/is-thorium-the-fuel-of-the-future-to-revitalize-nuclear/ (accessed July 23, 2020)

[403] J. Stephen Herring, Philip E. MacDonald, Kevan D. Weaver, and Craig Kullberg, "Low Cost, Proliferation Resistant, Uranium–Thorium Dioxide Fuels for Light Water Reactors," *Nuclear Engineering and Design*, vol. 203, issue 1, pp. 65–85, Jan. 1, 2001.

- Consequently, the ultimate waste stream of breakdown products can be a much smaller percentage than with uranium. Moreover, those products are more radioactive, but with a much shorter lifetime. That means that they decay and cool down within several hundred years rather than the tens of thousands of years from the products of uranium fuel. In fact, 90% of the thorium radioactive waste is gone within 30 years. Furthermore, the thorium daughter isotopes are simpler for reprocessing from the liquid fuel than the multitude of atoms similar to uranium (trans-uranium elements) with higher numbers in the periodic table, so that aspect can also be simpler and cheaper.

- Thorium is a much poorer material for bomb-making so it does not need the multiple levels of security of uranium/plutonium fuels. One short-lived daughter species produced in tiny quantities (after several steps) is thallium-208 that emits dangerous levels of gamma rays, so it would be dangerous to bomb makers and a tell-tale marker for any agencies hunting those bomb makers. (During the Cold War of the late 1940s until the early 1990s, thorium's poor bomb-making potential caused governments to avoid developing systems that used it; now poor potential for bomb-making is an advantage.)

- Another thorium daughter species is uranium-232, which is too radioactive to use in a practical fission bomb. (A group diverting fuel from a thorium reactor would need a major processing facility to separate the uranium-233 that they would want from that nasty uranium-232—they would do better processing the small amount of uranium-235 in widely available natural uranium ores to get bomb-making material.)

- This wild menagerie of isotopes is not a problem inside a reactor, but they would make diversion dangerous and easily detectable. Inside a potential bomb, their high radioactivity would harm electronics and reveal the bomb's location.

- Also, thorium reactors produce less plutonium than uranium reactors, and the plutonium produced is mostly plutonium-238, which releases a great deal of heat and alpha particles, but very few of the neutrons needed for a bomb detonation.[404] The high heat production and low neutron flux make plutonium-238 a highly prized material for outer space probes,[405] but that is another story.

---

[404] David Sylvain, Elisabeth Huffer, and Hervé Nifenecker, "Revisiting the Thorium-Uranium Nuclear Fuel Cycle," *Europhysics News*, vol. 38, no. 2, pp. 24–27, 2007. https://www.europhysicsnews.org/articles/epn/pdf/2007/02/epn07204.pdf (accessed June 24, 2920)

[405] David Szondy, "US Restarts Production of Plutonium-238 to Power Space Missions," *New Atlas*, Dec. 23, 2015. https://newatlas.com/ornl-plutonium-238-production-space/41041/ (accessed June 25, 2020)

- Conversely, the issues above would make it very difficult for any facility smaller than a major recycling center to separate uranium-235, plutonium-233, or plutonium-239 from the fuel mixture of a thorium molten salt reactor.

- Thorium is fertile rather than fissile. As opposed to fissile uranium, which spontaneously emits neutrons, thorium needs some neutrons to get started. There must be a starter, either an initiation rod of fissile material or an electrical accelerator that supplies particles. The advantage of that is that it affords much greater safety because the fission chain reaction is more easily stopped.

- A bonus advantage of developing thorium as a reactor fuel is that mining and refining thorium would also yield byproduct rare-earth elements. The rare-earth elements are crucial in small amounts for exotic high-tech applications such as display screens, catalysts, and super magnets. Being a company country, China has developed a near monopoly for production of rare-earth elements and has threatened to use that monopoly against the United States and other countries.[406,407] The short-term solution is for the federal government to begin a purchasing program of domestic thorium tailings and storing the ore for later use in thorium-based LFTR types of reactors. The Defense National Stockpile Center has maintained such stockpiles in the past for a number of other materials that were deemed to be strategic materials. For the longer term, the U.S. Congress could authorize ORNL to begin a program of cooperation with private industry to develop practical designs for thorium molten-salt reactors to replace the aging nuclear reactors in the U.S.[408]

Thorium has been used as part of the fuel mix for the fuel rods in conventional reactors. However, the advantages of thorium are much greater in molten salt reactors. Those advantages are particularly important when combined with the advantages of small modular reactors and molten salt reactors.

The molten salt reactor configuration would actually provide a benefit for transporting and then starting a small modular thorium reactor. A 5% uranium-235 portion in the thorium reactor fuel load would be enough to make a self-sustaining fission reaction, but it would only do so if the fuel were in the reactor chamber with the moderator.

If the fuel were to be released through the freeze plug into the drain pan below the reactor chamber, that fuel would be inert except for natural decay. Being widely spread in the drain pan, it would cool down without the input of powered cooling. Once cooled to a solid state, it could be

---

[406] David S. Abraham, *The Elements of Power: Gadgets, Guns, and the Struggle for a Sustainable Future in the Rare Metal Age*, Yale University Press, New Haven, Connecticut, 2015.

[407] Victoria Bruce, *Sellout: How Washington Gave Away America's Technological Soul, and One Man's Fight to Bring It Home*, Bloomsbury, New York, 2017.

[408] Mac MacDowell, "Rare Earth Minerals and Thorium," *American Thinker*, June 3, 2019. https://www.americanthinker.com/articles/2019/06/rare_earth_minerals_and_thorium.html (accessed July 23, 2020)

shipped to the installation site. For operation, a natural gas or oil heater would melt the fuel in the drain pan so that it could be pumped into the reactor chamber (with the moderator) where a self-sustaining reaction would commence. Conversely, draining down into the pan would allow the fuel to cool and solidify so that the reactor could be transported back to the factory for maintenance or final disassembly.

An experimental molten-salt reactor was run for 4 years, and thorium has been part of the mix of materials in the fuel load for several reactors. However, actually building and running a demonstration thorium (or partially thorium) molten-salt reactor, demonstrating success, getting design certification from regulators, selling investors on a manufacturing plan, securing sales, and delivering will be a big endeavor. One estimate is that the United States Government certification will require about $5 billion.[409]

There have already been business casualties. Transatomic Power, was founded in 2011 and closed down in 2018. Transatomic secured funding from venture capitalists and even received a U.S. Government research contract, but it was not enough. The good that came out of the venture was that Transatomic and its venture-capital supporters agreed to open source their design work so that somebody else would be helped to make the system work.[410,411]

Yet, other research and development efforts are proceeding. There are several national government programs including China (that has stores of thorium from its rare earth processing),[412]

---

[409] Sameer Surampalli, "Is Thorium the Fuel of the Future to Revitalize Nuclear?" *Power Engineering*, Aug. 13, 20019. https://www.power-eng.com/2019/08/13/is-thorium-the-fuel-of-the-future-to-revitalize-nuclear/ (accessed July 23, 2020)

[410] Jeff St. John, "Transatomic to Shutter Its Nuclear Reactor Plans, Open-Source Its Technology," *gtm*, Sept. 25, 2018. https://www.greentechmedia.com/articles/read/transatomic-to-shutter-its-nuclear-reactor-plans-make-its-technology-public (accessed July 20, 2020)

[411] Leslie Dewan, "Save the World with Nuclear Power," TEDxUniversityofRochester, June 25, 2019. https://www.youtube.com/watch?v=DoAcntoAVXE (accessed July 20, 2020)

[412] Alice Shen, "How China Hopes to Play a Leading Role in Developing Next-Generation Nuclear Reactors," *South China Morning Post*, Jan. 10, 2019. https://www.scmp.com/news/china/science/article/2181396/how-china-hopes-play-leading-role-developing-next-generation (accessed July 23, 2020)

India (with significant thorium deposits,[413] Canada,[414] and the United States (including work small modular reactors[415] and with chloride[416] instead of fluoride molten salt reactors).

Developing the quartet of thorium, molten-salt, modularized, and standardized reactors appears to be much like that of jet engines in the 1930s. Aeronautical engineers, and those who funded them, knew that the principle of the steam turbine had great promise for producing an aviation power plant that would be simpler, more powerful and faster than gasoline-fired piston engines. What they did not have was enough money to spend over years to make it work.

A gas turbine using combustion gases (a jet) could theoretically run hotter for greater efficiency and would not have the limitation of propeller blades that could only go so fast or be so great a length lest they go supersonic at tips. However, theoretical jets did not fly. The compressor pushing air and fuel to the igniting point had to be made lighter, more rugged, and more efficient. Likewise, the turbine and power shaft carrying some of that jet power back to the compressor needed those characteristics, plus they needed to withstand great heat.

This was the type of development that was too big for nimble entrepreneurs. This was a case where sustained government funding was required. In this case it was World War II during the first half of the 1940s. By the end of the war, hundreds of jet aircraft were flying.[417] Twenty years later, thousands of jet aircraft were flying, and the gas turbine had become a major type of power plant by itself or as an additional topping cycle expelling enough waste heat to run a steam-turbine cycle for another thirty percent more electricity from a heat source.

Commercializing the next generation of nuclear reactors will require a comparable effort, and it may yield even greater benefits.

8.3.2.5 Reprocessing and Transmutation—Vast Energy and No More under-the-Ruggism

Opponents of nuclear fission contend that there is not enough uranium and thorium fuel in the world to last more than a few decades and that they would leave behind dangerous amounts of

---

[413] Ritu Sharma, "Thorium Reactors Will Be a Reality in Few Years, Research Continues," *Nuclear Asia*, July 27, 2019. https://www.nuclearasia.com/news/thorium-reactors-will-reality-years-research-continues/3048/ (accessed July 23, 2019)

[414] Brian Wang, "Moltex Molten Salt Reactor Being Built in New Brunswick Canada," *NextBigFuture*, July 19, 2018. https://www.nextbigfuture.com/2018/07/moltex-molten-salt-reactor-being-built-in-new-brunswick-canada.html (accessed July 24, 2020)

[415] Dan Yurman, "DOE National Labs Kickstart Work on Micro and Small Modular Reactors at Two Sites," web page, The Energy Collective Group, Feb. 25, 2020. https://energycentral.com/c/ec/doe-national-labs-kickstart-work-micro-and-small-modular-reactors-two-sites (accessed July 25, 2020)

[416] Southern Company and TerraPower Prep for Testing on Molten Salt Reactor," web page, Office of Nuclear Energy, United States Department of Energy, June 28, 2020. https://www.energy.gov/ne/articles/southern-company-and-terrapower-prep-testing-molten-salt-reactor (accessed July 25, 2020)

[417] Matt Ridley, Chapter 3, "Transport," *How Innovation Works: And Why It Flourishes in Freedom*, Harper, pp. 80–111, May 19, 2020.

poisonous material. These fears are incorrect because there are two major ways to get more usable fuel and destroy unwanted byproduct materials: reprocessing and transmutation (also referred to as breeding). Together, they could be the keys to acquiring thousands of years' worth of fission energy, eliminating nuclear waste, and (of course for this book) reducing global warming.

First, reprocessing or recycling of unusable material from nuclear reactors has been done since the beginning of the nuclear power industry. Theoretically, fission reactors could use nearly all the potential energy in the uranium and thorium fuels (plutonium being an intermediate isotope transmuted from the uranium). In practice, this is not the case at present. The "reactor-grade" fuel is typically only 3 to 5 percent uranium-235 and/or plutonium-239 inside the long pipes called control rods. After 18 to 36 months, the "spent" or "used" fuel typically has about 1% uranium-235 and 0.6% fissionable plutonium isotopes bred during the reactor operation. It has about 95% nonfissile uranium-238 and 3% fission products and minor actinides. That 3% includes the xenon isotope that poisons the fission reaction.

When technicians remove the spent fuel, only about one twentieth of the potentially fissionable atoms have been used, so the so-called spent fuel still contains about 95% of its original energy. Plus, this unused material is hot both radioactively and in temperature. This is the material that the U.S; government presently expects to store for ten thousand years or more.

In contrast, a reprocessing regime (as operated in a number of countries other than the once-through-then-store system of the United States) sends the used fuel rods to a reprocessing facility. There, fission byproducts that could never become fissionable are removed, particularly the reaction-poisoning xenon. Fissile material is added to raise the active percentage to the required 3 to 5 percent, and the material is loaded into new fuel rods. That 95% nonfissile uranium-238 is recycled in the mix and eventually absorbs neutrons and breed up to other isotopes, many of which can function as fuel that is burned in reactors and destroyed. Ultimately, most of the fertile (but not fissile) uranium-238 is bred to fissile uranium-235 and burned in a reactor.

That brings us to breeding or transmutation of nuclear fuels. The medieval alchemists fervently hoped they could transform or "transmute" lead into gold with chemical processing and a pure heart. They never succeeded. However, breeder reactors have been routinely transmuting various elements into others for decades. The uranium-238 transformation to uranium235 discussed above is one of those transmutations.

Neutrons produced in a reactor, or by an accelerator, cause elements to split into lighter elements or bump up to slightly heavier elements. The most important transmutation reactions (or nonreactions) for this study are

- Uranium-238 transmuting into plutonium-239—plutonium-239 is a highly fissile isotope, meaning it is very useful for reactors or bombs. Both uranium-235 and plutonium-239 produce enough neutrons from natural decay that they can start and maintain a fission reaction if they have a sufficient mass of fuel in one place (called critical mass). They also transmute into the xenon isotope that slows down fission reactions, and small amounts of other trans-uranium elements (chiefly neptunium, americium, and curium) that keep nuclear waste dangerously radioactive for thousands of years.

- Thorium-232 transmuting (quickly via thorium-233 and protactium-233) into uranium-233—The uranium-233 can absorb neutrons at a wide range of energy levels. It breaks into a number of isotopes of smaller atoms (chiefly strontium and xenon) that are mostly unstable and break down further while producing more heat. A small amount of thallium 208 is also produced, and that isotope has another feature that would make it difficult to divert fuel to terrorists. Thallium 208 is a hard gamma-ray emitter, and those gamma-ray emissions would make it difficult for someone to hide or safely handle any stolen fuel from a thorium molten salt reactor. Thorium transmutation also produces the xenon isotope that is stable and must be removed because it slows nuclear reactions. (Handily, in a molten-salt reactor, the xenon simply bubbles out.)

These processes could be much easier with molten salt reactors, in which much of the byproduct unusable material can be removed on-site, replacement fissile material can be added on-site, and there are no costs for handling and processing fuel rods. However, there would still be some waste materials.

The thorium fuel cycle should have a much lower proliferation risk than the uranium/plutonium fuel cycle. Still, one danger claim has been made. There is an accusation that uranium-233 in thorium molten salt reactors does present a plausible proliferation issue. Even though uranium-235 and plutonium-239 are much better suited for making fission bombs, it is still theoretically possible to make bombs using uranium-233 produced from thorium via what has been called the protactium pathway described by Ashley et al.[418] The pathway and its issues can be summarized as follows.

- The uranium-233 that is major interim product from thorium and a major part of the fuel for a thorium reactor could be used for bomb making (1.6 metric tons [1.8 tons] of thorium radiated for a month could yield the critical uranium-233 mass for bomb making of 8 kilograms [18 pounds]).

- The good news is that there are difficulties in diverting the uranium-233 from a reactor. Uranium-233 in a reactor is also accompanied by the short-lived but highly radioactive uranium-232. Thus, separating out the uranium-233 with a half-life long enough to make a workable bomb must be done using remote techniques in heavily-shielded containment chambers (called hot cells) and a large reprocessing facility. These would be hard to hide.

- However, the intermediate isotope of protactium-233 is more tractable. It can be chemically separated from the uranium and thorium isotopes relatively easily. Then, after several 27-day half-lives of the protactium-233, the potential fuel diverters would have pure uranium-233 theoretically suitable for bomb making.

---

[418] Stephen F. Ashley, Geoffrey T. Parks, William J. Nutall, et al., "Thorium Fuel Has Risks," *Nature*, vol. 492, pp. 31–33, 2012. https://www.nature.com/articles/492031a (accessed June 21, 2020)

However, there have been no known instances of such bomb-making attempts being made.

- Reactors using conventional rods (tubes with fuel pellets) and high-temperature gas reactors using fuel balls would probably be much less of a danger for such a protactium pathway because a sudden decline in fuel rods or balls would soon become apparent to the fuel-supplying entity. Also, producing those fuel-holding items just to break them up would be exorbitantly expensive.

- Molten salt reactors might offer an opportunity for small-scale diversion to build up a supply of the uranium-233.

- Such diversion would probably require at least the resources of a small country rather than a terrorist group, and the separation techniques are still largely theoretical, but it is a concern.

- The suggested safeguards would be close monitoring of thorium production and use as well as continued monitoring of hot cells in which work can be done within a containment dome and in which materials can be remotely manipulated.

A happier use of transmutation (or breeding) is that the long-lived trans-uranic isotopes can be placed into a fast reactor and "burned" into smaller less radioactive daughter isotopes such as lead, bismuth, and helium. Both the integral fast reactors and the thorium molten salt reactors can burn so-called spent fuel into nonradioactive materials and materials with shorter half-lives.[419] Along the way, a more complex mix of isotopes would make also make diversion and processing of stolen bomb-making isotopes more complex and difficult.

How could the United States pay for nuclear waste disposal? For starters, the U.S. Department of Energy Nuclear Waste fund has already accumulated more than forty billion dollars for accepting and providing permanent disposal of spent fuel from decommissioned reactors.[420] The Waste Fund spent $11 billion dollars studying a possible long-term storage site at Yucca Mountain, Nevada, but Nevada politicians halted work at that site because of their concerns about storage … in their state. Meanwhile, the Waste Fund has not disposed of a single ounce of spent fuel.

The U.S. Government could allocate some of the Department of Energy Waste Fund money to subsidize fuel reprocessing and development of waste-burning reactor technologies, including integral fast breeders and thorium molten salt reactors. As a corollary, operators of reactors

---

[419] "Breed and Burn Reactors – Could Recycling Waste Redeem Nuclear Power? *Power Technology,* May 23, 2012. https://www.power-technology.com/features/featurenuclear-waste-disposal-reactor-technology-recycle-clean-energy/ (accessed June 23, 2020)

[420] "Nuclear Waste Burial Fund Grows to $43 Billion, But DOE Has Not Buried an Ounce of Spent Fuel, *Energy Central News,* Feb. 1, 2019. https://energycentral.com/news/nuclear-waste-burial-fund-grows-43-billion-doe-has-not-buried-ounce-spent-fuel (accessed June 23, 2020)

capable of waste burning should be paid for accepting and destroying nuclear wastes. This would be in addition to any revenue derived from electricity production.

Another suggested approach, probably further into the future, is to use a particle accelerator to hurl protons at a metal target, generating a neutron flux sufficient to generate localized fission, heat, and destruction of unwanted nuclear waste.[421] This approach would provide a major safety feature in that a loss of power would immediately stop the nuclear chain reaction.

However, proton accelerators powerful enough to drive a subcritical reactor have not been demonstrated, and the cost to get such power with presently available technology would be high.[422]

Still, the first steam engines had efficiencies of only a few percent and were barely practical. Technology evolves. There are technology paths that have not been explored. Theoretical physicists want the highest possible energy of a few particles for study with cost not being a major concern. Designers of a subcritical reactors would want the maximum number of particles at the lowest possible price.

8.3.2.6 Design Reactors to Be Inherently Safe

The Swedish PIUS reactor and the German-American small modular high-temperature gas-cooled reactor are inherently safe—that is, their safety relies not upon intervention of humans or of electromechanical devices but on immutable principles of physics and chemistry. A second nuclear era may require commercialization and deployment of such inherently safe reactors ...

Alvin Weinberg and Irving Spiewak, 1984[423]

The first designers of nuclear fission reactors were just happy to make them work. Only after reactors got larger did the issue arise of turning the fission chain reactor off and then cooling the reactor for hours or days lest equipment be damaged.

Then, safety features were added as afterthoughts, add-ons attached to the basic design. Unfortunately, the add-on safety features also added complexity and needed their own support. That was when several pioneers of fission energy began proposing designs with inherent safety.

The 1911 Fukushima accident provided a negative example of how add-on safety features can fail. Reactors 1, 2, and 3 were running, and they successfully shut down after the earthquake. They had water pumps for cooling a recently shut-down reactor. Because the reactor would not be making electricity for the pumps when it was shut down, the reactors had a grid connection to

---

[421] Motoharu Mizumoto, "Accelerators for Nuclear Waste Transmutation, *Proceedings of the 1994 International Linac Conference*, Tsukuba, Japan, 1994. https://accelconf.web.cern.ch/l94/papers/tu1-01.pdf (accessed June 21, 2020)

[422] Robert Hargraves, *Thorium: Energy Cheaper Than Coal*, CreateSpace Independent Publishing Platform, July 25, 2012.

[423] Alvin M. Weinberg and Irving Spiewak, "Inherently Safe Reactors and a Second Nuclear Era," *Science*, vol. 224, issue 4656, pp. 1398–1402, June 29, 1984.

get electricity and emergency generators if the grid was down. The grid went down, and the emergency generators (which were on lower ground than the reactors) were also put out of commission by the tsunami. Thus, a preventable disaster became inevitable.[424]

Temperatures rose in the three reactors. Partial core melt downs led to hydrogen explosions and venting of steam with radioactive materials from the three reactors that had been operating, and the nearby reactor 4.[425] Reactors 5 and 6, which had been down for maintenance were saved from serious damage.

The terms inherent safety and inherently safe have occasionally popped up in this discourse. These interchangeable terms refer to a condition whereby a reactor would shut down safely without overheating even if all human, electronic, and mechanical controls failed. Inherently safe features already developed or proposed include

- Staying with relatively small sizes of tens to only two- or three-hundred megawatts reduces risk of pressure and heat build-up

- Using fuel rods that swell an inhibit neutron flow at higher temperatures

- Using molten-salt fuels that expand when heated so that they have less concentration of the fuel, which causes the chain reaction to slow

- Maintaining an actively cooled freeze plug in molten salt reactors so that loss of power would cause the plug to melt and the fuel to drain into a pan below that would be spread out and would not have a moderator so the nuclear fuel would safely complete the cool down

- Adding more corrosion-resistant cladding (such as silicon carbide or silicon nitride) around fuel rods or fuel spheres to better resist oxidation (and steam generation in water-cooled reactors)

- Using heat-pipe wicking or convective air flow to take heat away from a reactor without any control or pumping power

- Using surrounding rock around a reactor sufficient to draw away enough heat for secure shutdown

- Using molten-salt fuels with the fuel metal (uranium, plutonium, or thorium) combined with fluorine (the most active oxidizing element) so that if the fuel were to leak out, it would not react with anything

---

[424] Robert Perkins, "Fukushima Disaster Was Preventable, New Study Finds," *USC News*, Sept. 21, 2015. https://news.usc.edu/86362/fukushima-disaster-was-preventable-new-study-finds/ (accessed July 18, 2020)

[425] "Chapter: 4 Fukushima Daiichi Nuclear Accident," *Lessons Learned from the Fukushima Nuclear Accident for Improving Safety of U.S. Nuclear Plants*, National Research Council of the National Academy of Sciences, 2014. file:///C:/Users/VALUED~1/AppData/Local/Temp/18294.pdf (accessed July 18, 2020).

A team led by Edward Teller, called the father of the American fusion bomb, provided a detailed proposal for one such inherently safe reactor.[426]

Inherent safety is becoming doubly important because, while the developed rich world has been building very few new reactors, the developing world is now increasingly the area building new reactors. The often-predicted nuclear renaissance is indeed happening, but it is in an area that is even more liable to nuclear accidents than the developed world has been.

A group of Chinese researchers remarked on this in a paper entitled, "Nuclear Safety in the Unexpected Second Nuclear Era,"[427] with an obvious homage to the earlier Weinberg and Spiewak paper regarding inherently safe fission reactors. Regarding the unexpected second nuclear era, they cited the figure of 55 fission reactors under construction with 47 of them being in the developing world.

They continue by noting that even the developed world with many technological and organizational advantages, experienced severe nuclear accidents, and nuclear power plants in the developing world have their own new set of dangers including

- New suppliers (such as India, China, and Korea) have had little experience in developing safety regulations.

- Many of the new countries starting to develop nuclear power have limited technical infrastructure and depth of trained nuclear personnel to combat an accident.

- Many of the new countries starting to develop nuclear power have low scores in corruption resistance and adherence to international law.[428]

- "Some regions are also stability-challenged, such as the Middle East." [I thought the British were the masters of understatement and diplomacy; I was so wrong.]

- It is not clear that the emerging suppliers would fully enforce safety standards of their country or international safety standards on recipient countries.

For those reasons alone, countries newly adopting fission nuclear power production should be encouraged to adopt next-generation reactor technology with inherent safety. If they fail to do so, they could experience one or more severe accidents with major loss of lives. Moreover, one or

---

[426] Edward Teller, Muriel Ishikawa, and Lowell Wood, "Completely Automated Nuclear Reactors for Long-Term Operation," summarized in Appendix B of Reese Palley, *The Answer: Why Only Inherently Safe, Mini Nuclear Power Plants Can Save Our World*, The Quantuck Lane Press, New York, 2011.

[427] Yican Wu, Zhibin Chen, Zhen Wang, et al., "Nuclear Safety in the Unexpected Second Nuclear Era," *Proceedings of the National Academy of Sciences of the United States of America*, vol. 116, no. 36, pp. 17673–17682, Sept. 3, 2019.
https://www.pnas.org/content/116/36/17673 (accessed July 19, 2020)

[428] *Evolving Nuclear Governance for a New Era: Policy Memo and Recommendations*, Global Nexus Initiative, April 2017. http://globalnexusinitiative.org/wp-content/uploads/2017/04/GNI-Policy-Memo-3.pdf (accessed July 19, 2020)

more such accidents could cause another world-wide retreat from fission technology and severely limited options for avoiding excessive global warming.

Major Things for Governments and Businesses to Do

- Government: Allocate some of the Department of Energy Waste Fund money to subsidize fuel reprocessing and development of waste-burning reactor technologies, including integral fast breeders and thorium molten salt reactors.

- Government: Support two nuclear initiatives of thorium molten salt reactors for the approach of using slow (also called thermal) neutrons for thorium reactors and fast neutrons for integral fast reactors.

- Government: In the short term, store thorium-containing rare-earth tailings for eventual processing.

- Government: Subsidize the return of rare earth-element mining and processing to the United States. These elements are crucial for the continued economic growth of the country and perhaps its survival. Besides thorium for power, tiny amounts of the other elements are increasingly vital for electronics, catalysts, high-strength magnets, and many other uses.[429,430]

- Government: Consider division of labor with other governments where one government might pursue one technology path, and another friendly government might pursue a second path.

- Government: Help developing countries willing to try something new as long as a certain amount of the funding stays in the United States. For instance, Indonesia, with many remote islands needing power, is considering a project to build the first commercial molten-salt reactor.[431]

- Business: Build small reactors for countries willing to experiment.

---

[429] David S. Abraham, *The Elements of Power: Gadgets, Guns, and the Struggle for a Sustainable Future in the Rare Metal Age*, Yale University Press, New Haven, Connecticut, 2015.

[430] Keith Veronese, *Rare: The High-Stakes Race to Satisfy Our Need for the Scarcest Metals on Earth*, Prometheus Books, Amherst, New York, 2015.

[431] Brian Wang, "Indonesia Consider Molten Salt Nuclear Reactor Cheaper than Coal Power," NextBigFuture, July 3, 2020. https://www.nextbigfuture.com/2020/07/indonesia-consider-molten-salt-nuclear-reactor-cheaper-than-coal-power.html (accessed July 6, 2020)

What You Can Do

- Support next-generation nuclear power startups because they are safer and probably the only energy source that can save the environment.

- Confront scare mongers with the fact that all energy production costs lives, and lack of energy costs the most energy of all.

8.4 Nuclear Fission Energy—Electricity, Heat, and Synthetic Fuels

8.4.1 Electricity, Space Heating, and Carbon Dioxide for Synthetic Fuel

Next-generation cheaper nuclear fission can provide revolutionary benefits. First, cheaper electricity makes all the uses associated with electricity more affordable. For the GND concern, roughly a quarter of present emissions of greenhouse gas agents come from electrical power generation from burning fossil-fuel. Increasing electricity production from solar and wind enough to replace all fossil-fuel power generation would probably not come within 12 years, if it comes at all.

Then, as the late-night infomercials say, there is more. Heat is a key cost component—and greenhouse emission source—for many key activities of society. Various types of reactors could be developed to provide different amounts and temperatures of heat and/or electricity for different uses.

First, what was old is new. The so-called pool reactors have been used for research for decades. They have fuel rods in pools of water. The water moderates neutron velocities so that there is a chain reaction, and it shields the surrounding area from stray radiation. Conversely, loss of water would cause the reaction to stop. For simplicity and cheapness of design, the pool reactors produce heat at less than the boiling point of water.

The pool reactors are inefficient for power generation because of their low temperatures, but they are sufficient for district space heating in cities with cold climates. Much of China has such a climate with district heating. and the common fuel for their district heating at present is coal, which produces greenhouse carbon dioxide and often harmful pollution. China National Nuclear Corporation has designed a pool reactor (and piping connections) to replace coal for district heating. With many units of a standard design and providing just low-grade heat, the Chinese developers hope to drive the price for heating energy to less than coal and much less than natural gas.

Cheap heat from pool reactors could also desalinate seawater, something that may be greatly needed in the future, with or without serious global warming. (Nuclear desalination has been a dream since the 1950s, but prices have been too high until now.)

If successful, this system may also be used in large areas of Russia, Northern Europe, and Canada.[432,433]

Pool reactors have been noted here because they may become a major energy source in those countries. They will probably not be widely used in the United States because this country uses very little district heating.

The opposite direction of benefit is that process heat can benefit electrical generation. If nuclear generation is a major portion of the power supply of an area, those plants need to follow the load. When local or sellable demand is low (such as selling to another region or country), operators must slow the reactors down (bad for the equipment) or run the excess electricity into the ground (a waste of potential income).

This has been the case of France, which generates more than two-thirds of its electricity using nuclear fission. To avoid waste of electricity when demand is low, there has been a proposal to use this extra electricity to electrolyze water to produce hydrogen. (Extra maintenance work required because of variations in electrolyzer settings is much cheaper than more frequent reactor maintenance due to variations in operating load.)

Then, the hydrogen generated can be combined with carbon dioxide collected from various waste streams such as from steel mills, paper pulping mills, and power plants. With buffering of the hydrogen and the carbon dioxide used by the chemical plant, there will be no wasted electricity and there will be additional production of some synthetic fuel such as methane, methanol, or dimethyl ether.[434]

There are additional research and development efforts in Germany, Britain, and the United States. These various nuclear-hydrogen initiatives would either use hydrogen directly (such as in steel mills) or combine it with carbon dioxide to generate some type of synthetic fuel.[435]

8.4.2 Process Heat Uses

A second major use of heat is in process industries that make the things we use from steel to plastics to the cement for concrete in our buildings. The next-generation of fission nuclear

---

[432] Yi-Xuan Zhang, Hui-Ping Cheng, Xing-Min Liu, and Jian-Min Li, "Swimming *Pool-Type Low-Temperature Heating Reactor: Recent Progress in Research and Application*," *Energy Procedia*, vol. 127, pp. 425–431, 2017. https://reader.elsevier.com/reader/sd/pii/S1876610217335968?token=B33752F2A31394B068315567D222C07B544BC880BF93946F1361764F1D068E114A749091F03B53E1797D358C3EA807D6 (accessed June 25, 2020)

[433] "CNNC Completes Design of District Heating Reactor," *World Nuclear News*," Sept. 7, 2018. https://www.world-nuclear-news.org/Articles/CNNC-completes-design-of-district-heating-reactor (accessed June 25, 2020)

[434] Jean-Marc Borgard and Michel Tabarant, "CO2 to Fuel Using Nuclear Power: The French Case," *Energy Procedia*, vol. 4, pp. 211-3 to 2120, 2011.

[435] Rachel Millard, "Nuclear looks to hydrogen in a bid to secure its future," *The Telegraph*, Nov. 29, 2020. https://currently.att.yahoo.com/att/nuclear-looks-hydrogen-bid-secure-150042845.html?.tsrc=daily_mail&uh_test=1_11 (accessed Nov. 30, 2020)

reactors can provide both heat from smaller, more accessible, sources and heat delivered at hotter temperatures than current reactors.

What are the additional uses available from cheap fission heat and electricity delivered in smaller—say truck-delivered—components. There are thousands of uses that are local and relatively small. A truck-borne-reactor heat source could replace coal or natural gas burners. Possible uses include

- Limestone Calcination to Calcium Oxide for Concrete ($CaCO_3$ to CaO) is by far the largest-tonnage human-performed geological process. The calcium oxide combines with water to make cement that holds aggregate (sand and/or pebbles) and some type of support rebar to be the reinforced concrete in everything from house foundations to highways and giant power dams. The cement to make concrete is made by baking limestone at 900°C (1652°F) and hotter to drive carbon dioxide from the rock, leaving only the lime of calcium oxide (CO) while emitting global-warming carbon dioxide.[436] The baking heat comes from burning coal or natural gas, which emits more carbon dioxide. The result is that 7% of the carbon dioxide entering the atmosphere comes from making cement.[437,438] Small modular reactors could supply 700°C of that heat (1,300°F) to preheat the limestone with the balance of the heat required supplied by fossil fuel (or electricity if volume production of SMRs drives electric prices low enough). Furthermore, as noted earlier, there would be byproduct carbon dioxide from the limestone. Next-generation reactors could provide heat and/or electricity to process this carbon dioxide into useful chemicals. Bottom line: There would be a respectable decline in carbon dioxide emission from cement production.

- Petroleum refineries burn roughly 15 to 20% of the chemical energy in the petroleum to boil the petroleum and distill it into the fractions ranging from the lightest of natural gas through gasoline and on down to asphalt, the heaviest. Thus, using fission heat for refining could move the same number of cars and trucks with a smaller carbon footprint. Furthermore, hydrogen is used in reforming heavier oil fractions to get lighter higher value products such as gasoline and diesel fuel. Using nuclear-cracked hydrogen from water could decrease the carbon footprint of vehicles even more.

---

[436] Antonia Moropoulou, Asterios Bakolas, Eleni Aggelakopoulou, "The Effects of Limestone Characteristics and Calcination Temperature to the Reactivity of the Quicklime," *Cement and Concrete Research*, vol. 31, issue 4, pp. 633–639, April 2001.

[437] Robbie M. Andrew, Global CO2 Emissions from Cement Production, Earth System Science Data, vol. 10, pp. 195–217, 2018. https://www.earth-syst-sci-data.net/10/195/2018/essd-10-195-2018.pdf (accessed Jan. 21, 2019)

[438] Vanessa Dezem, "Cement Produces More Pollution Than All the Trucks in the World," *Blomberg*, June 22, 2019. https://www.bloomberg.com/news/articles/2019-06-23/green-cement-struggles-to-expand-market-as-pollution-focus-grows (accessed June 24, 2019)

- Heat serves <u>agricultural production</u> in many ways. It transforms plant juices into dried and concentrated products such as sugar and alfalfa meal. Nuclear heat replacing burning of sugar cane fiber (bagasse) would allow its use for cloth or paper.) Heat dries grains for storage. It dries lumber and processes plywood. It cracks grains to allow fermentation and distillation for concentrated ethyl (such as vodka or bioethanol) alcohol. It boils vegetables for canning and provides hot water for cleaning of equipment. The leaf-protein concentrate mentioned earlier would be more profitable with cheaper heat for drying and cheaper mechanical energy for processing. The United States has thousands of agricultural facilities that could replace carbon-dioxide-producing heat with nuclear.

- The <u>biochar</u> and <u>charcoal</u> production mentioned earlier would become more profitable as the final stage of agricultural production and disposal of unused waste food and/or yard-waste material. When all other options have been used, transform it into loose charcoal (biochar) for application back in the soil where it can stay for centuries … and application into the soil eliminates tipping fees at the dump. Alternatively, in developing countries, compressed charcoal briquettes provide cleaner cooking heat without the toxic smoke of burning twigs and dung. An additional benefit is that those materials can then be incorporated back into the ground for soil building.

- <u>Water desalinization</u> using nuclear fission has been a goal since the first power reactors were being designed. People use more <u>water</u> than any other commodity. We use it for growing plants, processing food, drinking, bathing, industrial processes and more. Also, there is an infinite supply of water in the oceans; that water just needs de-salting or desalination. Unfortunately for great desalination hopes, water is still very cheap so it would be hard to justify expense except that desalination could provide a use during summer for unneeded heating energy when there is no need for district heating.

- <u>Water Splitting/Cracking to Get Hydrogen:</u> A key chemical for many processes is hydrogen (H_2), and it is as common as water. In fact, two hydrogen atoms are in every water molecule along with one atom of oxygen (H_2O). However, separating the hydrogen from the oxygen requires at least as much energy as the hydrogen can deliver as a fuel. Thus, a major study concluded that hydrogen from solar and wind would cost three times as much as natural gas.[439] Today, hydrogen producers generally get it more cheaply by cooking coal (C) or natural gas CH_4) with steam (gaseous (H_2O). Oxygen in the water goes to the carbon leaving hydrogen. However, it would be cheaper to just burn the coal or natural gas in the first place—the same amount of global-warming goes into the atmosphere. Next-generation nuclear fission will be able to split the hydrogen and oxygen in water with electrical splitting (electrolysis), more efficient high-temperature electrolysis, or straight

---

[439] National Academy of Engineering and National Research Council, *The Hydrogen Economy: Opportunities, Costs, Barriers, and R&D Needs*, 2004.

thermal decomposition at higher temperature. (Thermal decomposition of water can be done at 500°C [932°F], but the efficiency increases significantly as temperature increases.) The hydrogen produced can replace fossil fuels in a number of ways to be discussed further.

8.4.3 Using Mass-Produced Hydrogen and Heat

When affordable hydrogen, heat, and (if needed) carbon dioxide) are available, it can be used to synthesize fuels and process chemicals in many ways. Some of the most likely ways are for use of the following materials.

- Ammonia (NH_4) for fertilizer, or even liquid fuel, can be synthesized with heat in the Haber-Bosch process. Nuclear-produced hydrogen would replace 3 to 5% of world natural gas production—and the corresponding greenhouse carbon dioxide emissions. Ammonia synthesis is commonly done at 450°Celsius (842°Fahrenheit).

- Hydrochloric acid (HCl) is produced by burning hydrogen in chlorine and then dissolving the resulting hydrogen chloride in water. Hydrochloric acid is called a "workhorse chemical" because it is used for many products, including textiles, rubber, dyes, and plant fertilizers. The chlorine to make hydrochloric acid is produced by electrolysis from water and table salt (NaCl), respectively.

- Methane (natural gas, CH_4) can be synthesized from hydrogen produced by electrolysis or thermal cracking. Furthermore, carbon dioxide can be electrolyzed to yield just the carbon, which can more easily combine with hydrogen to make methane.

- Methyl alcohol (or methanol, CH_3OH) is synthesized from carbon monoxide (CO) and hydrogen. Methanol is an important material in chemical synthesis. Its derivatives are used in building up a vast number of compounds, including dyes, resins, pharmaceuticals, perfumes, solvents, and antifreezes. Methanol is also a high-octane, clean-burning fuel that is a potentially important substitute for gasoline in automotive vehicles, or it can be processed into dimethyl ether, a less caustic liquid than methanol and one that can directly replace gasoline.

- Ethyl alcohol (or ethanol, C_2H_6O, or more properly C_2H_5OH) has been fermented from grains and sugars for thousands of years, which is discussed elsewhere with the arguments for and against. It can serve as the previously mentioned biofuel, but it also has uses as a solvent and as a feedstock to produce a variety of chemicals. Ethanol can also be produced inorganically. One promising process uses electrolysis of water and carbon dioxide under low temperature and pressure to synthesize ethanol.[440] There are also improved processes to more efficiently

---

[440] Joseph E. Harmon, "Researchers Discover New Electrocatalyst for Turning Carbon Dioxide into Liquid Fuel," *Phys.org*, Aug. 5, 2020. https://phys.org/news/2020-08-electrocatalyst-carbon-dioxide-liquid-fuel.html (accessed Aug. 7, 2020)

separate carbon dioxide from waste streams such as from power plant combustion, calcining limestone for cement, and steel refining[441,442]

- Plastics production has many processes that require inputs of heat, hydrogen, and electrical power. Inputs of hydrogen can compensate for the many varieties of plastics that come back in the recycling stream.

- Heat serves agricultural production in many ways. It transforms plant juices into dried and concentrated products such as sugar and alfalfa meal. Nuclear heat replacing burning of sugar cane fiber (bagasse) would allow its use for cloth or paper.) Heat dries grains for storage. It dries lumber and processes plywood. It cracks grains to allow fermentation and distills for concentrated ethyl (such as vodka or bioethanol) alcohol. It boils vegetables for canning and provides hot water for cleaning of equipment. The leaf-protein concentrate mentioned earlier would be more profitable with cheaper heat for drying and cheaper mechanical energy for processing. The United States has thousands of agricultural facilities that could replace carbon-dioxide-producing heat with nuclear.

- The biochar and charcoal production mentioned earlier would become more profitable as the final stage of agricultural production and disposal of unused waste food and/or yard-waste material. When all other options have been used, transform it into loose charcoal (biochar) for application back in the soil where it can stay for centuries … and application into the soil eliminates tipping fees at the dump. Alternatively, in developing countries, compressed charcoal briquettes provide cleaner cooking heat without the toxic smoke of burning twigs and dung. An additional benefit is that those materials can then be incorporated back into the ground for soil building.

- Tar sands are petroleum deposits that have partially evaporated away over the eons, leaving deposits weighted toward the heavy asphalt end of the petroleum fractions. Still, these heavy oils can be fractured and reformed with heat and hydrogen into usable products, and the amount of available gasoline, diesel fuel, etc. is much greater than from conventional petroleum. There are great lakes of tar sands (such as areas in Athabasca, Canada delivering two million barrels per day and Lake Maracaibo, Venezuela capable of providing much more). Extracting these heavy deposits requires steam to flow to a refinery and more natural gas and/or coal to reform it into more useful materials such as gasoline and diesel fuel. These fossil fuels could themselves be fuels. How much better to supply nuclear heat without

---

[441] Mohammad Songolzadeh, Mansooreh Soleimani, Maryam Takht Ravanchi, et al., *Scientific World Journal*, vol. 2014, Article ID 828131, Feb. 17, 2014. https://www.hindawi.com/journals/tswj/2014/828131/ (accessed Oct. 29, 2020)

[442] David Chandler, "Engineers Develop a New Way to Remove Carbon Dioxide from Air," Techxplore.com, Oct. 25, 2019. https://techxplore.com/news/2019-10-carbon-dioxide-air.html (accessed Oct. 27, 2020)

- Another 1% of world greenhouse gas emissions comes from electroprocessing of aluminum ore into metal, roughly half from hydroelectricity and half from fossil-fueled power with growing use of coal. Furthermore, aluminum production is expected to double or triple by 2050.[443] Recycling aluminum requires only about 6% of the energy needed to refine aluminum from ore, and it is mostly heat rather than electricity. That is why the United States gets more aluminum from recycling than from primary production. Natural gas typically supplies the recycling processing heat of 650 to 750°Celsius (1,200 to 1,380°Fahrenheit). This is within the range of next-generation small modular reactors. Thus, aluminum recycling is a prime candidate for nuclear process heat.

- Steel production generates between 7% and 9% of direct carbon dioxide emissions from the global use of fossil fuel.[444] Steel mills produce global warming gases in two ways, heating iron oxide ore mix and carbon (coking coal) placed in the ore to remove oxygen from the ore. Regular coal supplies the heat for preheating air to between 540° and 870° C (approximately 1000° and 1600° F) to the furnace. Inside the furnace, coal with everything baked out but the carbon (coking coal) burns to carbon dioxide, drawing oxygen from the ore. The carbon dioxide goes to the atmosphere, and steel is left behind. Hydrogen-rich gas substituting for the carbon reduces the amount of carbon dioxide emitted. Natural gas (also called methane or CH_4) has already been incorporated in many steel mills. The next logical steps are nuclear-provided hydrogen replacing coking coal[445] and much of the pre-heat coming from nuclear heat.

the carbon dioxide emissions from burning the fossil fuels. Again, fission reactors can supply more petroleum products with less emissions of global warming agents.

In the 1950s, then President Dwight Eisenhower had a program called Atoms for Peace with the goal of sharing the power of nuclear fission. Promoters hoped reactors would make "electricity too cheap to meter." Nuclear reactors will probably never get that cheap, but the advanced concepts discussed earlier in this chapter can produce electricity cheaper than coal, solar, wind, and even natural gas … while leaving a smaller footprint of carbon, minerals, and land area.

---

[443] Ali Hasanbeigi, "Aluminum Industry: 10 Emerging Technologies for Energy-efficiency and GHG Emissions Reduction," *Global Efficiency Intelligence*, Nov. 21, 2017. https://www.globalefficiencyintel.com/new-blog/2017/technologies-energy-emissions-aluminum-industry (accessed Aug. 28, 2020)

[444] "Steel's Contribution to a Low Carbon Future," World Steel Association, Brussels, Belgium, 2019. https://www.worldsteel.org/about-us/who-we-are.html (accessed Sept. 15, 2019)

[445] Max Åhman, Olle Olsson, Valentin Vogl, et al., *Hydrogen Steelmaking for a Low-Carbon Economy: A joint LU-SEI Working Paper for the HYBRIT Project*, SEI working paper WP 2018-07, Stockholm Environment Institute, Sweden, Set. 2018. https://www.sei.org/wp-content/uploads/2018/09/hydrogen-steelmaking-for-a-low-carbon-economy.pdf (accessed Sept. 15, 2019)

Furthermore, the next generation of fission reactors can benefit from the advances of efficiency developed for fossil-fuel power stations and even proposed green energy systems. Those advances extend the range of heat harvested by two advances referred to as topping cycles and bottoming cycles. A topping cycle extends the higher (top) temperature by expelling hot gas through a gas turbine (much like a jet engine), which can run hotter than a steam cycle. Then, there is still enough heat remaining to run that steam turbine. Finally, another liquid that boils at a cooler temperature (such as ammonia) can be evaporated and the vapor expanded through a turbine—that being a bottoming cycle.

Adding a topping cycle and a bottoming cycle can double the efficiency of a thermal power plant from roughly a third to two thirds of the theoretical efficiency. Thus, the amount of available power generated roughly doubles. In total, the power that can be generated compared to existing fission reactors is roughly:

- 50 × existing uranium fuel using reprocessing & breeding = 50
- 3 × amount thorium than uranium (3 × 50) = 150
- 50 + 150 for the sum of those two resources = 200
- 2 × energy production with topping & bottoming = 400

In short, the amount of fission power available is roughly 400 times what was expected a few years ago. Thus, the available power is available for thousands of years, and the environmental impact of mining and processing per unit of power will be vastly reduced compared to today's once-through-and-store-for-millennia uranium/plutonium fuel cycle. That extra energy can more than pay for the expenses of reprocessing and breeding away most of the nuclear waste products.

Consequently, nuclear fission reactors could produce heat and electrical power for thousands of years. Furthermore, that heat and power would probably be cheaper than coal, solar, wind, and even natural gas.

Major Things for Governments and Businesses to Do

- Government: Expand research on generating synthetic liquid fuels of hydrocarbon and/or ammonia from hydrogen generated from surplus peak energy especially from low-load due to late night and mid-day peaks from solar.

8.5 Overchoice in Energy Economies—The Hydrogen, Methanol, Methane, Electricity, ... Ammonia Economy?

Caution: This section can only apply if more affordable fission (or much improved wind and/or solar) can provide cheap electricity and/or heat. Hopefully, it will be so. With that done, there will probably be an over-choice of ways to replace much fossil-fuels use with nuclear heat, nuclear-/solar-/wind-generated electricity and materials produced by nuclear processing.

The GND and related proposals for eliminating use of fossil fuels often refer to the hydrogen economy. We shall compare the previous energy economies, the proposed hydrogen economy, and several of the most likely alternatives.

First, we need some background. For thousands of years, we humans got our energy almost entirely from muscle power and wood burning; that might be termed the muscle and wood economy. During the Middle Ages, coal burning grew from occasional localized use to being a major commodity. By the late 1800s, coal was the predominant energy source heating homes, powering ships and trains, smelting metals, and providing chemicals. That era has often been described as the coal age and its economy the coal economy.

However, a rival energy source, petroleum (oil for short), was already rising because oil has many advantages over coal.

- Oil can be pumped up from the ground and transported more easily than coal.

- The various oil chemicals (fractions) in oil have a variety of uses available from the lightest propane and gasoline down to asphalt for roads) simply by separating out the different fractions.

- The liquid fuels derived from oil (such as bunker oil, diesel fuel, and gasoline) have much more energy per unit weight and per unit volume than coal. Oil is a mixture of hydrocarbon compounds with differing numbers of hydrogen and carbon atoms. Burning hydrocarbon fuels releases the energy of not only burning those two elements but also the energy of breaking the hydrogen–carbon bonds within those compounds.

- The liquid hydrocarbon fuels can also be delivered to engines and boilers with a simple fuel pump rather than a fireman shoveling coal or a mechanical worm drive carrying large amounts of coal. Thus, liquid hydrocarbon fuels are more practical than coal for small uses such as cars and house furnaces.

- The quality of hydrocarbon fuel can be maintained more uniformly than coal that may have different heating values, may have varying water content, and may have other minerals mixed in. Some of them are toxic (such as mercury, lead, and uranium).

- Liquid hydrocarbon fuels do not produce a residue of ash and clinkers that must be removed from the place where coal fuel was burned.

- Liquid hydrocarbons can be refined to burn cleaner than coal to produce much less soot and other pollution.

For all these reasons, oil was replacing coal in many uses during the first half of the 1900s. This was particularly so for transportation, in which the switch to oil meant greater range and/or more cargo. There was no question about fuel for the new markets of horseless carriages and airplanes; the power had to be from oil products.

Then, the energy economies got doubly complicated with the increasing use of natural gas (CH_4). Natural gas has the disadvantage of being low density compared to gasoline (but much denser than hydrogen). People driving cars with compressed natural gas generally have to sacrifice much of their trunk space for the compressed natural gas tank.

However, natural gas burns very cleanly, much cleaner than the liquid hydrocarbons. It does not even produce soot unless there is a malfunction of the burner. Hence, home heating, water heating, and cooking have largely gone from wood and coal to natural gas.

Taking coal, oil, and natural gas together, the world's transportation and chemical processing today functions mostly on those fossil fuels with most of transportation being a liquid-fossil-fuel economy, and transportation representing about 15% of greenhouse warming. There has been a string of periodic scares that fossil fuels would eventually become scarce and expensive, but each time geologists and energy companies have found new deposits and better ways to extract and use fossil fuels.

Now, of course, the new concern is excessive global warming resulting from burning of fossil fuels. Those warming agents include carbon dioxide, oxides of nitrogen, methane (from leaks), and soot. As noted earlier, solar and wind are presently expensive because of intermittency and limited in their practical locations. Hence, nuclear fission seems to be the best option available.

After two fission nuclear bombs ended World War II, it was immediately suggested that such power could and should be adopted for peaceful power generation instead of war. In the United States, there was a program called Atoms for Peace.[446] Fission power reactors were built throughout the world starting in the 1950s, and it was widely expected that reactors would dominate the economies of the world with nuclear-supplied electricity replacing oil and natural gas for water heating, house heating and cooling, and cooking. The best term for this would have been the nuclear–electric economy.

However, the nuclear–electric economy was slowed by several factors.

1. It is inherently inefficient to boil water and generate electricity at one-third efficiency, then lose a tenth of that third while transmitting electricity through a vast expensive grid of power lines just to heat bath water at the other end of the

---

[446] The rationale for the program was from *Isaiah 2:4 in the Bible: And he shall judge among the nations, and shall rebuke many people: and they shall beat their swords into plowshares, and their spears into pruninghooks: nation shall not lift up sword against nation, neither shall they learn war any more.*

grid. Even if the gas heater were to be only 90% efficient (with some warm smoke going up the chimney, you could get three times as much hot bath water in your house per unit of energy than with a gas heater than with electric heating. This is the key argument made by energy-efficiency proponent Amory Lovins in a string of publications including *Soft Energy Paths*.[447]

2. Nuclear–Electric power could not compete with liquid fossil fuels in powering cars and trucks because they would have required batteries better than any available at the time. Even the lithium batteries of the early 2000s are only competitive with liquid fossil fuels because of government subsidies. An electric car operates more efficiently and requires less maintenance, but it has a higher initial cost, so the electric owner may need 10 years to make a profit.[448] Furthermore, the major unnoted cost is that states charge heavy taxes on gasoline and diesel fuel, but electric-vehicle owners presently get a free ride. At some point, if electric vehicles grow into a major fraction of the vehicle fleet, cash strapped states will see their gasoline-tax revenues declining, and they will tap into a new revenue source— taxing car mileage. Ka-ching! In 2020, the state of Oregon began implementing a charge of 1.8 cents per mile (1.1 cents per kilometer) for electric vehicles.[449] It's the shape of things to come.

3. Fission reactors were associated (correctly) with atomic bomb technology. This caused suspicion and resistance to reactors, particularly from environmentalists and leftists. Political and legal resistance to every proposed power reactor increased costs and uncertainty for the investors.

4. As noted elsewhere, the first-generation reactors were touchier and more expensive to run than expected, causing overall prices to increase.

5. Meanwhile, those pesky competing fossil fuel power plants kept getting more efficient and more affordable.

6. The reactor makers responded by generating bigger and more complex designs, hoping to reap more efficient return per unit of input. Unfortunately, each of the new designs had to be run the gauntlet of government approvals and political challenges that drove costs constantly higher.

7. There were the major melt-down accidents at Three Mile Island in the United States, Chernobyl in the Ukraine (then a republic within the Union of Soviet Socialist Republics), and Fukushima Daiichi in Japan. Only Chernobyl had

---

[447] Amory B. Lovins, *Soft Energy Paths*: Towards a Durable Peace, (originally published by) Ballinger Publishing Co, Cambridge, Mass, 1977.

[448] "Electric vs Gasoline – Which is more cost effective?" *True Cost Blog*, Jan. 4, 2009. https://truecostblog.com/2009/01/04/electric-vs-gasoline/ (accessed Aug. 18, 2020)

[449] "OReGO: Oregon's Road Usage Charge Program," Oregon Department of Transportation website, undated. https://www.oregon.gov/odot/Programs/Pages/OReGO.aspx (accessed Sept. 25, 2020)

fatalities, but the public quite rightly got the impression that other fission reactors might fail spectacularly at any time.

8. As noted earlier (8.2.2 The Basic Lesson on Nuclear Fission Reactors), fears of nuclear materials led to resistance to long-term storage of nuclear waste, and that became another major expense and political concern in the United States.

The result of all these problems was a worldwide slowing of fission reactor construction, and a full-blown retreat from fission in many countries.

Meanwhile, the environmental movement exploded into prominence during the late 1960s. Besides opposing nuclear fission, the environmentalists have always opposed what they call "dirty fossil fuels." They did propose greater efficiency of end use, but efficiency could only do so much with a growing world economy. What was left?

On cue in 1970, chemist John O'Mara Bockris proposed the solar–hydrogen economy.[450] Solar-generated electricity (either from thermal collectors or photovoltaic panels) would break water into hydrogen and oxygen. Then the hydrogen would be distributed to run cars, heat bath water, and so forth. Burning hydrogen (H_2) with oxygen (O_2) would release only water (H_2O) rather than the global-warming carbon dioxide. Even better, hydrogen fuel cells could electrically combine hydrogen with oxygen—just the opposite of electrolysis—to generate electricity at two or three times the efficiency of internal-combustion car motors.

Environmentalists embraced the concept (often shortening it to just the hydrogen economy), and much of the nuclear fission community joined in with a nuclear–hydrogen variant. The hydrogen economy is a major, largely unspoken, component of the Green New Deal.

However, the hydrogen economy has its own major failings. As newspaper editorialist H. L. Mencken once said,

For every problem, there is a solution that is simple, elegant, powerful, ... and wrong.

The hydrogen economy may be just such a wrong solution. Both conservatives and many environmentalists complained about some serious practical problems detailed in books such as *The Emperor's New Hydrogen Economy*[451] and *The Hype about Hydrogen*.[452]

1. Hydrogen has incredibly low density. Remember natural gas tanks must be expanded into automobile trunk space, and natural gas has twice the energy per unit volume as hydrogen, so storing enough fuel to run a car would require some type of exotic storage or a very big tank. Such storage might include super-strong high-compression storage tanks and similarly strong transmission pipes and hoses to the engine (which is being done), cryogenic cooling (−253 °C, −423 °F) to keep the

---

[450] J. O'M. Bockris, The Origin of Ideas on a Hydrogen Economy and Its Solution to the Decay of the Environment," *International Journal of Hydrogen Energy*, vol. 27, issues 7–8, pp. 733–740, July–Aug. 2002.

[451] Darryl McMahon, *The Emperor's New Hydrogen Economy*, iUniverse, Inc., 2006.

[452] Joseph J. Romm, *The Hype About Hydrogen: Fact and Fiction in the Race to Save the Climate*, Island Press, New Edition, 2005.

hydrogen liquid (a car would need its cryo-refrigerator going when parked), or materials that could adsorb and release the hydrogen when needed. None of these hydrogen-storage systems are easy or cheap.

2. Hydrogen gas is a small molecule that is the most prone of all elements to leaking past seals or even through some metals.

3. Those small hydrogen molecules can intermingle with the atoms of many metal alloys causing embrittlement, which can lead to leaks and fires ranging from minor to catastrophic. This is particularly so for pipelines.

4. Last but most important, hydrogen is an energy storage medium or carrier rather than an energy source. Using some of these other energy carriers might be more efficient energy carriers (making them more profitable) than using them to make hydrogen—and then burning the hydrogen. Remember, they are all just energy carriers.

In fact, several other energy economies were being considered for the following reasons:

1. A resources computer simulation study sponsored by the prestigious Club of Rome suggested that supplies of affordable resources (including oil and natural gas) would be exhausted within a few decades.[453] Several analysts within the oil industry also forecast catastrophic shortages of oil and gas.

2. The oil market was heading toward one of its periodic peaks of high prices and panic, which has always been followed by new deposits and a new trough of low prices. This time it seemed to confirm the shortage forecasts of permanent exhaustion. Oil production in the area around Saudi Arabia started coming on after World War II, and oil from those countries was so cheap that the low oil prices depressed oil production elsewhere even while use of oil was increasing. An eventual price rise was inevitable.

3. Meanwhile, prices were so low that the oil producers felt they were being unfairly exploited by the oil-production companies and the developed-world customers. Consequently, many of the oil-producing companies formed a production cartel (the Organization of Petroleum Exporting Countries, OPEC) to drive prices up by agreements to limit their production. They had succeeded in getting some price increases.

4. In 1973, Arab countries around Israel made a coordinated set of surprise attacks that had the Israelis reeling toward defeat. The United States sent supplies that allowed Israel to repulse the invading armies, but the Arab oil producers were furious. They seriously reduced production, and world oil prices shot up. The price spike demonstrated not only the power of Arab oil producers but of the OPEC oil cartel in general. OPEC quadrupled oil prices around the time of "the Oil Crisis"

---

[453] Donella H. Meadows, Dennis L. Meadows, Jorgen Randers, and William W. Behrens, III, *The Limits to Growth*, 2nd ed.; Universe Books (N.Y., NY); 1974.

(now called "The First Oil Crisis") and was able to maintain high prices for several years. Those prices drove the economies of consumer countries into recession and stagnation.

5. In 1979, just as prices were drifting slowly back down, the government of Iran fell. The major oil supply from Iran ceased for some time, and the new government lobbied for more aggressive OPEC production limits. World oil prices doubled again. Economic pain in the developed world increased again during "The Second Oil Crisis."[454]

All these events suggested that something called "peak oil" was eminent. The peak-oil doomsayers predicted that as the last new oil fields were drilled, the amount of new production would decline. Meanwhile, older oil fields would produce less over time. These declines, would lead to rising oil prices. Worse, just as oil production had increased rapidly over the decades leading to that peak, the decline rate would match the rise with a corresponding rapid decline, much like a bell-shaped curve. The resulting production crash would be far worse than the First and Second Oil Crises.

M. King Hubbert was the first to codify this progression and plot it mathematically. In the 1950s and 1960s, He analyzed production in a number of oil-producing areas. Hubbert used his analysis to predict that oil production in the United States lower 48 (not counting Alaska and Hawaii) would reach a peak about 1965 or (as his less-likely "contingency case") 1970 and begin a downward trend. History proved him largely correct in that production in the lower 48 did peak in 1970 matching his backup 1970 prediction.

In 2001, Kenneth Deffeyes released *Hubbert's Peak: The Impending World Oil Shortage* in which he extended and updated Hubbert's predictions of U.S. peak oil production with more sophisticated mathematical analyses and more extensive updated data. He used those analyses to predict a world peak in oil production in about 2005.[455] That prediction meshed with world oil prices that were above $100 per barrel in much of the early 2000s. In 2006, Deffeyes followed up with another book that extended the resource-estimating methodology to the other energy resources that he suggested were running out—except uranium, which he concluded was much more available than commonly believed.[456]

In *Twilight in the Desert: The Coming Saudi Oil Shock and the World Economy*, Mathew Simmons, a prominent oil and gas market analyst, argued that Saudi Arabian oil production, particularly from Ghawar—the world's largest oil field—would peak in the near future (perhaps—

---

[454] Daniel Yergin, *The Prize: The Epic Quest for Oil, Money & Power*, (originally published by Simon and Schuster, 1992), Free Press; Reissue Edition. Dec. 23, 2008.

[455] Kenneth S. Deffeyes, *Hubbert's Peak: The Impending World Oil Shortage*, Princeton University Press, New Jersey, 2001.

[456] Kenneth Deffeyes, *Beyond Oil: The View from Hubbert's Peak*, Hukk and Wang, 2006.

there it goes again—2005) and that the Saudi government was conspiring to hide the impending production drop.[457]

These two expert pronouncements helped fuel any number of conspiracy-leaning and apocalyptic views of petroleum and natural gas futures. Chief among them was the "peak oil" movement with a number of books, newsletters, and even a Peak Oil website. The central tenet of the Peak Oil movement was that the impending drop in oil production would lead to a general collapse of civilization back to an 1800s level of technology with a resulting dieback of population. Peak oil blogs were rife with articles about plowing with mules, finding quiet villages that would not have large mobs of hungry ill-prepared neighbors, and (of course) fire arms with big ammo clips for defending against those same ignorant improvident slackers.

One resulting proposal was to replace the existing gasoline and diesel-fuel for cars and trucks with electricity. However, that would entail a huge additional energy cost and a huge financial cost of developing new infrastructure. Consequently, governments and industry pursued other options including synthetic fuel. The options that received the most government and investment support were:

1. Land-Based Biofuel (Ethanol, Ethyl Alcohol, C_2H_5OH) and plant oil for biodiesel: As noted earlier, ethanol is a major United States program for supplementing gasoline fuel for cars. Also as noted, there are two major limitations. First, the net energy return on investment (EROI) is low compared to petroleum. After the energy expenditures of plowing, planting, harvesting, transporting to the biofuel facility, malting, fermenting, and distilling the alcohol, corn bioethanol yields an EROI of about 1.3 for 1 unit of energy put in, put in while petroleum yields an EROI of 16. Likewise, biodiesel comes from soy beans in temperate climates and palm oil in the tropics. They have a better EROI than bioethanol at about 5, but that is still less than petroleum.[458] Second, the fuel comes at the expense of food that is processed into alcohol, so any food shortages could be made worse by biofuel production. As noted under food, the possible game changer for bioethanol is profitable technology to break the major structural plant molecules, such as cellulose and hemicellulose, into fermentable sugars. Another (but still theoretical) method employs a single catalyst to do multiple steps from food waste to chemicals

---

[457] Mathew Simmons, *Twilight in the Desert: The Coming Saudi Oil Shock and the World Economy*, Mathew Simmons, John Wiley & Sons, 2005.

[458] Jude Clemente, "Why Biofuels Can't Replace Oil," *Forbes*, June 17, 2015. https://www.forbes.com/sites/judeclemente/2015/06/17/why-biofuels-cant-replace-oil/#4247fbbaf60f (accessed Aug. 29, 2020)

8 Fixes from Energy and Material

or fuel.[459,460] It would be great, but like bringing fusion star power down to Earth, practical production always seems to be twenty years away.

2. Coal to Methane (natural gas, CH_4): From the 1800s to the mid-1900s, coal was heated with steam to make a mixture of carbon monoxide (CO) and methane. This synthetic fuel was the original "town gas" that was pipelined throughout cities for lighting and heating, so coal-to-gas would have been a replay of that previous technology. President Carter approved building of several coal-to-methane facilities. A major coal-to-gas program could have been done, but the lifting of gas price controls caused natural gas production to increase so much that synthetic methane gas could not compete. Then, even the cheaper natural gas fuel could not compete with gasoline and diesel fuel when oil prices collapsed back into one of their low-price troughs during the 1980s, so governments and businesses lost interest any kind of methane as a transportation fuel. Finally, from an GND perspective, coal-to-methane production on a national scale would have vastly increased global-warming carbon dioxide emissions into the atmosphere.

3. Coal and Methane to Methanol (wood alcohol, CH_3OH): Making methanol is just one chemical step farther from methane to make a liquid fuel that could substitute for gasoline (although more corrosive than gasoline). There were serious proposals for a methanol economy because liquid transportation fuel is much more practical and energy dense for vehicle use than natural gas.[461] Furthermore, flexfuel cars that can adjust to burn gasoline, ethyl alcohol, or methanol have been sold for years although not in large numbers.[462] The two biggest barriers were setting up all the pipelines and gas-station pumps for providing the methanol to cars and building cars that could run on methanol—two considerable investments. As with compressed natural gas, the oil price collapse of the early 1980s, prevented any major implementation of nationwide methanol fuel. And again, this other coal use would have also caused a major crease in carbon dioxide emissions.

4. Coal to Oil Fractions: Yes, the Germans made synthetic gasoline from coal during World War II. Similarly, South Africa made fuels from coal when their government was being boycotted by oil producers, and this production continued even after the

---

[459] Mark A. Isaacs, Christopher M. A. Parlett, Neill Robinson, et al. "A spatially orthogonal hierarchically porous acid–base catalyst for cascade and antagonistic reactions," *Nature Catalysis*, Oct. 26, 2020. https://www.nature.com/articles/s41929-020-00526-5#citeas (accessed Nov. 8, 2020)

[460] Brian Wang, "Breakthrough for Remaking Cooking Oil and Food Waste into Plastic or Biodiesel," *Nextbigfuture*, Nov. 5, 2020. https://www.nextbigfuture.com/2020/11/breakthrough-for-remaking-cooking-oil-and-food-waste-into-plastic-or-biodiesel.html#more-167482 (accessed Nov. 8, 2020)

[461] George Olah, Alain Goeppert, and Surya Prakesh, *Beyond Oil and Gas: The Methanol Economy*, Wiley, March 2006.

[462] Robert Zubrin, *Energy Victory: Winning the War on Terror by Breaking Free of Oil*, Prometheus, April 10, 2009,

boycotts were lifted. The South African company doing that synthesis has also added ammonia fertilizer to its production. Unfortunately, for this and the previous two options, using coal to replace a major fraction of transportation fuel would have exhausted conventionally mined coal within a few decades, as well as (one more time) seriously contributing to global warming.

5. Carbon Dioxide to C_2 Chemicals: If one were to have sufficient affordable energy, carbon dioxide would no longer be a waste product to be emitted. Instead, it would be a resource. Chemical catalysts have recently become available to electroreduce carbon dioxide (CO_2) with high efficiency to chemicals such as ethane (C_2H_6) that can be converted to many other chemicals such as ethanol, methanol, or ethylene.[463,464] One proposal is to separate carbon dioxide (CO_2) from the other atmospheric gases so that it could be processed it into some useful product except that there are two problems: (1) concentrating the carbon dioxide costs energy; and (2) converting it into something useful costs more energy. Energy equals money. The equipment and processing costs to get the carbon dioxide would be about $94 to $232 per ton ($85 to $210 per metric ton).[465] A more affordable way to get carbon dioxide would be to mine existing exhaust streams with high percentages of the substance (breweries, steel mills, cement calcining facilities).

6. Exotic fossil-fuel deposits of natural gas: During the two energy crises, it was recognized that vast potential gas resources might be tapped when technology evolved sufficiently. First, areas such as the Texas–Louisiana Gulf Coast have a type of often large natural gas deposits called geopressured methane.[466] These deposits are hot, salty, and under pressure—all of which can be harnessed for electrical power. Unfortunately, the triple energy comes at a triple chance of ripping a drill rig apart. The biggest exotic potential energy source is methane hydrates (or clathrates). Water and methane can freeze together under high pressure to hold vast quantities of methane and act as cap rock to hold additional methane

---

[463] Taehee Kim and Tayhas R. Palmore, "A Scalable Method for Preparing Cu Electrocatalysts That Convert CO2 into C2+ Products," *Nature Communications*, vol. 11, July 17, 2020. https://www.nature.com/articles/s41467-020-16998-9#ethics (accessed Aug. 15, 2020)

[464] Ayan Maity, Sachin Chaudhari, Jeremy J. Titman, and Vivek Polshettiwar, "Catalytic Nanosponges of Acidic Aluminosilicates for Plastic Degradation and CO₂ to Fuel Conversion," *Nature Communications*, vol. 11, July 31, 2020. https://www.nature.com/articles/s41467-020-17711-6#Sec2 (accessed Aug. 22, 2020)

[465] David W. Keith, Geoffrey Holmes, David St. Angelo, and Kenton Heidel, "A Process for Capturing CO₂ from the Atmosphere," *Joule*, vol. 2, issue 8, pp. 1573–1594, Aug. 15, 2018. https://www.cell.com/joule/fulltext/S2542-4351(18)30225-3 (accessed Sept. 16, 2020)

[466] Chacko J. John, Brian J. Harder, Reed J. Bourgeois, and Raymond Fortuna, "The Gulf Coast Geopressured-Geothermal Gas Resource: A Multipurpose, Environmentally Safe and Potentially Economic Reality in Today's Market?" *Gulf Coast Association of Geological Societies Transactions*, vol. 56, p. 303-307, 2006. https://kgs.uky.edu/kgsweb/geothermal/la/Louisiana_Document5.pdf (accessed Aug. 22, 2020)

and/or oil that may percolate up from hotter geologic sedimentary rocks below. (Much of the North Atlantic and South Atlantic deep waters [abyssal plains] could be oil and gas provinces.) Methane hydrates ("flammable ice") probably hold two or three times more energy than all other fossil fuels combined.[467] Once again, the pressure often associated with methane hydrates can send 8-inch-diameter (20-centimeter) drill stem flying high above a drill rig—dangerous stuff. Still, the Peoples Republic of China is pressing on, and reported nearly a million cubic meters of production from a test site in the South China Sea.[468]

7. Land-Based Methane Biofuel (Biogas): Biological breakdown of organic material in swamps and manure dumps has always produced methane as a byproduct. Through much of the 1900s, methane evangelists proposed major production of methane via breakdown of waste manure and waste plant matter. Such methane production is greenhouse neutral because burning that methane only releases carbon that had recently been drawn down from the atmosphere by plants. Methane-producing biodigesters do reduce toxic and water-eutrophying pollution (also known as green slime) in waters and rivers. However, they have not yielded a large enough reduction of net emissions of carbon dioxide to matter. Nor are they likely to do so because land-based growing of plant matter for biogas must compete with growing of food and fiber—both of which are worth more per unit mass than methane for burning.

8. Ocean-Based Biofuel Methane and/or Oil Production: More than twice the area of all land on Earth is ocean. As will be detailed in the next chapter, some of this area could be cultured for algae and/or seaweed production. These plants could be processed into methane and/or liquid transportation fuels. These would (again) be greenhouse neutral in that any fuels produced would be from carbon that marine plants drew down from the sky. Indeed, the lost byproduct organic matter falling to the ocean floor should make marine-derived fuel carbon negative. Similarly, food production associated with this fuel production would symbiotically benefit from the giant sea forests and help lower the costs of marine-derived foods. The delaying factor for this oceanic development is building the needed technologies, social structures, and expensive infrastructure in the oceans.

9. Ocean-Thermal Energy Conversion (OTEC) to Make Hydrogen and/or Ammonia: Another concept that was seriously considered in the energy crises of the 1980s was using the temperature difference (*delta-T*) between warm tropical surface waters

---

[467] Zheng Rong Chong, She Hern Bryan Yang, Ponnivalavan Babu, et al., "Review of Natural Gas Hydrates as an Energy Resource: Prospects and challenges," *Applied Energy*, vol. 162, pp. 1633–1652, Jan. 2016.

[468] Echo Xie, "China extracts 861,400 cubic metres of natural gas from 'flammable ice' in South China Sea," *South China Morning Post*, March 27, 2020. https://www.scmp.com/news/china/society/article/3077156/china-extracts-861400-cubic-metres-natural-gas-flammable-ice (accessed July 6, 2020)

and the colder deep waters in all oceans that are only slightly above freezing.[469] (Methanol as an energy carrier has been analyzed more extensively than ammonia,[470,471,472] but methanol production proposals entail shipment of coal to the OTEC site and significant carbon dioxide emissions when the synthetic methanol would be burned.) The potential energy from OTEC power plants would be several times the energy use of all humanity. However, a great disadvantage of the OTEC power proposals was that they needed to be in deep tropical waters, which are far from energy customers on land from deep water. Thus, OTEC proponents needed some sort of energy carrier (there is that term again). One of their proposed solutions was electrolyzing water to get hydrogen for shipment in super-cold liquid-hydrogen (cryogenic) tankers, but chilling liquid-hydrogen tankers was very expensive. A cheaper proposal was to use that hydrogen to synthesize ammonia (NH_3), which can be compressed to a liquid at room temperature. Transporting energy via ammonia would thus be much cheaper running cryogenic-hydrogen tankers. Upon delivery on land, the ammonia could be used for fertilizer or fuel. Again, the U.S. OTEC research was largely dropped after oil prices dropped in the 1980s.

10. The Ammonia (NH_3) Economy: Industrial ammonia production combines hydrogen with nitrogen to produce a gas that can be kept liquid at room temperature under mild compression. As such, it carries much more energy per unit volume than hydrogen and with much less storage cost; thus, it is a much better energy carrier, and there could be an "ammonia economy." Such an ammonia economy was demonstrated during World War II when Germany occupied most of Europe, including Belgium. The opposing forces of Great Britain, the United States, and others blockaded continental Europe, so Belgium had limited access to oil-derived

---

[469] W. H. Avery, D. Richards, and G. L. Dugger, 'Hydrogen Generation by OTEC Electrolysis, and Economical Energy Transfer to World Markets via Ammonia and Methanol," International Journal of Hydrogen Energy, vol. 10, issue 11, pp. 727–736, 1985. https://www.sciencedirect.com/science/article/abs/pii/0360319985901089 (accessed Nov. 5, 2020)

[470] Subhashish Banerjee, Md. Nor. Musa, Abu Bakar Jaafar, Economic assessment and prospect of hydrogen generated by OTEC as future fuel," International Journal of Hydrogen Energy, vol. 42, issue 1, pp. 26–37, Jan. 5, 2017. https://www.sciencedirect.com/science/article/abs/pii/S0360319916334036 (accessed Nov 5, 2020)

[471] C. B. Panchal, P. P. Pandolfini, and W. H. Kumm, *Ocean Thermal Plantships for Production of Ammonia as the Hydrogen Carrier*, ANL/ESD/09-6, Argonne National Laboratory, Oak Ridge, Tennessee, Aug. 2009. https://publications.anl.gov/anlpubs/2009/12/65627.pdf (accessed Nov. 5, 2020)

[472] C. B. Panchal, "Case Study of Ammonia Production in the Island States Using Ocean Thermal Energy Conversion," *Ammonia Energy Conference 2019*, Orlando Florida, 2019. https://www.ammoniaenergy.org/wp-content/uploads/2019/08/20191113.1623-E3Tec-614b-AIChE-Annual-2019-NH3.pdf (accessed Nov. 7, 2020)

transportation fuels. The Belgian response was to keep vehicles running by burning ammonia instead of gasoline.[473] Ammonia has recently been proposed as an alternative to oil to power ships.[474,475] Ammonia is one of the major chemicals produced in the world with extensive production and pipeline-distribution facilities, so an ammonia economy could be a much more practical than the hydrogen economy.[476,477] Ammonia could be burned directly, decomposed onsite into hydrogen for use in a hydrogen fuel cell, or used directly in an ammonia fuel cell.[478,479] Even if just burned as a gasoline or diesel substitute, the benefit would be that there would only be oxides of nitrogen to deal with in catalytic converters; there would be no partially burned carbon compounds, which are a major cause of smog. Of course, as with hydrogen, ammonia is only a carrier of the energy source used to synthesize it. How much does that source cost, and what effect does that source energy have on climate?

11. Nuclear Fission Electrical or Thermal Breaking of Water to Make Hydrogen and Combining It with Carbon or Nitrogen: Nuclear cracking of water has been discussed for decades; however, it was not pursued, possibly because of the growing distrust of fission, the growing cost of the first-generation fission reactors, and the temperature limitations of materials at the time. When breakthroughs in affordable fission energy occur, the synthetic fuel options of hydrogen, methane,

---

[473] Robert Hargraves, "Chapter 7, "A Sustainable World," *Thorium: Energy Cheaper Than Coal*, CreateSpace Independent Publishing Platform, July 25, 2012.

[474] "New Research Shows Benefits of Ammonia as Marine Fuel," *The Maritime Executive*, July 18, 2020. https://www.maritime-executive.com/article/new-research-shows-benefits-of-ammonia-as-marine-fuel (accessed July 18, 2020)

[475] Niels de Vries, *Safe and Effective Application of Ammonia as a Marine Fuel*, Delt University and C-Job & Partners B.V., June 12, 2019. https://repository.tudelft.nl/islandora/object/uuid:be8cbe0a-28ec-4bd9-8ad0-648de04649b8?collection=education (accessed July 18, 2020)

[476] Dave Borlace, "Hydrogen Energy Storage in AMMONIA: Fantastic Future or Fossil Fuel Scam?" *Just Have a Think*, No. 133, available in *YouTube*, Dec. 20, 2020. https://video.search.yahoo.com/yhs/search;_ylt=AwrVrN5jsvJfanEAawEPxQt.;_ylu=Y29sbwNncTEEcG9zAzEEdnRpZAMEc2VjA3Nj?p=just+have+a+think&fr=yhs-symantec-ext_onb&hspart=symantec&hsimp=yhs-ext_onb#id=2&vid=5a78e4fb14e3027b7e096bb2848923ed&action=view (accessed Jan. 5, 2021)

[477] Robert F. Service, "Ammonia—a renewable fuel made from sun, air, and water—could power the globe without carbon," *Science*, July 12, 2018. https://www.sciencemag.org/news/2018/07/ammonia-renewable-fuel-made-sun-air-and-water-could-power-globe-without-carbon (accessed Jan. 5, 2021)

[478] Ahmed Afif, Nikdalila Radenahmad, Quentin Cheok, et al., "Ammonia-Fed Fuel Cells: A Comprehensive Review," *Renewable and Sustainable Energy Reviews*, vol. 60, pp. 822–835, July 2016.

[479] A. Valera-Medina, H. Xiao, M. Owen-Jones, et al., "Ammonia for Power," *Progress in Energy and Combustion Science*, vol. 69, pp. 63–102, Nov. 2018.

methanol, and (of course) ammonia become much more practical. Hydrogen or ammonia would decrease carbon dioxide emissions the most, but even synthetic hydrocarbon fuels would yield twice the work for each molecule of carbon dioxide emitted.

12. Electric Batteries: Many of the first horseless carriages in the late 1800s and early 1900s were powered by batteries—quiet, clean, and no fumes. However, the internal combustion engines got steadily more powerful per unit weight. Electrical pioneer Thomas Alva Edison received an apologetic letter from Henry Ford saying that he could not see a way to make electrics work, so he was leaving Edison's employment to found … what later became the Ford Motor Company. And, despite many attempts, batteries lagged far behind internal-combustion engines for a century. Then, in the early 2000s, Elon Musk's Tesla Motors began using newly available lithium batteries that more closely approached the power density of internal-combustion engines. Electric cars are still more expensive per mile than internal-combustion cars, and limitations on lithium supply might keep electric prices from dropping much more.[480] However, one battery revolution suggests that there could be another—say batteries using sodium, calcium, or magnesium instead of lithium. That would be vital because one or more of them might yield considerably more power density. One more battery revolution could be a game changer. Still, the barriers would be a need for a massive new infrastructure of charging stations and a massive increase in the electrical grid to support those stations.

Surprisingly, the correct answer for maintaining adequate energy was hydrofracturing (fracking) for oil and gas. Peak-oil Hubbert, his disciple Deffeyes, and many other energy experts, did not seriously consider the revolutionary advances possible from fracking.[481] By the late twenty-teens, surging American oil and gas production drove the price of both those quantities down, and new expert opinions began estimating centuries of world supply.

The good news from an energy standpoint is that we have many times greater reserves of oil and natural gas than we thought even a decade ago. That means that the coal-to-transportation fuels proposed of the 1970s are definitely not needed. In fact, the ongoing natural gas replacement of much coal and oil use is giving us more time to find and develop technology alternatives that put lesser amounts of greenhouse agents into the atmosphere.

The bad news from a global-warming standpoint is that steadily rising use of oil and natural gas will eventually more than compensate for decreasing use of coal.… By the way, coal use is

---

[480] James Temple, "Why the Electric-Car Revolution May Take a Lot Longer Than Expected," *MIT Technology Review*, Nov. 19, 2019. https://www.technologyreview.com/2019/11/19/65048/why-the-electric-car-revolution-may-take-a-lot-longer-than-expected/ (accessed June 27, 2020)

[481] Robert Rapier, "What Hubbert Got Really Wrong About Oil," *Forbes*, Sept. 8, 2016. https://www.forbes.com/sites/rrapier/2016/09/08/what-hubbert-got-really-wrong-about-oil/#720b66492a3b (accessed Aug. 9, 2020)

still increasing in the developing world. Thus, slowing global warming requires adoption of one or more new sources and one or more new energy carriers to get usable power to the consumers.

The good–bad news is that there are a bewildering number of choices for that energy carrier or carriers. Some choices are not even known today. Choices among various developing technologies are often murky and often don't work out.

Crash programs often do just that—crash. Remember, coal-derived gas and oil were part of the crash program in the 1970s. Another crash development of that era was the molten sulfur battery, which also faded away. The Wankel rotary engine was another hoped-for revolution in vehicle engines great power density and smooth running. Unfortunately, it had lower efficiency, more emissions, and quicker wear-out of seals. Hundreds of millions of dollars later, no more Wankel rotaries are being manufactured.[482,483] The lighter-than-air dirigibles of the early 1900s that seemed destined to rule the sky ultimately dwindled to the Goodyear and Fuji blimps televising sporting events and providing mobile billboards.

Conversely, the lithium-ion battery was only invented in 1980, and it only evolved into something that could become a major power source after another three decades. The molten-salt reactor was invented and tested in the 1950s and 1960s and then ignored for a half century. For the past several years, many researchers and entrepreneurs have been attempting to start a battery revolution. From the 1990s through 2019, more than 300,000 battery-related patents have been filed.[484] As with many inventors' dreams throughout history, they can be supported and encouraged, but ultimately, they must stand on their own.

Even if they fail, other energy revolutions will still evolve in ways that cannot be guessed today. There will undoubtedly be other game changers besides those mentioned in this book.

We need to be patient enough to allow energy sources to evolve to the point that they can affordably supply the electricity and/or sufficient temperatures for producing noncarbon fuels or at least decreasing the net carbon footprint of fuels production. Only then, will it be practical to begin developing one or more [fill-in-the-blank] economies.

Major Things for Governments and Businesses to Do

- Government: Support research and analyses on possible alternatives to the present fossil-fuel economy. However, do not help destroy existing industries first. Remember that President Jimmy Carter in the 1970s was pushing toward a heavily

---

[482] "The Problem with Rotary Engines: Engineering Explained," *Car Throttle*, 2015. https://www.carthrottle.com/post/engineering-explained-why-the-rotary-engine-had-to-die/ (accessed Sept. 25, 2020)

[483] "4 Reasons Why the Rotary Engine Is Dead," *YouTube*, Jan. 13, 2016. https://www.youtube.com/watch?v=v3uGJGzUYCI (accessed Sept. 25, 2020)

[484] Neil Gladstone, "Lithium-ion is just the beginning. Here's a peek at the future of batteries," *digitaltrends*, April 26, 2020. https://www.digitaltrends.com/features/the-future-of-batteries-energy-storage-technology/ (accessed Oct. 30, 2020)

coal economy—sort of forward to the past. Now, government experts want to destroy the coal industry.

- Government: Definitely, support the technical research on splitting water for hydrogen either thermally or by electrolysis. If business manages to develop cheaper nuclear, wind, and/or solar, any or all of them could reduce global warming by either repurposing carbon dioxide wastes (for synthetic methane, gasoline, and diesel) or totally avoiding carbon use with ammonia.

- Government: Support basic research and offer \prizes for continued advances in battery performance. The long-awaited battery revolution is happening, and it may grow into a game changer.

- Business: Assemble a business group and offer to put Iceland on the ammonia economy so that the hydrogen energy could be stored and transported much more practically.

What You Can Do

- I grew up in farm country, and I appreciate welfare for farmers, but do not make an extra effort to buy the biofuels. Bioethanol drives up the price of corn. Biodiesel drives up the price of soybeans and subsidizes destruction of rain forests to make palm-oil plantations. After the greenhouse emissions from fertilizer, farming, and processing, there is very little climate benefit.[485]

8.6 Controlling Fire Emissions with Fountains into the Sky

Humanity's greatest single invention, fire, has many side effects. People immediately noticed and dealt with the bad side effects of choking smoke and of fire-starting cinders. That was an issue with open-air camp fires. It was a greater issue when humans brought fire inside tents and huts with smoke going out a hole in the ceiling. Fire safety in larger structures evolved into fireplaces with chimneys so that less smoke entered the living spaces of the structure and burning cinders had more time to burn out before landing on the fields and thatched roofs of nearby neighbors. Industrial combustion for factories and mills led to steadily larger chimneys so that the large amounts of toxic smoke could disperse before material dropped onto nearby areas … or at least be somebody else's problem far away.

This was especially so because the coal that was the main fuel of the Industrial Revolution had varying amounts of sulfur. Burning that coal also yields sulfur dioxide (SO_2). The sulfur dioxide forms aerosol particles that reflect light. Hence, it is an anti-greenhouse agent. Indeed,

---

[485] Mario Loyola, "Stop the Ethanol Madness," *The Atlantic*, Nov. 23, 2019. https://www.theatlantic.com/ideas/archive/2019/11/ethanol-has-forsaken-us/602191/?gclid=EAIaIQobChMIwOWZ1_q56gIVfz6tBh3eewjkEAAYBiAAEgL9_PD_BwE (accessed July 6, 2020)

some commentators in the 1950s and 1960s predicted that what they called "global dimming" would contribute to colder weather ... perhaps even a new ice age.

Fears of a new ice age faded away in the late 1970s as the new theory of global warming became widely accepted. However, a second concern was not just theoretical but painfully real. The SO_2 of sulfur dioxide combines with the H_2O of water in the atmosphere to make the H_2SO_4 of sulfuric acid. In a lesser but similar manner, fire makes oxides of nitrogen and oxides of carbon, which become nitric acid and carbonic acid when they meet water. The combined acidic dragon breath can be a killer when weather inversions push the smoke down to the ground. The Great Smog of London in 1952 killed about 4,000, with another 6,000 or more related deaths in the subsequent several months.[486]

Progressively taller smokestacks carried the acid fogs into the trade winds where the fogs could disperse as they drifted far downwind, but there was so much acid rain that it was damaging, even when dispersed. Often the fogs found their way in acid rain to pine forests in Scandinavia, New England, and Central Europe. Pine forests naturally produce acidic soils, and the acid rain pushed that acidity to more harmful levels. The acidity led to forest diebacks, especially when forest managers tried to increase timber production with nitrate fertilizer.

European and North American governments used a cap-and-trade system so that power plants could install sulfur-cleaning in their emissions or buy credits from other power plants that did install or just shut down. Over a number of years, these countries did decrease their acid rains. That was a victory for the green movement.

However, it was not a total victory. To make profits, owners of the obsolete sulfur-emitting power plants sold them. Then Asian buyers had them dismantled, shipped to their countries, and then restarted to spew sulfur in those areas. That is part of the reason China and India have terrible smog conditions.

Furthermore, the cap-and-trade markets have been defrauded on multiple occasions.[487,488] This means that the monies spent have much less effect on sulfur pollution and global warming than advertised. Worse, exposed frauds are recurring stimuli for growing cynicism about efforts to mitigate global warming.

Largely replacing coal fuel will require a more cost-effective (that is cheaper) energy source, which was described in 8.6 The Return to Atom Splitting Power with Better Next-Generation Nuclear Fission Reactors.

Meanwhile, the continued growth in energy use has caused the noticeable growth of another fire byproduct—waste heat. Thermal engines in power plants work based on the difference

---

[486] Kacey Deamer, "Scientists determine cause of London's 1952 'killer fog'," *CBS News*, Dec. 12, 2016. https://www.cbsnews.com/news/londons-1952-killer-fog-cause-revealed/ (accessed Dec. 12, 2016)

[487] McKenzie Funk, "Cap and Fraud," *Foreign Policy*, pp. 32–39. Jan./Feb. 2015.

[488] Jude Clemente, "Cap-And-Trade Is Fraught with Fraud," *Forbes*, Oct. 1, 2015. https://www.forbes.com/sites/judeclemente/2015/10/01/cap-and-trade-green-climate-fund-are-fraught-with-fraud/#6fcafb649405 (accessed Aug. 27, 2020)

in temperature between the heat source (coal, gas, nuclear, concentrating solar, etc.) and something to cool it, a heat sink. Typically, water from a local body of water supplies the cooling and then returns as 10–20°F (6–10°C) warmer water to that body of water. This can cause fish die-offs in rivers and lakes due to hot water.

That leads to a question. What is the largest use of water in America? Is it showers and toilette flushes? No. Is it irrigating farm land? That's a better guess, but no.

The largest single use of water in America and much of the world is for power-plant cooling.[489] Most large power plants in the world use water to remove heat from the power plant in one of three ways.

1. Direct or "Once-Through" Cooling: If a power plant is next to the sea, a big river, or large inland water body, the operators can simply pump a large flow of water through the plant to cool the steam condenser. That water returns to the source water body a warmer. That is the simplest and cheapest method. The water may be salt or fresh.

2. Evaporative Cooling. Most commonly, the power plant does not have access to abundant water, so the operators use an evaporative cooling tower, where an updraft of air through water droplets cools the condenser water. The cooling tower evaporates 3% to 5% of the flow, and that percentage must be continually replaced. Along the way, it leaves a brine too salty for other use.

3. Dry Cooling: A few power plants are cooled simply by air, without relying on the physics of evaporation. This may involve cooling towers with a closed circuit, or high forced draft air flow through a finned assembly like a car radiator. This avoids water loss, but it costs more in lowered power-plant efficiency.

The greater the temperature difference (*delta-T*) between the heat source and the cooling source, the greater the power plant's mechanical efficiency. Hence, there is great advantage in siting a power plant alongside a large body of cold water, and this advantage grows in cold weather months. (As discussed in the next chapter, this advantage applies in the ocean where it is always winter a kilometer [two-thirds of a mile] or more from the surface.)

There are two disadvantages with water cooling. First, once-through cooling may release water so warm that it harms local water plants and animals. Second, evaporative cooling uses water that is in particularly short supply in the American Southwest. The American Southwest is fast approaching peak usable cooling water, and other areas have seasonal shortages when the source cooling water is less available and/or is warmer—hence less useful for cooling a power

---

[489] Cheryl A. Dieter, Molly A. Maupin, Rodney R. Caldwell, et al., *Estimated Use of Water in the United States in 2015*, Circular 1441, U.S/ Geological Survey, Reston, Virginia, 2018. https://pubs.usgs.gov/circ/1441/circ1441.pdf (accessed Sept. 6, 2020)

plant.[490] Those hot problem times coincide with the times of highest air-conditioning use, which is what caused California's rolling blackouts in August 2020.

Thus, dry cooling is becoming the wave of the future. As with a fan blowing on a car radiator, the much larger fans for dry air cooling for power plants has a larger "parasitic loss" than water-to-water radiators or evaporative cooling towers. Increasingly, it will be the only option available.

One of the alternate-energy designs proposed in the desperate days of the two energy crises in the 1970s was the vertical-vortex wind machine or tornado wind turbine patented by James Yen.[491] This is different from a conventional windmill, which has a propeller at the top of a tower spinning on a horizontal axis parallel to the ground. Such conventional windmills require progressively larger towers and progressively larger propellers at the top of the tower to get more power. Those giant towers require massive amounts of steel and massive amounts of foundation concrete to anchor the towers.

A vertical-vortex wind turbine does not have propeller at the top of a tower high above. Instead, the turbine is a turbine at ground level, and it is in a fixed position rather than rotating to face the wind. Slats in a silo-shaped structure are closed on the lee side from the wind and are open on the windward with a twist in the slats so that the air swirls upward in a tornado-like vortex. Thus, the entire side area of the silo captures wind energy.

The vertical vortex can harvest additional power if warm air is exhausted from the base into the vortex. As the warm air rises and cools, it strengthens the vortex and provides more flow to the turbine at ground level. Thus, Yen's machine can be both a heat engine and a wind collector.

The heat engine aspect of Yen's machine has another important purpose. It anchors the vortex. The artificial tornado will not escape and run away across the countryside like a regular weather tornado because that heat source does not move. If the vortex were to begin drifting away from the rising heat column, the vortex would lose power, and a new vortex that would form over the heat source.

However, Yen's tornado machine has two major weaknesses. First, the giant slotted silo around the vortex, a skyscraper-sized silo-like structure, is a major expense. Second, that large structure is susceptible to failure during high winds.

---

[490] Uisung Lee, Joseph Chou, Hui Xu, et al., "Regional and Seasonal Water Stress Analysis of United States Thermoelectricity," *Journal of Cleaner Production*, vol. 270, Oct. 10, 2020. https://www.sciencedirect.com/science/article/pii/S0959652620322812 (accessed Sept. 5 2020)

[491] James T. Yen, *Tornado-Type Wind Turbine*, United States Patent 4070131, Jan. 24, 1978.

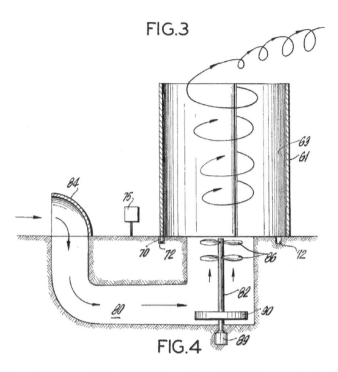

Image of vertical vortex tornado turbine from James T. Yen patent. Warm air is injected at the base. It expands and cools while rising up and driving the vortex.

Louis Marc Michaud suggested an alternate approach, which he described as an aerovortex engine. His aerovortex engine avoids the cost and complexity of a single turbine at the base. Instead, multiple turbines guide the warm air in flows angled for tangential inflows toward the direction of the desired vortex flow.[492,493]

---

[492] Louis Marc Michaud, Atmospheric Vortex Engine, United States Patent 7,086,823 B2, Aug. 8, 2006.
Louis Marc Michaud, *Enhanced Vortex Engine*, United States Patent 7,938,615 B2, May 10, 2011.

[493] Louis Michaud and Erick Michaud, "Harnessing Energy from Upward Heat Convection," *Power*, March 1, 2010. http://www.powermag.com/issues/features/Harnessing-Energy-from-Upward-Heat-Convection_2511.html (accessed Dec. 28, 2012)

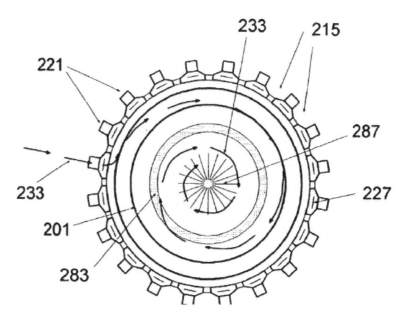

Tangential flow into base of aerovortex engine in Louis Michaud's U.S. Patent 7,086,823 B2.

The aerovortex engine does not need a giant silo surrounding the vortex because the vortex itself maintains the integrity of the spinning flow. The vortex is essentially an artificial tornado that holds the rising air together. Without the limiting costs and structural-integrity concerns of the surrounding silo, designers can send bigger vortices higher into the sky to harness the greater temperature difference between the warm exhaust and the greater chill of the higher altitudes. This is the same principle that powers dust devils going across fields and likewise the much larger tornadoes. While the vortex rises, it sucks air through the turbines at ground level. The turbines can help pull cooling air past radiators; if there is extra air flow (say, on a cool evening), the turbines can generate additional electricity.

Air-flow cooling from the aerovortex engine will produce slightly more net energy than conventionally water-cooled thermal power plants. Vastly more important, it will conserve water for those other tasks of agriculture, bathing, etc. The urgency of saving water will only increase with a warmer climate in the U.S. Southwest matching or perhaps significantly exceeding the Medieval Warm Period.

Yet, there is one more service the aerovortex engine can provide. As will be discussed in the next chapter about sea and sky, there have been climate crises of both heating and chilling. Both have caused widespread destruction. These climate crises have been caused by unusual levels of certain materials getting material above the weather of the troposphere and on up to the stratosphere. The concerns regarding gradual long-term global warming from such materials have been widely discussed.

There are also phenomena that cause the opposite effect of global chilling. As noted in "9.1.7 Earthquakes, Volcanoes, Giant Waves, Large Objects from the Sky, and Nuclear-Bombs," large volcanoes have often been observed to send aerosols of sulfur dioxide and dust high into the sky above the weather layer of the stratosphere. Above the weather, the aerosols have different sizes and different effects. The larger dust particles are a greenhouse agent that allows visible to pass through but which absorbs infrared (heat waves)—increasing warming. But the dust particles fall out more quickly than the smaller sulfur dioxide aerosol particles. The smaller sulfur dioxide aerosol particles have the opposite effect, global chilling by reflecting visible light while allowing infrared heat waves through. The 1815 eruption of Mount Tambora in Indonesia caused "the year without a summer."[494] The much larger Mount Toba eruption (also in Indonesia) about 74,000 years ago is associated with the start of the most recent ice age and may have driven humanity almost to extinction.[495]

A war with nuclear weapons might cause a comparable chilling of a nuclear winter.[496] In addition to the chilling, a nuclear winter's aerosol load in the stratosphere would include tiny radioactive particles atomized so small that they would give maximum breathing exposure to people over s couple years as they rained down.

Not as common would be a major asteroid or comet strike.[497] Such an object striking Earth directly, if large, would have effect like an explosive volcanic eruption.

A large version of any of these global chilling events might be more dangerous than global warming because such events happen suddenly with no time to adjust, but both could be serious threats.

Either warming or chilling could be a threat to humanity. Hence, people have looked for climate fixes, such as Cesare Marchetti when he invented the term geoengineering in 1976.[498]

There have been speculations about ways to deal with warming, chilling, droughts, storms, and more for centuries, but humans have never had any capabilities to deliberately modify the atmosphere until recently. That started to change in the 1800s when self-proclaimed rainmakers generated various smokes to induce rain (although there is no evidence that their methods actually worked). In the 1940s, people began dropping frozen carbon dioxide (dry ice) or other chemicals to provide nucleation sites for moisture to condense on and form rain drops. Hence, the term

---

[494] William K. Klingaman and Nicholas P. Klingaman, *The Year Without Summer: 1816 and the Volcano That Darkened the World and Changed History*, St. Martin's Press, New York, 2013.

[495] Donald R. Prothero, *When Humans Nearly Vanished: The Catastrophic Explosion of the Toba Volcano*, Smithsonian Books, Oct. 16, 2018.

[496] Paul R. Ehrlich, Carl Sagan, Donald Kennedy, et al., *The Cold and the Dark: The World After Nuclear War*, W. W. Norton & Company, 1985.

[497] Peter Bobrowski and Hans Rickman, editors, *Comet/Asteroid Impacts and Human Society: An Interdisciplinary Approach*, 2007 edition, Springer, Jan. 8, 2007.

[498] Cesare Marchetti, "On Geoengineering and the CO_2 Problem," RM-76-17, International Institute for Applied Systems Analysis, 1976. https://core.ac.uk/display/33892139

8 Fixes from Energy and Material

applied was cloud seeding. That has had some success leading to proposals, and virulent environmentalist resistance, involving geoengineering.[499,500,501,502]

However, dealing with the existential threats of global warming and global chilling would generally require transporting large amounts of material high into the sky. Such proposals have often included copying the volcanic injection of sulfates or carbonates into the stratosphere (for long-term effect) or simply spraying seawater to make reflective low-lying clouds such as over endangered coral reefs (for cheaper, but shorter-term effects).[503,504]

The stratospheric injection would entail large expenses for delivery by airplanes, balloons, cannons, or rockets. Developing such capability would best be done over many years of gradual warming met with years of developing a transportation capability. All that investment would yield no return until and unless a major need suddenly appeared.

In contrast, power stations with aerovortex engines could provide return on investment whether or not they were ever needed for transportation. Then, if there were to be a crisis of warming or chilling, they could transport geoengineering materials into the stratosphere more cheaply than the other methods proposed so far.

There have already been natural analogs observed of this stratospheric delivery method. Large forest fires have long been known to generate fire whirls that are several thousand feet high and swirl burning embers in a tornado fashion. Larger fire whirls have been called firenadoes. A comparable firenado happened in Bardstown, Kentucky in 2003 when lightning hit a Jim Beam warehouse causing the so-called "bourbonado" that lit the night sky as a pillar of flame.[505] The firenadoes waft smoke and water vapor from the combustion thousands of feet into the sky and forming thunderhead-like pyro-cumulonimbus clouds, which chill as they rise—often causing any moisture from the flames to freeze and fall back down as hail, and sometimes extinguishing the

---

[499] James Rodger Fleming, *Fixing the Sky: The Checkered History of Weather and Climate Control*, Columbia University Press, New York, 2010.

[500] Eli Kintisch, *Hack the Planet: Science's Best Hope - or Worst Nightmare - for Averting Climate Catastrophe*, John Wiley & Sons, Hoboken, New Jersey, 2010.

[501] Jeff Goodell, How to Cool the Planet: Geoengineering and the Audacious Quest to Fix Earth's Climate Mariner Books; 1st Edition, April 6, 2011.

[502] Steven D. Levitt and Stephen J. Dubner, Chapter 5, "What Do Al Gore and Mount Pinatubo Have in Common?" *Super Freakonomics: Global Cooling, Patriotic Prostitutes and Why Suicide Bombers Should Buy Life Insurance*, Harper Collins, New York, 2009.

[503] Justin McClellan, David W. Keith, and Jay Apt, Cost Analysis of Stratospheric Albedo Modification Delivery Systems, *Environmental Research Letters*, vol. 7, 2012. http://iopscience.iop.org/1748-9326/7/3/034019/pdf/1748-9326_7_3_034019.pdf (accessed Dec. 20, 2012)

[504] Christopher Flavelle, "As Climate Disasters Pile up, a Radical Proposal Gains Traction," *New York Times*, Oct. 28, 2020) https://www.nytimes.com/2020/10/28/climate/climate-change-geoengineering.html (accessed Oct. 28, 2020)

[505] Steve Huff, "The Bizarre Tale of the Kentucky Bourbon Fire Tornado," *Maxim*, Sept. 4, 2015. https://www.maxim.com/maxim-man/kentucky-bourbon-fire-tornado-2015-8 (accessed Nov. 27, 2019)

fire below. However, in the last several decades, some forest-fire firenadoes have been observed to rise higher, actually punching up high enough to reach the stratosphere.[506]

However, global warming is a threat that would tend to grow over many years so that countervailing measures could be taken over years or decades. Based on geological history, greater and more frequent dangers have come from sudden unexpected chilling events. These have volcanic winters from mega-volcano-eruptions, and a significant nuclear exchange could do the same as well as spreading poisons in the atmosphere. Both of these threats cause damage by injecting particles up into the stratosphere—volcanic eruption by anti-greenhouse particles hanging high in the stratosphere and nuclear war by both cooling particles in the stratosphere and by a prolonged rain of poisonous radioactive particles.

The chilling effect only lasts for several years, but such events can have dire results. The 1815 Tambora volcanic eruption was followed the next year by the 1816 "year without a summer" and many famines throughout the world. The Toba volcanic eruption 74,000 years ago apparently caused a several-year mini-ice-age.

There is one powerful chemical that can remove these deadly agents from the stratosphere and mitigate global winters. That chemical is water. Very light particles can hang for months or years in the stratosphere just above the weather-filled troposphere because the stratosphere is drier than the Sahara Desert. Water vapor condenses on aerosol particles, and the combined heavier particles sink down into the troposphere where they accrete more water and are soon rain all the way to the ground. While in the stratosphere, water can mitigate volcanic winter and/or nuclear winter because water vapor is twice as strong a greenhouse gas as carbon dioxide—The greenhouse effect or radiative warming for water is around 75 watts/square meter while carbon dioxide only contributes 32 watts/square meter (90 and 38 watts/square yard).[507]

There is already a precursor technology for washing pollution from the sky. The Chinese have developed truck-borne "water cannons" to spray atomized water mists into dust clouds above

---

[506] Michael Fromm, Andrew Tupper, Daniel Rosenfeld, René Servranckx, and Rick McRae, "Violent pyro-convective storm devastates Australia's capital and pollutes the stratosphere," *Geophysical Research Letters, Atmospheric Science*, vol. 33, L05815, doi:10.1029/2005GL025161, March 15, 2006. https://agupubs.onlinelibrary.wiley.com/doi/epdf/10.1029/2005GL025161 (accessed Nov. 26, 2019)

[507] J. T. Kiehl and Kevin E. Trenberth, "Earth's Annual Global Mean Energy Budget," *Bulletin of the American Meteorological Society*, vol. 78, no. 2, Feb. 1997. https://journals.ametsoc.org/bams/article/78/2/197/55482/Earth-s-Annual-Global-Mean-Energy-Budget (accessed Aug. 30, 2020)

construction areas or just areas of intense smog.[508] India[509] and Thailand[510] have also applied similar mist cannons.

Injecting particles as far as the stratosphere is more difficult, but an aerovortex engine might be the most economical way to do so. An aerovortex engine could be boosted to function as a fire tornado, a fountain, carrying water up to the stratosphere.

Just as nature pumps material into the stratosphere using the heat of volcanoes and forest fires, people can pump more energy into a fire tornado vortex and make the vortex grow stronger and taller to reach the stratosphere. More importantly, it could carry more material with it to the stratosphere. Facility operators need only feed additional energy into a vortex as it is rising so that it reaches higher altitude. Some potential ways to feed and maintain the energy of a vortex include:

- Run the air turbines at the aerovortex-engine base as motors rather than generators. A more powerful vortex can maintain itself higher into the sky.

- Increase the temperature of heat input just above the base of the vortex. This produces an artificial firenado. As with a natural firenado, this reduces the density of the gas in the vortex, which increases the buoyancy and increases the spinning of the vortex to achieve greater height before dissipating.

- If desperate, insert flammable solid materials into the vortex that there will be continued burning and continued heat supplied far into the sky. During World War II in the 1940s, the German air-defense forces experimented with a *Windkanone* or "Whirlwind Cannon" using burning coal dust to maintain and generate vortices high enough to attack Allied bombers flying overhead.)[511]

- Switch on microwave beams around the periphery of the vortex with a frequency range matching an absorption band of water or any other chemical being lofted to the stratosphere. Any vortex target material straying into the periphery microwave beam would be reheated and redirected back toward the vortex core. The greater heat and the thrust back into the vortex would both help increase the final altitude that the vortex could reach before it dissipated. The practicality of such microwave boosting will soon be radically increasing because more efficient (and longer

---

[508] "Unique ways to deal with China's air pollution," *YouTube*, May 28, 2014. (accessed Sept. 14, 2020)
https://www.youtube.com/watch?v=jy9EpaOLf1whttps://www.youtube.com/watch?v=4cOFf962PDc (accessed Sept. 14, 2020)

[509] "Anti-Smog Gun," *Current Affairs Plus*, July 15, 2020.
https://neoiascap.com/2020/07/15/anti-smog-gun/environment/interventions/ (accessed Sept. 14, 2020)

[510] "Bangkok uses water cannons to fight air pollution," *Global News*, Jan. 14, 2019.
https://globalnews.ca/video/4846846/bangkok-uses-water-cannons-to-fight-air-pollution (accessed Oct. 3, 2020)

[511] Jack Beckett, "Greatest Mysteries of WWII – Nazi Germany's Top Secret Weapons," *War History Online*, Dec. 12, 2015. https://www.warhistoryonline.com/whotube-2/greatest-mysteries-of-wwii-nazi-germanys-top-secretweapons.html (accessed Aug. 30, 2020)

lasting) solid-state microwave transmitters may soon replace the magnetrons invented for radars in World War II. These solid-stat transmitters will probably also improve microwave ovens by increasing efficiency, providing more even heat, and extending service life.[512,513]

- Other possible materials for a vortex to carry might include sulfur dioxide or calcium carbonate for cooling[514])

- Another type of material to inject would be nucleating agents (such as bismuth tri-iodide (BiI_3) into the upper troposphere or lower stratosphere so that it would cause the small ice of cirrus clouds to clump into bigger ice particles that would fall out of the sky sooner. The tiny ice crystals of cirrus clouds are another global-warming agent, so reduction in such clouds would cause cooling. The nucleating agent would be in tens of parts per billion so there would be no health effect on people.[515]

Thus, there are a number of ways for aerovortex engines to function as aerial fountains carrying material to the stratosphere for dealing with sudden global warming or global chilling.

Major Things for Governments and Businesses to Do

- Government: Support research and analyses on aerovortex engines or other vertical-vortex power systems for reducing water use in cooling and increased efficiency in energy production.

- Government: Support research and analyses on aerovortex or other vertical-vortex power systems for injecting materials into the stratosphere if need be for washing pollutants down to the ground.

---

[512] Bill Schweber, "LDMOS Transistor Seeks to Displace Consumer Microwave-Oven Magnetron," *Electronic Design*, Nov. 22, 2020. https://www.electronicdesign.com/power-management/whitepaper/21147904/electronic-design-ldmos-transistor-seeks-to-displace-consumer-microwaveoven-magnetron?oly_enc_id=1869D3889512F1V (accessed Dec. 3, 2020)
LDMOS is an acronym for laterally-diffused MOSFET (metal-oxide semiconductor field-effect transistor)

[513] Mark Murphy, "Here Comes the Solid-State RF Energy Evolution," *Electronic Design*, March 9, 2018. https://www.electronicdesign.com/power-management/article/21806218/here-comes-the-solidstate-rf-energy-evolution (Dec. 4, 2020)

[514] Ribooga Chang, Semin Kim, Seungin Lee, Soyoung Choi, Minhee Kim, and Youngjune Park, "Calcium Carbonate Precipitation for CO2 Storage and Utilization: A Review of the Carbonate Crystallization and Polymorphism," *Frontiers in Energy Research*, vol. 5, article 17, July 10, 2017.

[515] David L Mitchell and William Finnegan, "Modification of cirrus clouds to reduce global warming," *Environmental Research Letters*, doi:10.1088/1748-9326/4/4/045102, vol. 4, Oct.–Dec. 2009. https://iopscience.iop.org/article/10.1088/1748-9326/4/4/045102/fulltext/ (accessed Dec. 1, 2019)

- Business: Invest in air-cooled aerovortex engines dry-cooling systems for thermal power plants, and (most importantly) include an open-space area for possible swapping in of a small modular reactor when government certification is achieved.

What You Can Do

- Speak up for research on research for climate modification/geoengineering. If there were to be a ~~include sulfur~~ serious and rapid chilling or a warming event, it would be too late to research and develop such safeguards.

- Speak up for NOT using geoengineering unless a rapid warming or cooling event was definitely proved to be happening. Remember, there was a serious 1960s proposal to air drop soot onto the Arctic Ocean ice to stop the expected onset of anew ice age.[516]

8.7 Key Toolmaking Technologies

Fortunately, just as the climate needs grow, the technological development options are also growing. The following toolmaking technologies particularly help in energy production, but they help other technology areas as well. The following nine areas are probably the most important such keys, and they are detailed in the following subsections.

1. Computer Power for Calculation, Simulation, Analysis, and More
2. The Range from Servocontrols Leading to Robotics
3. Advanced Materials Processing Such As 3D Printing (Additive Manufacturing)
4. Prefabricated and Modular Construction
5. Supercritical Carbon Dioxide Turbines—Another Supporting Energy Revolution
6. Electronic Remote Connection—The Internet, Webinars, and Offices Anywhere
7. Green-Energy Storage
8. Materials for Absorbing. Reflecting, Generating, and Using Energy
9. Putting It All Together: Hybrid Systems of Combustion, Ultracapacitors, ... And Batteries

---

[516] P.M. Borisov, "Can we Control the Arctic Climate?", Bulletin of the Atomic Scientists, March, 1969, pp. 43-48.

8.7.1 Computer Power for Calculation, Simulation, Analysis, and More

Computer power is the gateway to almost every advance coming for humanity. From at least the building of the pyramids in Egypt until today, people have needed steadily more and better calculations and analysis. In fact, the so-called drunken pyramid in Egypt could not be finished because of inadequate calculation, and the lopsided ruin is a monument to that failure.

Mechanical calculators in the late 1800s led to the codebreaking and artillery-plotting vacuum-tube computers of World War II in the 1940s. Then, the 1950s saw the solid transistor finding its way into calculators and computers. Single transistors morphed into integrated circuits with several transistors fabricated as a unit. Then, integrated circuits morphed into large scale integration with hundreds of circuits fabricated in one chip. Now, they are simply called microcircuits with millions of circuits crammed into correspondingly smaller spaces. Meanwhile, memory storage has gone from refrigerator-sized units down to easily misplaced data sticks. One shorthand description is that computer capability increased a trillion-fold between 1956 and 2015.[517]

Furthermore, computer power will continue to grow. A rough approximation called Moore's Law says that microcircuit capabilities (and thus computer power) double roughly every 2 years as circuit sizes shrink. Although classical Moore's Law shrinking is nearing its end as circuits are approaching a single nanometer, circuits computer power will continue growing because of stacking chips up sideways for a third dimension, parallel computing where software splits many tasks into pieces that can be processed my many processors, quantum computing, and returning to the efficient coding employed when early computers had limited capacity. This continued increase in computing power is important because computer calculations do the following things, and more are coming.

- Designing large construction projects such as chemical plants, power plants, high-rise buildings, and aircraft to maximize performance

- Generating weather and climate forecasts

- Processing geological data for oil exploration and production. Precise mapping of thin oil-bearing layers made the fracking revolution possible because the drills could follow the oil-bearing layer exactly.

- Processing and decoding genetic data to improve human health and help breed those better crops in Section 7.3 Plant Breeding Miracles the Neo-Prairie, and the Ten-Foot Moon Shot.

---

[517] Nick Routley, "Visualizing the Trillion-Fold Increase in Computing Power," *Visual* https://www.visualcapitalist.com/visualizing-trillion-fold-increase-computing-power/ (accessed Dec. 6, 2019)

- Modeling chemical processes to more quickly find and use better alloys, catalysts, and processes.[518] These research advances are key to nuclear-fission related advances described shortly.

- Simulation for training, predicting results, avoiding the expense of physical experiments, including just fun video games, (from which, computer programming and equipment advances have repeatedly fed back into scientific simulations.

- Modeling and simulating nuclear processes for safety of civilian reactors and to minimize testing of real bombs by the military. These researches will eventually lead to that seemingly unattainable goal of practical fusion reactors.

Conversely, the same On/Off binary switches that do those calculations can activate the motions of and robots to be discussed in 8.7.2, the next section.

8.7.2 The Range from Servo Controls Leading to Robotics

A servocontrol or servomechanism is a device with a sensor that monitors some mechanical system parameter and a controller to modify the behavior of that device based on that monitoring. For instance, a thermostat maintains the temperature of a house, a governor on a car motor maintains the air and fuel inputs to the motor at the desired level, a program on a machine tool makes cuts or welds on a part per preprogrammed instructions. Sensor inputs in a chemical plant (such as a manufacturing plant or a petroleum refinery) may have thousands of servomechanisms controlling a comparable number of parameters in the plant.

The development of servocontrols was a major part of the technical advances of the 1900s. Some have even called servocontrols the Second Industrial Revolution. Without servocontrols, myriads of people would have been needed for all the tasks run by servocontrols.

The early 1900s had controls that were mostly pneumatic or hydraulic. The late 1900s and beyond have increasingly used electronic controls. Electronics allow the servocontrols to respond faster, in more complex ways. Eventually, those more complex servocontrols can be called robots. They are not thinking entities, such as Robby the Robot in the film *Forbidden Planet*, but affordable widespread computing power to serve roles everywhere. They can be gigantic, miniscule, or even distant. For example:

- Machine tools for manufacturing are increasingly robotic. Computer-aided design (CAD) generates an electronic instruction set for computer-aided manufacturing (CAM) by various machine tools. For factory production, a set of these tools operate an assembly line with faster and higher production rates than human

---

[518] Marco Annunziata, "Mind over Matter: Artificial Intelligence Can Slash the Time Needed to Develop New Materials," *Forbes*, Dec. 3, 2018.
https://www.forbes.com/sites/marcoannunziata/2018/12/03/mind-over-matter-artificial-intelligence-and-materials-science/#4572719fe9db (accessed Dec. 29, 2019)

workers. The largely deserted factory in the movie *Minority Report* comes to mind.[519]

- Fabrication of those increasingly tiny microelectronic systems requires work at microscopic levels, which is difficult for humans but just a design parameter for robotic fabrication systems.

- Autonomous (self-driving) cars and trucks are now on the verge of commercial breakthrough. Millions of such vehicles will be a revolution equivalent to Henry Ford's mass-produced Model T in the early 1900s.[520] Cheap autonomous cars operating as taxis[521] could reduce the number of cars in the car fleet and their energy costs.

- Thermostats already reduce energy use in buildings, but that is only a start. The greenhouses, vertical farms, fungi-growing chambers, and minilivestock areas in Section 7 need exquisite control of temperature, water, nutrient levels in the water, lighting, and humidity. Energy Crises of 1970s came at the same time as robotic controls became more capable and more affordable. The technologies have advanced to where a "smart home" can save energy by switching heating and cooling on when people are arriving and using appliances to match time-of-day electricity pricing. However, that is only a start.

- Similarly, the robo farm workers of Sections 7.5 and 7.6 are evolving to produce the detailed judgement and labor to allow revolutionary farming advances. There are robo weeders,[522] moving robots for transferring heavy loads, robots for transplanting sprouts,[523] and water-knife harvesters to get lettuce (and maybe sugar cane) with much less effort than arguably the hardest job on Earth.[524] There will be thousands of other robotic systems to bring greater agriculture productivity at lower cost. Robots can do agricultural chores more cheaply especially where it is expensive, difficult, and/or hazardous to get human labor.

- Even more so, ocean agriculture, discussed in the next chapter, is already benefitting from robotic workers that are not bothered by hypothermia from

---

[519] Steven Spielberg, director, *Minority Report* is a 2002 science fiction detective movie set in 2054. One chase-and-fight scene is set in a futuristic-looking car assembly plant that seems to be entirely robotic. The film makers shot their images in an existing car assembly plant.

[520] Michael E. McGrath, *Autonomous Vehicles: Opportunities, Strategies and Disruptions: Updated and Expanded Second Edition*, independently published, Dec. 2, 2019.

[521] The 1990 science fiction adventure film, *Total Recall*, had robotic Johnny Cabs.

[522] Tharran Gaines, "The Future of Robotic Weeders," *Successful Farming*, Nov. 29, 2018. https://www.agriculture.com/technology/robotics/the-future-of-robotic-weeders (accessed Dec. 8, 2019)

[523] Michael Liedtke, "Meet the farmers of the future: Robots," *Phys.org*, Oct. 3, 2018. https://phys.org/news/2018-10-farmers-future-robots.html (accessed Dec. 8, 2019)

[524] Matt Simon, "Robots Wielding Water Knives Are the Future of Farming," *Wired*, May 2017. https://www.wired.com/2017/05/robots-agriculture/ (accessed Dec. 8, 2019)

working in cold water or danger of decompression sickness (the bends) from going back to the surface too quickly. Furthermore, human workers require heroic technologies (such as armored suits or saturation diving) beyond roughly a hundred feet (thirty meters). Thus, work beyond that depth can be done much more practically by robo workers (such as Big Geek and Little Geek in the 1989 science fiction adventure movie, *The Abyss*.[525]

- Of course, robotic guidance and remote control were developed for aerospace exploration and (eventually) exploitation, and developments are continuing. One evening during the fall of 2018, this author was on a restaurant patio in South Pasadena, California. A little bright light rose in the western sky. Then it split into three. One continued up, and two went back down. The two dropping lights were boosters that landed under their own control and later flew again. Similarly, there are already robotic orbiting radio repeating stations that transmit millions of phone, video, and data messages—all without hands-on human control. Robots already operate autonomously in space Ultimately, there will also be robotic equipment fabrication, assembly, and repair in outer space. Advances in robotic technology are the only things that will allow commercial conquest of space.

- Back on Earth, that hydrofracturing (fracking) revolution made possible by detailed geological mapping also needs robotic controls to guide the drill bits to stay in those thin oil-bearing layers. Together, the computation for mapping and the steerable horizontal drilling have made an industry providing tens of billions of dollars of oil and natural gas each year. For good and ill, those liquid fossil fuels that were expected to dry up within a couple decades could last centuries.

- For the nuclear fission reactors in this chapter, robotic assembly, maintenance, and emergency cleanup can be done remotely without risk to human life.[526,527] Indeed, remote manipulators, also known as telemanipulators, or Waldos were invented in 1945 for manipulating nuclear materials, and they were inspired by a science fiction short story.[528] After more than eighty years, electronics allow such telerobotics to have remote operators hundreds (or even thousands) of miles away. Better yet, many tasks could be performed by largely autonomous robotic action.

---

[525] James Cameron, writer and director, *The Abyss*, movie, 1989.

[526] Sharon Gaudin, "Robotics Industry Learns from Successes and Failures at Fukushima," *Computerworld*, March 15, 2017. https://www.computerworld.com/article/3181407/robotics-industry-learns-from-successes-and-failures-at-fukushima.html (accessed July 14, 2020)

[527] Thomas Hornigold, "How Fukushima Changed Japanese Robotics and Woke Up the Industry," *SingularityHub*, April 25, 2018. https://singularityhub.com/2018/04/25/how-fukushima-changed-japanese-robotics-and-woke-up-the-industry/ (accessed July 14, 2020)

[528] Robert Heinlein's "Waldo" in a 1942 *Astounding Science Fiction*, August 1942.

Unfortunately, there are two concerns, weaknesses, in the Robotics Revolution. They both involve increasingly tiny parts of microcircuits being increasingly sensitive to radiation: first long waves, often called electromagnetic pulses and radiation from just particles and very short waves (such as gamma rays) from reactors and from background radiation in outer space.

Long waves (also known as electromagnetic pulses) first demonstrated themselves in September 1 and 2, 1859 when the Sun had a solar storm that burped a billion tons of material into space, much of it in the form of positively charged hydrogen ions, and (crucially) Earth happened to be directly in the path of those ions—small target in a big sky, but occasionally it happens.

When the hydrogen ions hit the atmosphere, the effect was something like a series of giant lightning bolts. Accounts from Europe and North America mentioned northern lights (aurora borealis) bright enough to read a newspaper throughout the entire night. At the same time, the only electric facilities were the telegraph offices. Sparks flew off the telegraph keys, and there were a few nearby stacks of papers caught fire.

However, there was very little major damage, and the telegraphs were operating the day after the solar storm ended. An astronomer named Richard Carrington connected the electrical phenomena with a large number of sun spots, so the "Carrington Event" is named after him. That is a footnote in history.

However, since then, electronics have shrunk to sizes in the nanometers (a nanometer is about 4 times 10 to the minus 8th power of inches or 0.00000004 inch). Such small spaces in the electronic parts make them much more susceptible to damage from electrical sparking. A direct hit on Earth from a solar flare, like that in 1859 might could cause "extensive social and economic disruptions" with costs as great as $2 trillion.[529,530]

This all just a large version of one of the greatest reliability issues for electronic parts— electrostatic discharge (ESD), little lightning bolts inside the electronic components. Finished parts must be increasingly sealed from ESD and be shielded with circuit protection devices.

Hostile governments could cause similar electronic problems deliberately using much less power than a flare from the Sun but an instantaneous short intense electromagnetic pulse (EMP). The EMP would come from some mechanical device in a local area or a nuclear bomb in space above the target area. Such an EMP could overload or disrupt at a distance numerous electrical systems and high technology microcircuits across large swathes of territory. Both the United States and the Soviet Union (now Russia) tested such high-altitude bombs in the 1950s and 1960s. The largest United States' test, Starfish Prime, was 1.4 megatons at an altitude of 250 miles (400 kilometers) above Johnston Atoll in the Pacific Ocean. The vacuum tubes of that age were larger, cruder, and hence, more rugged regarding electromagnetic pulse than the electronics of

---

[529] Christopher Klein, "A Perfect Solar Superstorm: The 1859 Carrington Event," *History.com*, original March 14, 2012, updated Aug. 22, 2018. https://www.history.com/news/a-perfect-solar-superstorm-the-1859-carrington-event (accessed Aug. 29, 2020)

[530] Chris Gebhardt, "Carrington Event Still Provides Warning of Sun's Potential 161 years Later," *NASA Spaceflight.com*, Aug. 28, 2020. https://www.nasaspaceflight.com/2020/08/carrington-event-warning/ (accessed Aug. 29, 2020)

today; so, there was no serious damage to equipment at that time beyond blown fuses that knocked out 300 street lamps on Oahu Island about 900 miles (1,500 kilometers) away, and several satellites failed prematurely due to an increase in the particle levels in the Van Allen Radiation Belts around the Earth.[531]

However, the increasingly capable electronics are becoming increasingly susceptible to EMP. Russia used tactical non-nuclear EMP weapons to suppress communications of Chechen rebels in the early 200s and against Ukrainian forces in the border war of the twenty-teens. Both Russia and China have military doctrines of using nuclear EMP attacks.[532],[533] There was even a casual comment by a Russian diplomat in a 1999 negotiation regarding the Serbian conflict at the time.

> *If we really wanted to hurt you, with no fear of retaliation, we would launch an SLBM [submarine-launched ballistic missile] from the ocean, detonate a nuclear weapon high above your country, and shut down your power grid and your communications for six months or so.*
>
> Deputy Duma Chairman Vladimir Lukin[534]

Another problem environment for electronic parts is nuclear fission reactors, especially reactors needing emergency repairs, produce large amounts of damaging radiation in the form of neutrons and short-wavelength electromagnetic radiation (such as gamma rays). That was the problem at the Fukushima Daichi reactors in 2011. After the earthquake and tsunami, robots were sent into the damaged reactors to reconnoiter and perform repairs. Unfortunately, the robots lost electronic function within days.[535]

The nuclear-repair robots apparently did not have the more expensive radiation-hardened (rad-hard) electronics used by the military. This is a concern in that robotic maintenance of reactors offers the possibility most reactor operations being done without risk to humans and being more easily watched to prevent diversion of nuclear materials. However, this option would be

---

[531] Mark Wolverton, *Burning the Sky: Operation Argus and the Untold Story of the Cold War Nuclear Tests in Outer Space*, The Overlook Press, New York, 2018.

[532] Ariel Cohen, "Trump Moves to Protect America from Electromagnetic Pulse Attack," *Forbes*, April 6, 2019. https://www.forbes.com/sites/arielcohen/2019/04/05/whitehouse-prepares-to-face-emp-threat/#655d99fce7e2 (accessed Aug. 29, 2020)

[533] Strategic Primer: Electromagnetic Threats, American Foreign Policy Council, Washington, D.C., vol. 4, winter, 2018. https://www.afpc.org/publications/special-reports/strategic-primer-electromagnetic-threats (accessed Aug. 29, 2020)

[534] Statement of Representative Roscoe Bartlett (Republican from Maryland), *Congressional Record*, June 9, 2005. The quote was cited in Chapter 17, "The Fire of Damocles," in Wolverton's *Burning the Sky*.

[535] "The Robots Sent into Fukushima Have Died," *Newsweek*, March 9, 2016. https://www.newsweek.com/robots-sent-fukushima-have-died-435332 (accessed July 14, 2020)

much less practical if available electronics were to be either incapable of surviving a high-rad environment or fabulously expensive.

Electronics in outer space experience both pulses of energy from solar flares and still-significant levels of radiation as part of the natural background. For different reasons than fission reactors on Earth, certain space facilities operate totally without any human crew for years. Proposals for major space facilities can only be practical if affordable rad-hard electronics are available.

Major Things for Governments and Businesses to Do
- Government: Implement the recommendations for making the electrical grid more robust against sabotage and electromagnetic pulses due to solar storms or electromagnetic-pulse weapons.
- Government: Fund a long-term development effort for radiation-hardened (rad-hard) electronics for use in fission-reactor maintenance and repair and in long-term outer-space applications.
- Government: Generate a long-term set of regulations for gradually increasing the minimum levels of static resistance and radiation hardening for electronic parts.

What You Can Do
- Increase the protection on all your computers and other electronics with surge-protector power strips and other precautions.

8.7.3 Advanced Materials Processing

Advancing materials have made key improvements for better human living at least since fire hardening of pointed sticks for spears to weaving of fabrics. For energy, it continued on with brass and steel for steam engines—then special alloys and composites for jet engines.

Likewise, advanced materials are helping make the next-generation and small modular reactor revolution. A 2020 *Forbes* article entitled "3D Printing Has Entered the Nuclear Realm" noted that Siemens had installed the first commercial 3D printed part in a nuclear reactor, that the part had been in successful operation since then 2017, and that other 3D nuclear parts are coming.[536,537]

---

[536] Victor Anusci, "Siemens Sets Milestone with First 3D Printed Part Operating in a Nuclear Power Plant," *3D Printing Network*, March 9, 2017.
https://www.3dprintingmedia.network/siemens-sets-another-milestone-first-3d-printed-part-operating-nuclear-power-plant/ (accessed June 26, 2020)

[537] James Conca, "3D Printing Has Entered the Nuclear Realm, *Forbes*, May 9, 2020.
https://www.forbes.com/sites/jamesconca/2020/05/09/3d-printing-has-entered-the-nuclear-realm/#5ccf7ed67d34 (accessed June 26, 2020)

However, the article soon commented that 3D printing is just one of the advanced materials that are being applied or researched for nuclear power production. The nuclear industry is smaller than aviation, and has been moving slower because reactor construction has been largely stalled for three decades—but the advanced materials are coming.[538,539] Some of the most important advanced materials are:

- Additive manufacturing (usually called 3D printing), the best-known family of advanced materials
- High-temperature alloys
- Sintering and pressure on metal powders
- Laser welding and electron-beam welding
- Ceramics and composites

Additive manufacturing (AM or 3D printing) is actually another robotic application, but it gets its own topic because it is a very powerful advance that will impact every aspect of technology. 3D printers build parts up using a very thin layer at a time instead of the traditional machining modality of cutting or grinding material off bigger pieces (subtractive manufacturing) and then attaching them to other pieces. Microwave, laser, or electron beams cook (fix) jets of powders into 3D positions much like toner cartridges in printers fix letters in 2D patterns of text on pages of documents.

3D printing in the late 1980s was in plastics, and the plastic products were only suitable for plastic toys and prototypes. Since then, 3D printing has been extended to a progressively wider range of metals, ceramics, and composites. Businesses are working to increase speeds, apply multiple materials at the same time, and (most of all) lower costs so that 3D printers can become ubiquitous—used everywhere in technology. 3D printing processes can:

- Reduce the cost and time needed to start a new product line (those prototypes again)
- Reduce the cost and time for modifying a product because a design modification is only a change in the computer-aided-manufacturing software (and as always, test it under load!)
- Reduce waste production by as much as 90% because it adds material rather than milling off material as waste

---

[538] *Roadmap for Regulatory Acceptance of Advanced Manufacturing Methods in the Nuclear Energy Industry*, Nuclear Energy Institute, Washington, D.C., May 13, 2019. https://nei.org/CorporateSite/media/filefolder/resources/reports-and-briefs/Report-NEI-Advanced-Manufacturing-Regulatory-Roadmap-Final-2019-05-13.pdf (accessed June 26, 2020)

[539] Xiaoyuan Lou and David Gandy, "Advanced Manufacturing for Nuclear Energy," *JOM* (Journal of The Minerals, Metals & Materials Society), vol. 71, pp. 2834–2836, June 26, 2019.

- Reduce the cost and space requirements for different kinds of machine tools and jigs for particular jobs because an increasing number of jobs can be run with just one 3D printing machine
- Produce more useful shapes that cannot be produced with conventional production processes
- Reduce part counts by producing complex parts that would ordinarily be produced in multiple pieces and then fastened together; for example, a jet engine developed by GE and Snecma Safran replaced twenty welded parts with a single fuel nozzle that is 25% lighter, is five times more durable, and reduces fuel use by 15% compared to the nozzle made by welding the twenty pieces together[540]
- Reduce assembly time and possible failure points from incorrect fastening by building unitary parts,
- Along with prefabricated modules, radically reduce the time (and time is money!) in building residential and commercial structures.

All these 3D printing advantages are increasing the speed of developing new technologies while decreasing cost and energy use. The applications extend from as small as nanometers in fabricating microcircuits, to high temperature ceramics of jet and rocket engines, to large concrete structures.[541] In particular for this chapter, The Oak Ridge National Laboratory is working to develop a 3D stainless-steel printed reactor core with the goal of 3D printing an entire reactor by 2023.[542,543]

Note: This reactor will be a Transformational Challenge Reactor (TCR). It will be an experimental reactor built with many materials and techniques not according to building codes. Lessons from building and operating this TCR will be used for applying to get those materials approved.

The most important implication for a GND, and for progress in general, is that 3D printing will spread more capable industrialization everywhere. A local machine shop with a few 3D printers will be able to make new equipment and/or repair parts in hours rather than ordering shipments from out of town or even out of country. A small local factory will be able to develop a revolutionary product and sell the manufacturing details worldwide rather than paying the

---

[540] Steven Brand, "5 Major Benefits of Additive Manufacturing You Should Consider," CMTC Manufacturing Blog, California Manufacturing Technology Consulting® (CMT), 2019. https://www.cmtc.com/blog/benefits-of-additive-manufacturing (accessed Dec. 15, 2019)

[541] Benedette Cuffari, "Additive Manufacturing of Concrete," *AZO Build*, July 8, 2019. https://www.azobuild.com/article.aspx?ArticleID=8333 (accessed Jan. 1, 2020)

[542] David Szondy, "Oak Ridge developing 3D-printed nuclear reactor core," *New Atlas*, May 11, 2020. https://newatlas.com/science/oak-ridge-3d-printed-nuclear-reactor-core/ (accessed June 27, 2020)

[543] Sonal Patel, "Nuclear Reactor with 3D-Printed Core Slated for Operation in 2023," *Power*, May 12, 2020. https://www.powermag.com/nuclear-reactor-with-3d-printed-core-slated-for-operation-in-2023/ (accessed June 27, 2020)

money, energy, and pollution costs of building, inventorying, and shipping physical product. 3D printing will enable the other technologies described in this book, plus many more, to stop or reverse global warming.

Improving metal alloys has been a key technology throughout the ages from steel swords, to steel plows, to aluminum light enough for metal airplanes, to the turbine alloys for jet aircraft and nuclear reactors.

The alloy Inconel 617 is one such key advance. The Inconel family of alloys is a combination of nickel, chromium, cobalt, and molybdenum. Inconel 617 was first developed for use in high-temperature gas reactors, but it can also be applied to molten salt and liquid metal reactor designs. The new metal offers significant improvements over previously approved alloys in the code and can withstand operating temperatures of 1,750°F (350°C)—nearly 400°F (200°C) hotter than the next-best material. The expanded operating range could also open up new market opportunities for the nuclear industry by using its thermal heat to directly thermal crack water for hydrogen and retort limestone into cement by driving off carbon dioxide.

Nuclear reactor builders in the United States must use only metal alloys in the American Society of Mechanical Engineers code. Getting an alloy approved to be placed in the code is a lengthy process and requires significant amounts of data. The Department of Energy invested $15 million over 12-years to make Alloy 617 available in the fall of 2019 for supporting demonstration and deployment of advanced reactor concepts.[544] It is the first high-temperature material cleared for commercial use since the 1990s.

This, oddly, leads to another argument for small modular reactors and the even smaller micro-reactors. The materials code writers often limit the first code approvals to smaller sized parts with the reasoning that smaller parts experience less stress. Use of new materials in small units could come much sooner and would provide some of the data for eventually securing approval of larger units.

Sintering uses heat (below the melting point) and pressure to bond and partially fuse particles together. It is often referred to as powder metallurgy, but it can also use powdered ceramics.

The fuel pellets in high-temperature gas reactors are often sintered to increase density. Then, some corrosion-resistant material is applied as a coating layer, and this can be done with 3D printing.

Advances in welding for nuclear reactors and other high-pressure high-temperature systems are in two tracks. First, developers of laser and electron-beam welders want the maximum strength. Second, they want to do the operations as remotely as possible. This tele-operation is another aspect of the robotics discussed earlier.

---

[544] "New Alloy Material Approved for Use in High-Temperature Nuclear Plants," web page, Department of Energy, Washington, D.C., May 19, 2020. https://www.energy.gov/ne/articles/new-alloy-material-approved-use-high-temperature-nuclear-plants (accessed June 26, 2020)

Metal workers have used refractory ceramics for centuries to smelt ores because the ceramics could withstand heat and chemical attack better than the metals being processed. Ceramics have also worked well in the high-temperature structures of gas turbines for jet aircraft and power plants.

In both those uses, the ceramics are often limited by brittleness and weakness in flexing. In those cases, the ceramic is mixed with some type of fiber to provide strength. These fibers may be carbon,[545] silicon carbide,[546,547] silicon nitride,[548] and others.

The combinations of plastics, metals, ceramics, and more are called composites. We know composites in fiberglass and concrete. Those are workhorse fabrication materials, but much stronger, lighter, and high-temperature versions of them are on the high-performance edge of technology … without being on *The Limits to Growth* edge of resource exhaustion.

There is great potential in these advanced materials, and many of them will be realized and will fulfill their potential. However, there is an important delay factor that is longer than the GND, 12 years to do or die. This delay factor can be seen in the case history of electric motors, which provide an historical analog to 3D printing. Electric motors contain a cluster of technological revolutions making steadily smaller, yet more capable, electrical motors that have spread throughout industry starting in the 1880s and continuing to this day. These revolutions include direct current, alternating current, generators for both, utilities to distribute power, and (most recently) batteries for portable power.

Before the spread of electric motors (and the electrical grids to support them), factories tended to be huge affairs with a central power source and a variety of complex power-transmission devices connected to the actual work stations. These transmission devices included various combinations of drive shafts, belts, pulleys, pneumatic lines, and hydraulic lines. They were inefficient and dangerous to workers. Worst of all, they confined production to limited areas and slowed updates because changing the power-transmission linkages was a major effort in itself.[549]

---

[545] Ramani Venugopalan, D. Sathiyamoorthy, and A. K. Tyagi, "Development of Carbon / Carbon Composites for Nuclear Reactor Applications," *BARC Newsletter*, issue 325, pp. 16-20, April 2012. http://www.barc.gov.in/publications/nl/2012/2012030409.pdf (accessed June 27, 2020)

[546] Yutai Katoha and Lance L. Snead, "Silicon carbide and its composites for nuclear applications – Historical overview," *Journal of Nuclear Materials*, vol. 526, 151849, Dec. 1, 2019.

[547] Leda Zimmerman, "Nuclear: Can Silicon Carbide Fuel Rod Cladding Improve Safety, Performance? *Energypost.eu*, June 17, 2020. https://energypost.eu/nuclear-can-silicon-carbide-fuel-rod-cladding-improve-safety-performance/ (accessed June 27, 2020)

[548] A. Rueanngoen, K. Kanazawa, M. Akiyoshi, et al., "Effects of Neutron Irradiation on Polymorphs of Silicon Nitride and SiAlON Ceramics, *Journal of Nuclear Materials*, vol. 442, Issues 1–3, Supplement 1, pp. S394–S398, Nov. 2013.

[549] Warren D. Devine, "From Shafts to Wires: Historical Perspective on Electrification," *Journal of Economic History*, vol. 43, no. 2, p. 356, June 1983.

The anarchocommunist, Peter Kropotkin, had the technical vision to see the powerful advantages that electric motors could provide.[550] His somewhat libertarian philosophy of communal groups cooperating together (without central-government or capitalist control) for production fit perfectly with individual workstations that could be powered with a cord to one of many plug-ins so that a workshop could be small, cheap, and easily reconfigured. There could be a much better and more productive world.

Kropotkin was correct about the technological potential for electric motors and that they would spread across the world. Electric motors helped people build such an advance from the smallest workshops to giant factories (which also grew more efficient). Unfortunately, for Kropotkin's hopes, no anarchist societies managed to make that revolution work. Instead, both capitalists and big-government communists embraced the technology and thrived. Electric-motors helped the Soviet Union (Russia plus the former Russian Empire conquered countries) evolve from an agrarian to an industrial economy, helped Europe and Japan to recover from the devastation of World War II, and helped the United States to maintain its lead as the greatest economic power in the world.

Yet, the thousands of innovations that made the electric motor revolution took several decades to become a major part of the industrial base. The original technology was largely direct current. Then, it went to alternating current. Materials got progressively better, and fabrication for components got steadily better for both larger and smaller motors, as well as the generators and utility grids to support them. But it didn't come easy, and it didn't come fast.

Likewise, 3D printing and the other advanced material technologies have many obstacles to overcome. The benefits of fine control come at the expense of sensitivity to slight variations in electrical current pumping the input powder and fixing powder with laser or electron beam. Quality control for all those exquisitely fine layers is much more complex than for simple rolled or stamped materials. Quality control must extend to the feed powders, recycling (or excessive recycling) of unused feed powder, and many more factors.

Sintered materials must be heated just enough to make a solid metal with the desired properties, but not heated enough to liquify and lose the properties. New welding technologies, just like the old welding technologies, must work at the correct balance of melting to join materials without being too strong and blowing material away from the intended weld. Composites have two or more materials combined in the correct amounts and the correct array. The rebar of a concrete structure must be in all the right places to support its structure—even more so, carbon fiber must be in the right places and orientation within a composite to make a turbine blade or other high-strength structure.

---

[550] In *Fields, Factories, and Workshops*, 1899—republished a number of times, Peter Kropotkin theorized about the technical advantages of using electric motors and what he thought this would do to the social milieu. Kropotkin, was from the kinder gentler school of anarchism—he wrote more about technical analyses of concepts rather than the bomb throwing he championed in his youth.

Advanced materials will definitely be a major factor in a successful GND, but many of those advanced materials will not come into play for several years. Materials development is another area where patience could yield great rewards.

8.7.4 Prefabricated and Modular Construction

Ikea furniture best illustrates the concept of prefabricated (prefab) construction on a small scale. One or more boxes contains all the parts. This kit of parts can be easily transported to the buyer. The buyer, in turn, uses a small set of common tools to assemble the dresser, book shelf, bed or whatever.

Similarly, prefabricated structures and kits have been widely used since at least the 1830s when London Carpenter John Manning dispatched kit houses to Australia, South Africa, and various places in Great Britain. From the early 1900s to 1940, Sears–Roebuck and Company in the United States sold half a million mail-order house kits.

In the late 1940s, after World War II, manufacturers filled much of the demand for housing with prefabricated mobile homes (trailers) that could become permanent, prefab modules that could be put together to make larger structures, and kit structures. As with kits starting in the early 1800s, structures with these building techniques have many advantages.

- Expensive on-site construction time can be greatly reduced.

- Shorter construction times decreases project risk.

- Factory production of modules, or even entire buildings, allows the factory advantages of mass production, inventory at the factory site, use of specialized tools (such as powder coating to reduce corrosion without paint or lacquer), and fabrication in a controlled (weather-protected) site.

- Factory production has less scrap material, and the scrap can often be recycled to other uses.

- Prefab and modular construction can be done with fewer highly skilled laborers. (This is particularly important in countries such as the United States and Japan where there are increasing shortages in many construction skills.)

- If modular, the modules can often be repurposed when needs change.

Meanwhile, the associated technology advances in 3D printing, robotics, and computer-controlled production are making prefab and modular construction both cheaper and more flexible. For example, modules and prefabs were built with jigs, but computer controls can often change design features with only changes to a computer program. Likewise, 3D printing can be used onsite, and can be adjusted for different tasks.

Taken together, these advances could make construction tremendously cheaper in another decade or two—just the sort of thing that would be useful if many people needed to relocate to higher ground and/or to more poleward ground. (Quebec could become the new Adirondacks).

8 Fixes from Energy and Material

The biggest barrier is developing uniform building code enforcement. As with many other countries, the United States recognizes the International Building Code as a model code for new construction. However, the individual states have modifications. Furthermore, the administration of the state codes is done by local authorities who have their own variations on forms, procedures, computer software, and other details. Cynics have suggested that these layers of complexity were developed to protect local contractors from outside competition. Others have suggested that the local bureaucracies just grew up before better transportation and communication allowed widespread construction.

Whatever the origin, the crazy quilt of construction enforcement multiplies the cost and complexity for any progress toward standardized high-volume construction, the kind that would be needed for a major reconstruction of the built environment to radically increase energy efficiency and generation of alternate energy. The needed improvement is to merge the standards of local enforcement with the procedures of the model code with clearly noted exceptions for local conditions (such as extra wind resistance in areas with hurricanes or extra structural bracing for earthquake-prone areas).

8.7.5 Supercritical Carbon Dioxide Turbines—Another Supporting Energy Revolution

Most heat engines making electricity use the 1884 steam turbine of Charles Parsons with some minor variations plus the gas turbine that evolved from the steam turbine. Parsons' first 7.5-kilowatt engine evolved into units rated as high as thousands of megawatts today. The pressure also increased by using supercritical steam, which has so much heat and pressure that it is neither a liquid nor a vapor.

The next major development for power cycles is supercritical carbon dioxide (sCO_2).[551] The sCO_2 power cycle uses sCO_2 as the working fluid medium in a closed or semi-closed Brayton thermodynamic cycle. This is comparable to supercritical steam that has provided higher performance than ordinary steam for some decades except that the sCO_2 power cycles have significantly higher cycle efficiency and smaller equipment size; therefore, they have lower capital cost. An sCO_2 power cycle might be only 5% of the mass and volume of a steam-cycle turbine while delivering 10 to 20% better efficiency.[552] Furthermore, supercritical carbon dioxide is gentler on the turbine materials than supercritical steam.

Finally, sCO_2 has its greatest advantage at the lower end of the temperature range of steam cycles, so there might be an advantage in a stacked system with a gas turbine or very high temperature steam cycle topping a supercritical CO_2 cycle. These advantages are sparking an sCO_2

---

[551] Sonal Patel, "What Are Supercritical CO2 Power Cycles?" *Power*, April 1, 2019. https://www.powermag.com/what-are-supercritical-co2-power-cycles/ (accessed July 7, 2020)

[552] Václav Dostál, Pavel Hejzlar, and Michael J. Driscoll, "The Supercritical Carbon Dioxide Power Cycle: Comparison to Other Advanced Power Cycles," *Nuclear Technology*, vol. 154, no. 3, pp. 283–301, June 2006. https://www.researchgate.net/publication/285014511_The_Supercritical_Carbon_Dioxide_Power_Cycle_Comparison_to_Other_Advanced_Power_Cycles (accessed July 7, 2020)

revolution in the power industry, and that revolution would also be useful for solar-thermal power production.

8.7.6 Electronic Remote Connection—The Internet, Webinars, and Offices Anywhere

The electronic remote complex of cell phones, personal computers, the Global Positioning System, and the internet in the late Twentieth and early Twenty-First Century is and industrial revolution comparable to the Industrial Revolution of steam power and mass production. Furthermore, it may even be the single most important development for a working GND.

This electronic complex consists of thousands of hardware and software innovations for instantaneous transmission connecting much of the world, personal computers that can generate or receive massive amounts of data, and software that can connect everything together as well as control all the connections.

The economic and social changes this electronic remote complex has already brought and that are coming include the following:

- Millions of people are part of a growing trend toward telecommuting to work without the carbon footprint of driving or mass transit. Furthermore, the 2020 corona-virus pandemic has increased this trend.[553,554]

- Hundreds of millions of transactions each year are done remotely through companies such as Amazon—one delivery van in a day replaces hundreds of carbon footprints of people driving to "brick-and-mortar" stores.

- The carbon footprint of many of brick-and-mortar stores is shrinking. There are fewer stores with names such as Sears, K-Mart, and J.C. Penny. Millions of people download music and movie files rather than driving to the stores and buying or renting the physical storage disks. Furthermore, the selection is much larger. The last Blockbuster video store was in Alaska; it may still be there.

- Millions of people each year participate in virtual (or remote) meetings rather than adding to the carbon footprints or driving, plane flights, and motel rooms. This trend is growing as the software becomes more user friendly and more capable of providing useful services such as meeting transcripts, audios, and videos.

- The 2020 pandemic has spurred a surge in remote working in general. "As COVID-19 forced many workers across the U.S. to move from downtown office towers to

---

[553] Rob Enderle, "Coronavirus (and 5G) Will Boost Telecommuting, Change Our Tech Future," *Computerworld*, Feb. 28, 2020. https://www.computerworld.com/article/3529957/coronavirus-and-5g-will-boost-telecommuting-change-our-tech-future.html (accessed Aug. 31, 2020)

[554] Baruch Feigenbaum, "Telecommuting Helps Fight Coronavirus, Will Likely Outlive the Panic," *The Orange County Register*, March 9, 2020. https://www.ocregister.com/2020/03/06/telecommuting-help-fight-coronavirus-and-beyond/ (accessed Aug. 31, 2020)

spare rooms and kitchen tables, their commutes shrank from an average of almost 30 minutes (often in bumper-to-bumper traffic) to a few steps down the hall. A May survey of 2,500 Americans found that 42 percent were teleworking full-time," which has also caused a major reduction in the carbon footprint.[555]

- At the height of the corona-virus pandemic (and we hope it was the height), 1.2 billion students throughout the world were participating in remote learning.[556] With course material that can be downloaded at any time with little or no carbon footprint for travel, brick-and-mortar educational facilities will probably continue declining except for proctored examinations and final projects. Even those might be largely done by virtual presence and/or satellite education centers.

- The 3D printing of parts mentioned earlier will lessen transportation costs and carbon footprints as well as reducing the costs of down time.

- In the ancient days, people might do research in a great library, such as in Alexandria or Bagdad. Now, people around the world have access to a world's worth of knowledge many times larger than the Library of Alexandria and often millions of times faster … while staying in their own homes.

- More than anything, the remote electronic complex is transforming the entire planet into one gigantic virtual city. Many commentators have noted that innovation increases immensely as cities grow.[557] What wonders might come from the virtual world city via Internet?

8.7.7 Green-Energy Storage

Futurist and inventor Arthur C. Clarke once said that it is be hard to beat the energy storage capacity of gasoline. It stores a tremendous amount of blazing energy per unit mass, and it can flow into the combustion chambers of vehicles.[558] Of course, gasoline is one of the hydrocarbon fossil fuels that produce global-warming emissions. Now, because of global-warming concerns,

---

[555] Ainslie Cruickshank, "COVID Pandemic-19 Shows Telecommuting Can Help Fight Climate Change," *Scientific American*, July 22, 2020. https://www.scientificamerican.com/article/covid-19-pandemic-shows-telecommuting-can-help-fight-climate-change/ (accessed Nov. 27, 2020)

[556] Cathy Li and Farah Lalani, "*The COVID-19 Pandemic Has Changed Education Forever. This Is How*," World Economic Forum, April 29, 2020. https://www.weforum.org/agenda/2020/04/coronavirus-education-global-covid19-online-digital-learning/ (accessed Aug. 31, 2020)

[557] Edward Glaeser, Triumph of the City: *How Our Greatest Invention Makes Us Richer, Smarter, Greener, Healthier, and Happier*, Penguin Press, New York, 2011.

[558] Arthur C. Clarke, Profiles of the Future: An Inquiry into the Limits of the Possible, Arthur C. Clarke, Harper & Rowe, February 1963.

we want to develop noncarbon, or at least decreased carbon, ways to gather and store energy for later use. These noncarbon energy storage methods fall under four broad categories.

1. Breaking water to get hydrogen for fuels
2. Improving batteries and other electrical storage devices
3. Developing more affordable fuel cells to use hydrogen or hydrogen-containing fuels (such as methane, methanol, or ammonia)
4. Hybrid systems of combustion, ultracapacitors, and batteries[559]

8.7.7.1 Breaking Water to Get Hydrogen for Fuels

Life on Earth separates hydrogen from water and then stores it in various compounds of hydrogen and carbon—hydrocarbons. GND technologies could do the same or just use the hydrogen as is. Such an undertaking would create synthetic fuel that would release much less greenhouse-warming carbon dioxide into the atmosphere. If the carbon for the synthetic hydrocarbon were derived from waste carbon dioxide, such fuel might be nearly carbon neutral.

But there is no free lunch. The difficult part is separating the hydrogen and using it in whatever form at an affordable price. The hydrogen (H_2) and oxygen (O_2) of water (H_2O) are together because combining them was an exothermic process that released usable energy. Conversely, one must apply energy (an endothermic process) to break them apart so that the stored energy of the hydrogen can be then released for some useful task elsewhere. That energy can be sunlight for making wood or other biofuels, heat from nuclear fission or concentrating solar collectors, or breaking water with an electrical current (electrolysis) that can come from many alternate-energy sources. The costs are in getting energy to break the water to get hydrogen fuel (or hydrogen combined with carbon), the efficiency loss of actually breaking the water bonds, any losses in reforming hydrogen to make more convenient fuels (such as ammonia [NH_4] or methane [CH_4]), and the storage and transmission costs.

The sums of the costs and efficiency losses have kept synthetic low-carbon or noncarbon fuels much more expensive than fossil fuels. Biofuels come in return for the input costs of land, tending plants, harvesting the plants, and processing the harvest into some usable fuel. Concentrating solar, nuclear fission, or any other heat source comes at the capital costs of building those facilities, running them, and purchasing the land on which to site them.

The higher temperatures and/or cheaper electricity available from advanced nuclear-fission technologies, advanced materials to build them better, and the robotics to run them are making all energy systems more efficient.

However, enough of these advances to become significant will take longer than 12 years. New fission reactor designs need to get their permits and operate successfully for several years

---

[559] CSIRO Australia (2008, January 18). "UltraBattery Sets New Standard for Hybrid Electric Vehicles," *ScienceDaily*, Jan. 18, 2008. http://www.sciencedaily.com¬/releases/2008/01/080118093341.htm (accessed Jan. 20, 2008)

before they can be widely adopted and designs can go down the learning curve to cheaper prices, less down time, and greater safety. Likewise, windmills and solar facilities need more years of the same type of improvements to reduce their costs.

Only then can entrepreneurs and government entities commit billions in funding to industrial-scale water breaking. There have been decades of theoretical analyses and lab-scale experimental processes for catalysts and process steps for more efficient heat breaking of water, but no demonstrator plants the size of small refineries. Likewise, there has been no industrial-scale electrolytic breaking of water.

The dirty little secret of proposals for a hydrogen economy is that all industrial hydrogen production systems use the reforming of natural gas (CH_4) or byproduct gases from coal gasification. Steam at 700–1100 °C (1300–2000 °F) strips away the carbon as carbon dioxide (CO_2) leaving hydrogen behind. The natural gas is essentially burned with an emission of carbon dioxide; plus, there is additional combustion to raise the high-temperature steam so more carbon dioxide is produced than by simply burning the natural gas. Furthermore, the hydrogen product is more expensive to store and transport because of its lesser density and its greater tendency to escape from containers.

Yet, there is a serious proposal from the city of Los Angeles, California to issue utility contracts for hydrogen from solar and wind power and then using the hydrogen to replace natural gas in power plants.[560] These contracts would be based on processes and equipment still to be developed. This does sound a bit like Mao Tse-Tung's Great Leap Forward, in which it was fervently believed that peasant farmers would develop methods to make high-quality steel in backyard blast furnaces.

8.7.7.2 Improving Batteries and Other Electricity Storage Devices

The seemingly mundane world of batteries and other electricity storage devices are key to any hope for a GND.

Intermittent and undependable green energy sources (such as wind, solar, and waves) need backup power to be actually profitable instead of being used because of government mandates and subsidies. At present, these intermittent sources are competitive (or nearly competitive) when available, but their total cost multiplies when the utility must buy peaking power, such as from gas turbines, when the sun doesn't shine and/or when the wind speed drops too low or rises too high.

Calling that back-up power expensive is a bureaucratic subterfuge to hide the expense and continued global-warming emissions of those backup power sources. Many governing entities have mandated that utilities must purchase green energies at premium rates; that adds to electricity

---

[560] Sammy Roth, "Los Angeles Wants to Build a Hydrogen-Fueled Power Plant. It's Never Been Done before," *Los Angeles Times*, Dec. 10, 2019.
https://www.latimes.com/environment/story/2019-12-10/los-angeles-hydrogen-fueled-intermountain-power-plant?utm_source=Daily%20on%20Energy%20121119_12/11/2019&utm_medium=email&utm_campaign=WEX_Daily%20on%20Energy&rid=9580 (accessed Dec. 13, 2019)

rates. Furthermore, the utilities must make these purchases even if there is more electricity available than needed. To compensate for the required electricity purchases, the utilities shut off other power sources, which then become "back-up power." Because the back-up power plants deliver less power compared to their fixed costs, they appear more expensive, and that becomes an argument for more green energy.

Unfortunately, more green energy and less "expensive" back-up power makes the electrical grid more expensive and more prone to failure as was predicted for California in 2013[561] and as California demonstrated in 2020.[562]

Back at electricity, it is a much more useful form of energy than just heat. It can run motors, televisions, computers, and much more. Typically, about two thirds of the energy going into a power-generating facility is lost—leaving a third. Using electricity to make and store hydrogen might lose half of that third—leaving a sixth. Then, using hydrogen to generate electricity might only get another third of the energy as regenerated hydrogen would deliver one-eighteenth of the original energy. Some efficiencies might be better, so we could charitably say roughly one-sixteenth or 6%. If that generating plant is located in an area with significant wind, the electricity delivered might have long-distance losses, dropping delivered electricity down to 5% of the original energy.

That is an inefficient process. Thus, batteries and other electrical storage devices are a key part of any GND.

In the future, a revolution in electricity storage devices could make portions of the GND much more practical—or at the very least, make the California electrical grid less of a disaster. Batteries slowly developed in the early 1800s with just enough energy for small power loads such as running telegraph systems. Then, in the late 1800s and early 1900s, several rechargeable battery types were developed that could both produce chemically derived electricity and then reverse the process if electricity was supplied to the battery (such as from a car's generator). For the first several years of the car age, electric cars dominated the market with small cars gliding quietly around towns.

Then, combustion engines improved to the point that they could deliver significantly more power and range than batteries, all without several hours of recharging before the next leg of the trip. After that, car batteries only served auxiliary functions such as the starter and the lights. The most affordable was the lead-acid battery of sulfuric acid and lead plates, which is the battery type powering most automobiles and trucks to this day. The low price of materials and easy recyclability

---

[561] Rebecca Smith, "California Girds for Electricity Woes: Increased Reliance on Wind, Solar Power Means Power Production Fluctuates," *Wall Street Journal*, Feb. 26, 2013. https://www.wsj.com/articles/SB10001424127887323699704578328581251122150 (accessed Sept. 20, 2013)

[562] Michael Shellenberger, "Why California's Climate Policies Are Causing Electricity Blackouts," *Forbes*, Aug. 15, 2020. https://www.forbes.com/sites/michaelshellenberger/2020/08/15/why-californias-climate-policies-are-causing-electricity-black-outs/#f8fb46c1591a (accessed Sept. 20, 2020)

of lead acid batteries allowed them to hold onto the car and truck market into the 2000s. (The environmental effects of leaking lead compounds are another issue.)

For most of a century, batteries were stuck in the role of auxiliary power. Many thought that batteries would never improve enough to be a prime power source for cars and trucks. Batteries lacked sufficient power to carry even a small car more than about 70 miles (112 kilometers) compared to internal-combustion cars that typically had a range of at least 250 miles (400 kilometers) and could be quickly refueled at the seemingly infinite number of gas stations.

Then, technology changed again. Lithium-ion batteries appeared in the late 1980s. At first, they were small units only good for cell phones and laptop computers, but that changed in the early twenty-teens when car makers scaled them up to run cars. By 2019 several electric cars had ranges around that 250 miles (400 kilometers) under optimal conditions. (Buyer beware: Optimal means one should subtract for cold weather, air conditioning, heating, lights, and so forth.) Still, the lithium-ion batteries brought electric vehicles back to being nearly competitive with combustion engines.

For the future, lithium-ion batteries are becoming a mature technology that might or might not have significant potential for more improvements. For instance, once range has increased, how many cycles can the battery withstand before performance degrades. This is a significant because, although an electric car avoids much of the high parts count and downstream maintenance costs of the internal combustion engine, but the battery is a major expense (maybe six-thousand dollars). As that battery charges and discharges, its capacity begins to decline to the point that electric-car owners must often make that expensive purchase again a few years after buying the car.

However, the revolution of lithium-ion batteries suggests that there may be more lithium-ion improvements. First, better fabrication techniques might lead to a "million-mile battery," which decrease the vehicle lifetime cost significantly.[563,564] Second, there might be improvements to lithium-ion batteries with slight additions of other materials, and/or entirely new families of batteries that might be developed with potential power-per-unit mass of two, three, or even four times that of lithium-ion batteries. These might be fabricated using lithium with various other materials[565] such as lithium sulfur[566] or lithium iron phosphate ($LiFePO_4$). There might be batteries substituting magnesium for the lithium—with magnesium having two valence electrons instead of

---

[563] Tim Mullaney, "Tesla and the Science Behind the Next-Generation, Lower-Cost, 'Million-Mile' Electric-Car Battery," *CNBC*, June 30, 2020. https://www.cnbc.com/2020/06/30/tesla-and-the-science-of-low-cost-next-gen-ev-million-mile-battery.html (accessed Oct. 31, 2020)

[564] Vanessa Bates Ramirez, "New Record-Crushing Battery Lasts 1.2 Million Miles in Electric Cars," *Singularity Hub*, June 11, 2020. https://singularityhub.com/2020/06/11/road-trip-new-record-crushing-battery-lasts-1-2-million-miles-in-electric-cars/ (accessed Oct. 31, 2020)

[565] Priscila Barrera, "6 Lithium-ion Battery Types," *Investing* News, May 13th, 2019. https://investingnews.com/daily/resource-investing/battery-metals-investing/lithium-investing/6-types-of-lithium-ion-batteries/ (accessed Dec. 22, 2019)

[566] John Voelcker, "Advances in Lithium-Sulfur Batteries Offer Promise for Electric Cars," *Green Car Reports*, Feb. 14, 2015. https://www.greencarreports.com/news/1096683_advances-in-lithium-sulfur-batteries-offer-promise-for-electric-cars (accessed Dec. 22, 2019)

the one of lithium for twice the capacity, minus a decrease in capacity per unit weight because magnesium is slightly heavier.[567]

Such improvements also may be crucial for a new mineral resource limits to growth. Lithium and of cobalt are two elements in lithium-ion batteries that might experience shortages if production of those batteries increases faster than supplies. Lithium production has environmental issues, and the best cobalt deposits to date have been in the unstable Republic of the Congo.[568]

Recycling of lithium and cobalt has been considered too difficult and expensive to pursue, but that may need to change.[569] Lithium itself is not considered a serious pollutant, but lithium-ion batteries do contain cobalt, nickel, and manganese, which are heavy metals—not as poisonous as lead, but poisonous nevertheless.[570]

Even without new battery types, incremental (*kaizen*) improvements could significantly produce cheaper and higher performance batteries. Some of these probable incremental improvements include developing a major battery supply chain[571] and applying new production techniques, such as robotics and 3D printing.[572]

Just a doubling in battery capacity would extend the range of electric cars—Los Angeles, California to the San Francisco Bay Area rather than recharging in the Central Valley (and fast recharging comes at the cost of a faster degradation of battery performance). Such a doubling in battery capacity will also facilitate the robotics and distributed manufacturing previously mentioned and even enable electric airplanes, which are described in Section 8.9 in part to show the incredible novelty that innovation (and better batteries) can make common place.

---

[567] Alexander Hellemans, "Taiwanese Researchers Report Progress Towards a Magnesium-ion Battery," *IEEE Spectrum*, posted Nov 7, 2014. http://spectrum.ieee.org/energywise/semiconductors/materials/taiwanese-researchers-report-progress-towards-a-magnesiumion-battery (accessed Dec. 22, 2019)

[568] David L. Chandler, "Will metal supplies limit battery expansion?" MIT News Office web page, Massachusetts Institute of Technology, Cambridge, Massachusetts, Oct. 11, 2017. https://news.mit.edu/2017/will-metal-supplies-limit-battery-expansion-1011 (accessed Oct. 31, 2020)

[569] Chantelle Dubois, "A Looming Shortage of Lithium and Cobalt? Depends on Electric Cars, Politics, and Battery Chemistry," *All About Circuits*, June 20, 2018. https://www.allaboutcircuits.com/news/looming-shortage-lithium-cobalt-electric-cars-politics-battery-chemistry/ (accessed Nov. 1, 2020)

[570] Robert Rapier, "Environmental Implications of Lead-Acid and Lithium-Ion Batteries," *Forbes*, Jan. 19, 2020.

[571] *A Vision for a Sustainable Battery Value Chain in 2030: Unlocking the Full Potential to Power Sustainable Development and Climate Change Mitigation*, World Economic Forum, Geneva, Switzerland, Sept. 2019. http://www3.weforum.org/docs/WEF_A_Vision_for_a_Sustainable_Battery_Value_Chain_in_2030_Report.pdf (accessed Dec. 3, 2019)

[572] "Is 3D Printing the Future of Battery Design?" *MIT Technology Review*, Dec. 22, 2019. https://www.technologyreview.com/2019/12/22/131439/is-3d-printing-the-future-of-battery-design/ (accessed Sept. 28, 2020)

Changing the focus up from cars to the larger utility scale, there are many civil-engineering-sized ways of storing electricity or ways electricity can be cheaply regenerated to store power on the scale of electric utilities.[573]

- Pumped hydroelectric storage (PHES) is the most mature and most used storage technology for large amounts of electricity. Water is pumped uphill from one reservoir to a higher reservoir. Releasing water back down can provide roughly 80% returned electricity. The prime limitation to this type of storage is finding locations with both a dependable water supply and two basins that have significantly different altitudes. Also, such places tend to be highly scenic, and the locals do not always appreciate reservoirs that can change from a lake to a mud hole on any given day.

- lithium-ion batteries and variants (as noted earlier) have been rapidly growing in capability since electric cars returned to the car market. They have an advantage in that battery innovations can be developed for small uses and then scale up just as lithium-ion batteries evolved to serve the light electronics market and then worked up to powering cars.

- Industrial-sized reduction–oxidation (redox) flow batteries cause certain electrolytes to oxidize or reduce when they flow through a membrane either when a current is applied to the battery or produced by the battery. The batteries do not store as much electricity per unit mass as lithium ions, but the units can be many tons, and they may last 10 to 20 years. Redox batteries return about 80% of the electricity that went in. Functioning units have stored tens of megawatt hours, and a unit of 800 megawatt-hour capacity is being installed in Dalian, China.[574,575] (A megawatt hour is a million watts, or a thousand kilowatts, of electricity for one hour.) Redox batteries could largely mitigate the intermittency problem of solar and wind. However, redox batteries have their own mineral-shortage concern with

---

[573] Seamus D. Garvey and Andrew Pimm, Chapter 5, "Compressed Air Energy Storage," *Storing Energy: with Special Reference to Renewable Energy Sources*, Trevor H. Letcher, editor, Elsevier; May 11, 2016.

[574] Tom Lombardo, "Massive 800 MegaWatt-hour Battery to Be Deployed in China," *Engineering.com*, June 5, 2016. https://www.engineering.com/ElectronicsDesign/ElectronicsDesignArticles/ArticleID/12312/Massive-800-MegaWatt-hour-Battery-to-Be-Deployed-in-China.aspx?e_src=relart (Dec. 27, 2019)

[575] Z. Gary Yang, "It's Big and Long-Lived, and It Won't Catch Fire: The Vanadium Redox-Flow Battery," *IEEE Spectrum*, Oct. 26, 2017. https://spectrum.ieee.org/green-tech/fuel-cells/its-big-and-longlived-and-it-wont-catch-fire-the-vanadium-redoxflow-battery

vanadium, and several development teams are attempting to produce redox batteries that substitute iron and other materials for the scarce vanadium.[576,577]

- Flywheel energy storage uses a rotating wheel spun at high speed much like the flywheel in a car only larger. The flywheel spins up to speed as a motor using electricity and produces electricity as a generator as it slows down. Flywheels have been used for uninterruptible power systems (UPS) with sizes into several hundred-kilowatt hours. They hold high capacity because of the high efficiency of magnetic bearings and partial vacuum in the flywheel chamber. Their price has been about half that of a battery UPS, and there will probably be a major technology-development competition between flywheels and batteries. Flywheel upgrades being applied or proposed include going from a steel wheel to composite, running the flywheel in vacuum, using better magnetic bearings to reduce friction, and going from a conventional generator to a high-efficiency generator using rare-earth elements or even going to a superconducting generator.

- Capacitors hold electricity like a battery, but there is no chemical change. That allows them to receive and give back energy quickly but not as much total energy—batteries are the long-distance runners while capacitors are the sprinters. Larger capacitors, called ultracapacitors or supercapacitors, can buffer batteries from the surge loads of rapid acceleration or the surge input electricity of electric brakes putting electricity back in (regenerative braking) and for load balancing of utilities. The greater surface area of nanomaterials (specifically carbon nanotubes) has allowed more powerful ultracapacitors.[578]

- Compressed air energy storage (CAES) puts compressed air (say 100 times atmospheric pressure) in underground geologic formations, and then regenerates power when the air is released. The formation used must be something like a salt dome with pockets of open space to keep the air trapped. (Caution: the formation must not have any hydrocarbons or even some potentially flammable minerals because many things burn in the presence of high-pressure air.) There are two operating plants. One in McIntosh, Alabama can produce 110 megawatt hours for 26 hours (2,860 megawatt hours total) at 54% efficiency. Another (in Huntorf, Germany) can hold air enough to produce 300 megawatts for 3 hours or

---

[576] Robert F. Service, "New Generation of 'Flow Batteries' Could Eventually Sustain a Grid Powered by the Sun and Wind," *Science*, Oct. 31, 2018. https://www.sciencemag.org/news/2018/10/new-generation-flow-batteries-could-eventually-sustain-grid-powered-sun-and-wind (accessed Nov. 1, 2020)

[577] Bo Yang, Advaith Murali, Archith Nirmalchandar, et al., "A Durable, Inexpensive and Scalable Redox Flow Battery Based on Iron Sulfate and Anthraquinone Disulfonic Acid," *IOP Science*, April 9, 2020. https://iopscience.iop.org/article/10.1149/1945-7111/ab84f8 (accessed Nov. 1, 2020)

[578] Rob Matheson, "New Applications for Ultracapacitors," *MIT News*, Sept. 7, 2016. http://news.mit.edu/2016/new-applications-ultracapacitors-drilling-aerospace-electric-cars-0907 (accessed Dec. 27, 2019)

900 megawatt hours. CAES is theoretically just a mechanical pumping process, but the air heats up in compression (like in a bicycle pump), and the decompressing air gets colder (like in a refrigerator). The system must moderate these temperature differences (including burning some natural gas), so it is more expensively complex than it would seem. Nevertheless, the two facilities have operated reliably for some years.[579]

- Underwater compressed air energy storage would pump compressed air down into the ocean or deep lakes. Water pressure would contain the air in tanks anchored to ocean or lake floor. No drilling or particular geologic formations would be required, so the potential storage is theoretically unlimited. Unfortunately, building and maintaining underwater facilities would be expensive, and no such facilities have been constructed.

- Rail energy storage would have a short electric railroad with one or more lines carrying weighted cars upslope to store the potential energy of the mass on the rail cars with regenerating of electricity rolling back downslope. The potential energy of traveling technology might get an 80 to 90% return on electricity. The Advanced Rail Energy Storage (ARES) company hopes to build a demonstration facility with 200 megawatt hours. The disadvantage is that rail storage requires a great deal of land area with largely unused sloping terrain—available in places like the American Southwest but much less so elsewhere.[580]

- Energy vault would stack giant concrete blocks into a skyscraper to save energy and would lower the blocks back down when dispatchable power was needed. Energy Vault's advantage over other rail storage is that it does not require a large area of sloped ground. Its advantage over compressed air energy storage is that the permits and environmental concerns are less.[581,582] Cost estimates based on the

---

[579] A major proposed site in Utah has an existing coal-fired power plant that is scheduled to shut down in 2025, a high-voltage transmission from that site to Los Angeles, and a nearby salt dome (Sammy Roth, "A clean energy breakthrough could be buried deep beneath rural Utah, Los Angeles Times, Aug. 8, 2019 [accessed August 11, 2019].)

[580] "Advanced Rail Energy Storage," *All Sustainable Solutions*, Jan. 2020. https://allsustainablesolutions.com/advanced-rail-energy-storage-system/ (accessed Aug. 31, 2020)

[581] Julian Spector, "Can Newcomer Energy Vault Break the Curse of Mechanical Grid Storage?" *Green Tech Media*, Nov. 14, 2018. https://www.greentechmedia.com/articles/read/energy-vault-stacks-concrete-blocks-to-store-energy (accessed Nov. 30, 2020)

[582] "Energy Vault Named Technology Pioneer by World Economic Forum as It Offers an Economic Way to Store Clean Energy and Deliver Dispatchable Power," *Business Wire*, June 16, 2020. https://www.businesswire.com/news/home/20200616005363/en/Energy-Vault-Named-Technology-Pioneer-World-Economic (accessed Nov. 30, 2020)

Energy Vault design specifications indicate that it could store electricity cheaper than pumped hydro storage.[583]

8.7.7.3 Developing More Affordable Fuel Cells

Fuel cells have been a great alternate energy hope since about 1970 because they can provide electricity with far greater efficiency than a thermal engine. While burning fuel in an engine might deliver power at thirty-percent efficiency, a fuel cell might chemically produce electricity at efficiencies as great as eighty percent while producing very few emissions. Moreover, fuel cells might also work in small units such as power generation for individual houses and (of course) long-range electric cars.

That would be using hydrogen, which is the ultimate noncarbon fuel. However, at a small penalty in efficiency, a more conveniently available fuel might be used. These could include ammonia (NH_4, also can be noncarbon), methane (CH_4, available from the existing natural-gas-distribution network), methanol (CH_3OH, storable as a liquid), or others. It was hoped that there would be a fuel-cell revolution using this cleaner, greener, more efficient energy system.

Fuel cell use has grown, but it has not been a revolution because the fuel cells have generally been too expensive. The catalysts that make fuel cells work use platinum and other elements in the precious-metals group. Those expensive catalysts have kept prices high. Chemists have sought cheaper catalysts through experimentation, the power of increased computer calculation capabilities, and attempts to find nanotech catalysts, but progress has been slow.

Now, hopes have risen again. Futurist Garry Golden described it as, "Back to the Future: The Latest Fuel Cell Revolution," which might become a major force within the decade.[584] Once again, the vision is that fuel cells can replace vehicle internal combustion engines, and fuel cells could provide both electricity and hot water for a combined efficiency of 75 to 90%. It would be wonderful if those hopes were correct … this time.[585]

As a side note, a revolution of fuel-cell cars might render electric charging stations obsolete. Should we be rushing into that multi-billion-dollar investment? Yes, it would be a game changer, but no, it would not be a good investment unless someone manages to develop a working affordable catalyst, and they have been trying for fifty years.

---

[583] "Energy Vault Named Technology Pioneer by World Economic Forum as It Offers an Economic Way to Store Clean Energy and Deliver Dispatchable Power," Business Wire, June 16, 2020. https://www.businesswire.com/news/home/20200616005363/en/Energy-Vault-Named-Technology-Pioneer-World-Economic (accessed Nov. 30, 2020)

[584] Garry Golden, ""Back to the Future: The Latest Fuel Cell Revolution," *NEMA Currents*, Jan. 25, 2019. https://blog.nema.org/2019/01/25/back-to-the-future-the-latest-fuel-cell-revolution/ (accessed Dec. 25, 2019)

[585] Fuel cells were developed by the U.S. National Aeronautics Administration to support crewed missions such the Apollo missions to the Moon and the International Space Station. There were great hopes in the 1970s for widespread commercial use of such cells, but the catalysts required always kept them too expensive to be profitable.

The bottom line is that batteries, though expensive compared to the cost of generating electricity would still reduce the wind and solar costs of wasted power when the supply is too great and buying replacement power when the supply is too little.

8.7.8 Materials for Absorbing. Reflecting, Generating, and Using Energy

Techniques such as computer materials analysis, 3D printing, and nanotechnology (processing with very tiny portions of materials) have been producing materials with revolutionary new capabilities. These new materials facilitate new processes and business models described elsewhere in this book. The following are just some of the most important of these materials.

8.7.8.1 Photovoltaics, Thermovoltaics, and Photo-Emitters

Photovoltaics, materials that convert light into electricity, have been the great green hope for changing the world's energy system for sixty years. For many years before that, photovoltaic-generated electricity cost thousands of dollars per watt of electricity—worthwhile only for motion-sensing burglar alarms and safety sensors for garage-doors. Then, after *Sputnik* launched into orbit in 1957 as the first artificial satellite, photovoltaics became a key competing power source for space probes. Even though fabulously expensive, solar cells had no moving parts, required no maintenance, and worked well in the cloudless realms outside Earth's atmosphere. Space program researchers developed cheaper and more efficient solar cells, and futurists began to suggest that solar might be the power source of the future.

Then, the two Energy Crises of the 1970s and warnings that fossil fuels were nearly exhausted led to government programs throughout the world seeking to develop industrial-scale solar energy that would extend individual cells into panels that could be as large as billboard signs. These programs focused on:

1. Reducing the price of highly pure electronics-grade silicon and getting silicon-based photovoltaics to the present efficiency of about 30 percent.

2. Developing techniques to fabricate silicon cells as large numbers of cells connected together in panels rather than individual cells (often produced at high temperatures in long tubes that were then sawn into wafers), which were then connected into panels.

3. Developing photovoltaic materials using entirely new materials (such as gallium arsenide or indium sulfide) that might be easier to fabricate and/or might use different parts of the light spectrum.

4. Developing solar cells with multiple layers of silicon and some of those different types of photovoltaic materials so that one layer would stop a certain light band with the remaining light continuing on to be used by one or more additional layers below to harvest different frequency bands to get more energy per unit area. Multijunction cells used in outer-space application can attain efficiencies of about 40 percent. However, manufacturing them cheaply enough for competitive

electrical generation on Earth will require continued evolution of robotics and nanotechnology.

5. Reducing cost by standardization and manufacturing in volume.

6. Increasing efficiency and panel lifetimes with evolving fabrication improvements over time.

7. Developing affordable energy storage technologies for inconvenient hours of, clouds, low-sun, and no-sun.

In addition to the cost of storage, the second cost barrier for photovoltaics is that two thirds of the cost now consist of the installation, and only a third is in the modules themselves. Consequently, the next step in conventional silicon panels is to develop modules that are cheaper to install … and cheaper for removal and disposal. (As with coal and nuclear, spent photovoltaics contain toxic wastes.) The continued solar revolution depends on manufacturers and installers making those incremental improvements needed to stay ahead of their competitors. That is the slow-moving revolution to transform (solar) photovoltaic panels from a government-subsidized luxury into a bottom-line winner.

This revolution can happen first in hot desert urban areas such as Los Angeles, California; Phoenix, Arizona; most of North Africa; and much of the Middle East. Such areas could get all their electricity from photovoltaic panels—if they had some type of affordable storage.

The simplest approach to providing that desert solar energy is strings of giant "solar farms" of several square miles out in largely thinly settled desert areas distant from any urban area so the land price is cheap. Such farms usually have one size of panels with one set of power inverters to convert the direct current into alternating current, and one set of transformers tying into the high-power line with a few hundred megawatts. A series of such farms along the power line would supply the urban areas with tens, hundreds, or thousands of megawatts. Maintenance crews could drive through at all hours for maintenance and replacement of panels and transformers. All these features would provide a number of efficiencies.

However, a crucial disadvantage is that the locals do not want to sacrifice major fractions of their nearby lands for powering cities elsewhere. It may be desert, but it is home to them. It is also home to numerous endangered species that environmentalists have fought and continue to fight bitterly to protect. Furthermore, the long stretches of high-power transmission lines are another major financial and environmental cost.

A simpler approach would be using the urban roofs that cover sufficient area to potentially provide all the electricity used in the urban areas. Moreover, generating electricity in the urban area would allow tying into the local electrical grids and would avoid the costs of giant solar receiver farms in distant locations and the high-power transmission lines to bring that power to urban areas.

Unfortunately, urban roofs also have the most expensive costs for installing and maintaining. In contrast to the standardized solar farms, the urban photovoltaic complexes would have tens of thousands of smaller units of different sizes, placements, and installation times. Each unit would require local permitting and inspection. Each unit would require its own power inverter and transformer to tie into the local grid. (An alternate approach would include individual storage in each building rather than going to the grid, but that would have its own additional expenses and complexity.) All maintenance would have to be scheduled with the building owners. In residential areas maintenance could only be done during certain daylight hours as prescribed by city ordinances.

Last but not least, including photovoltaic panels on an existing structure costs much more than on new structures because it entails risk and possible reconstruction. Can the roof support the panels? Conversely, can the roof hold down rows of solar panels that have some similarities to the sails of clipper ships? Does the building plan on file with the city planning office have sufficient information to even make those decisions? Many installations on existing structures would require consultation with a structural engineer and special approval from the local planning commission.

For all those reasons, developing urban photovoltaic power on existing structures (as with the proposed upgrades to energy efficiency of all building proposed by the GND), would be impractically expensive in the short term of 12 or even 22 years. Major photovoltaic installations are expensive, but also has advantages for new buildings or major rebuilds. Thus, affordable use of all available roofing areas is something that should be done over a number of decades, although it could be expedited with some technologies still under development.

One such new technology under development is photovoltaic materials that absorb only infrared and ultraviolet and are otherwise transparent—thus, "Solar Energy That Won't Obscure Your View."[586] Furthermore, these materials can be thin, and (not working in the visible band) they can be coatings on windows, walls of buildings, or even roofs and panels of cars. Even though they might be less efficient, they could cheaply cover large additional surface areas that cannot be used by conventional photovoltaic materials.[587] This is another material that has just recently evolved in lab-scale performance.[588] and could be another revolution, but only after

Another approach is not photovoltaic of visible light but thermophotovoltaic of heat, cells geared to converting heat difference into electricity. Mirrors concentrating all the light frequencies

---

[586] "Solar Energy That Won't Obscure Your View," *Engineering.com*, Aug. 26, 2014. https://www.engineering.com/DesignerEdge/DesignerEdgeArticles/ArticleID/8329/Solar-Energy-That-Wont-Obscure-Your-View.aspx?e_src=relart (accessed Feb. 13, 2020)

[587] Stephen Battersby, "The Solar Cell of the Future," *Proceedings of the National Academy of Sciences of the United States*, vol. 116, no. 1, pp. 7–10, Jan. 2, 2019. https://www.pnas.org/content/pnas/116/1/7.full.pdf (accessed Sept. 3, 2020)

[588] Yongxi Li, Xia Guo, Zhengxing Peng et al., "Color-Neutral, Semitransparent Organic Photovoltaics for Power Window Applications," *Proceedings of the National Academy of Sciences of the United States*, vol. 117, Sept. 1, 2020. https://www.pnas.org/content/117/35/21147 (accessed Sept. 3, 2020)

onto highly absorbing surfaces (more nanotechnology) should be able to get at least 50% conversion efficiency and possibly more.[589]

There is a surprisingly new angle from nanotechnology, radiative cooling panels. Panels pointing straight up into the sky can tap a vast freely available heat sink only slightly above absolute zero. On clear nights, heat radiates out to the cold of outer space. The radiative panels emit more energy in infrared (heat waves) than they absorb in visible. Their emission frequency band is optimized to be in the band that best penetrate the atmosphere so that energy can dissipate out into space. This is the same process as heat radiating out on a clear winter night, but it also works in the day. Radiative cooling has a great potential to dissipate the enormous waste heat we generate every day on Earth.[590]

A polymer layer embedded with microspheres, backed with a thin layer of silver provides passive radiative cooling. The result is an easy-to-manufacture material near the theoretical limit for radiative cooling both night and day. The translucent and flexible film can be made in large quantities for a variety of energy technology applications.[591]

1. "Self-cooling Solar Cells Boost Power, Last Longer."[592] A cooling layer could be added to the surface of ordinary photovoltaic panels. Cooling the panels would allow them to operate more efficiently, and less heat would slow the decreased performance caused by heat, thus extending service life of the panels.

2. Sky-pointing cooling panels would increase the efficiency of air conditioners and industrial refrigeration, thus decreasing electricity needed. (In winter, panels might need covering to prevent loss of heat much as screen windows are replaced with storm windows in the fall.)

3. For poorer regions where people would tolerate some discomfort, cooling panels could eliminate air conditioning entirely.

4. Greenhouses are usually thought to be for protecting plants from cold, but hot desert greenhouses could be kept cooler by a radiative layer and/or a thin power generating layer on the glass that stopped some infrared while using those frequencies to power

---

[589] Kevin Bullis, "Better Thermal Photovoltaics: A New Way to Convert Heat into Electricity Could Lead to More Efficient Solar Power," *MIT Technology Review*, Jan. 21, 2009.

[590] Yao Zhai, Yaoguang Ma, Sabrina N. David, et al., "Scalable-manufactured randomized glass-polymer hybrid metamaterial for daytime radiative cooling," *Science*, vol. 355, issue 6329, pp. 1062–1066, March 10, 2017.
https://science.sciencemag.org/content/355/6329/1062 (accessed Jan. 2, 2020)

[591] Yao, Zhai, Yaoguang Ma, Sabrina N. David, et al., "Scalable-manufactured randomized glass-polymer hybrid metamaterial for daytime radiative cooling," *Science*, vol. 355, issue 6329, pp. 1062–1066, March 10, 2017.
https://science.sciencemag.org/content/355/6329/1062 (accessed Jan. 2, 2020)

[592] "Self-cooling Solar Cells Boost Power, Last Longer," *Engineering.com*, July 23, 2014.
https://www.engineering.com/DesignerEdge/DesignerEdgeArticles/ArticleID/8100/Self-cooling-Solar-Cells-Boost-Power-Last-Longer.aspx?e_src=relart (accessed Feb. 13, 2020)

the greenhouse equipment.[593,594] Meanwhile, with lesser heat levels, the greenhouses could be kept largely sealed to save water and protect against pests.

5. Widespread cooling from radiative cooling could eventually slow greenhouse warming directly without any changes in greenhouse-agent emissions.

The three principles of photovoltaic, thermovoltaic, and radiative cooling are, indeed, a set of revolutions that will greatly reduce the need for burning fossil fuels. Yet here again, their development is not ready for a government crash program. They still need a couple decades of relatively small research grants, testing in the market place, and eccentric entrepreneurs to succeed or fail in that endeavor.

8.7.8.2 Composites

Composites are two or more materials combined so that they provide capabilities better than any of them separately. Plant lignin binder holds plant cellulose fibers in a composite matrix. Wood is a naturally grown composite that has been one of humanity's most important materials since prehistory.

The ancient Egyptians and Greeks cut out thin veneers of wood and put them back together in layers with the wood grain at different angles to increase strength. In the 1800s, sawmill owners redeveloped and improved the process with powered rotary lathes to cut the layers (or plies) for similar wood layers; hence, they made plywood. Development of such engineered wood (or reengineered wood, depending on the reference) has continued with engineered wood infused with plastic to better resist rot and termites.

But much more is coming. Processing of wood to reduce the amount of lignin binder to just the right level can increase the strength of wood to levels as strong as titanium but lighter.[595] This strengthened wood and multiple plies has been used for cross-laminated timber (CLT) high

---

[593] Eshwar Ravishankar, Ronald E. Booth, Carole Saravitz, et al., "Achieving Net Zero Energy Greenhouses by Integrating Semitransparent Organic Solar Cells," *Joule*, Jan. 29, 2020.

[594] Elinor P. Thompson, Emilio L. Bombelli, Simon Shubham, et al., "Tinted Semi-Transparent Solar Panels Allow Concurrent Production of Crops and Electricity on the Same Cropland," *Advanced Energy Materials*, vol. 10, 2001189, 2020.
https://onlinelibrary.wiley.com/doi/epdf/10.1002/aenm.202001189 (accessed Sept. 14, 2020)

[595] "Super Wood Could Replace Steel," *Science Daily*, Feb. 7, 2018.
https://www.sciencedaily.com/releases/2018/02/180207151829.htm (accessed June 27, 2020)

rises, including a ten-story office building in Brisbane, Australia.[596] Meanwhile, methods have been developing for infusing flame retardants into engineered wood.[597]

This leads to the question, "Engineered Wood – the new concrete on the block?" because the recent advances in engineered wood mean that it can provide better structural integrity than steel and concrete. [598] It leads naturally to the details of, "How to Build a Skyscraper out of Wood,"[599] which is entirely feasible although worrisome to safety inspectors.

For an GND, wood contributes less to global warming because it allows faster construction (less energy and cost in construction), it has good thermal insulation properties (less energy use throughout the life of the building), and it is more sustainable because it really does grow on trees.

Belted tires have Kevlar, steel, nylon, or other fibers that allow them to carry their vehicles better and achieve three or four times the mileage that tires achieved a half century ago. Similarly, structural composites have straight fibers (such as carbon, glass, or high-strength fiber) embedded within a polymer matrix (such as phenolic, polyester, or epoxy), which are deposited layer-by-layer. As with plywood, layers can be oriented in different directions to have strength in all directions.

Fiberglass (invented in 1936) at first had only glass fibers embedded in a polymer matrix. Applications, such as car and aircraft body parts, have been developed with higher performance plastics and stronger fibers, such as carbon fiber and aramid. Polymer production exceeded that of steel production in the late 1970s. Often, this was because of direct substitution of composites for steel, and the same substitution has also happened for aluminum.

Moreover, composites of various sorts have made turbine blades capable of handling greater stresses and higher temperatures. Ceramics with embedded fibers help overcome brittleness and sensitivity to thermal shock in the ceramics. That allows higher performance of

---

[596] Max Opray, "Tall Timber: The World's Tallest Wooden Office Building to Open in Brisbane," *The Guardian*, June 20, 2017. https://www.theguardian.com/sustainable-business/2017/jun/21/tall-timber-the-worlds-tallest-wooden-office-building-to-open-in-brisbane (accessed Jan. 2, 2020)
"The Tallest Timber Tower in Australia Opens in Brisbane," *ArchDaily*, Nov. 23, 2018. https://www.archdaily.com/906495/the-tallest-timber-tower-in-australia-opens-in-brisbane/ (accessed Jan. 2, 2019)

[597] "New highly effective, eco-friendly flame retardant," *Science Daily*, Jan. 5, 2015. https://www.sciencedaily.com/releases/2015/01/150105170241.htm (accessed Jan. 2, 2019)
Shan He, Yichen Guo, Tehila Stone, et al., "Biodegradable, Flame Retardant Wood-Plastic Combination via in Situ Ring-Opening Polymerization of Lactide Monomers," *Journal of Wood Science*, vol. 63, pp. 154–160, 2017. https://link.springer.com/content/pdf/10.1007%2Fs10086-016-1603-2.pdf (accessed Jan. 2, 2020)

[598] Leo Ronken, "Engineered Wood – The New Concrete on the Block? Gen Re, Cologne, Germany, Dec. 2016. http://www.genre.com/knowledge/publications/pmint1612-en.html (accessed Jan. 2, 2019)

[599] Jeff Spross, "How to Build a Skyscraper out of Wood," *The Week*, Jan. 14, 2019. https://theweek.com/articles/816653/how-build-skyscraper-wood (accessed Jan. 2, 2019)

rocket nozzles and jet turbine blades in the air and power-plant turbines on the ground. For instance, ceramic matrix composites (CMCs) have been called a revolution because their lighter weight and higher operating temperatures increase the efficiency of turbines from 60 to 65%.[600] This is a carbon-saving and money-saving revolution for nuclear, solar, and even fossil-fuel power plants.

Without this trend toward greater use of composites, humanity's carbon footprint would be much larger. Furthermore, this trend toward greater efficiency will continue.

8.7.8.3 Greener Concrete

Concrete is another composite, and it is by far the one that humanity uses most—approaching 4 billion metric tons (5 billion short tons Americans use) per year. Making that concrete emits roughly an equal weight of global-warming carbon dioxide into the atmosphere, and that is about 5% of world carbon dioxide emissions.

So, if concrete generates all that carbon dioxide, why do people use it? They use it because it is one of the cheapest, most fire resistant, and most durable construction materials available. Many concrete structures built by the ancient Romans are still standing. Considering those advantages, people would doubtless resist the loss of their concrete building materials. Also, sad to say, builders are seriously averse to more expensive "green" concretes.[601] However, there are a number of ways to decrease cost, decrease the carbon footprint, and increase the capabilities of concrete.[602]

- Most concretes are held together by some type of cement or other binder. The typical cement is made up of lime baked out of limestone, some hard and durable material (such as sand or gravel—called aggregate), and a little clay. As noted earlier in Chapter 7, the carbon footprint can be much less if nuclear fission reactors provide some or all of the heat for baking the carbon dioxide out of the limestone and then processing the carbon dioxide baked off into some useful product.

- Recycled concrete from structural demolition also reduces the carbon footprint and landfill costs from the demolition.

---

[600] Ivan G. Rice, "CMCs Will Revolutionize Aero and Land-Based Gas Turbines," Turbomachinery International, Nov. 5, 2015. https://www.turbomachinerymag.com/ (accessed Nov. 24, 2018)

[601] Vanessa Dezem, "Cement Produces More Pollution Than All the Trucks in the World," *Bloomberg*, June 22, 2019. https://www.bloomberg.com/news/articles/2019-06-23/green-cement-struggles-to-expand-market-as-pollution-focus-grows (accessed Dec. 29, 2019)

[602] Prathyusha Yadali, R. Selvaraj, D. Neeraja, "Alternate Binders for Construction, Repair and Rehabilitation," *International Journal of Civil Engineering and Technology (IJCIET)*, vol. 9, issue 4, pp. 1220–1228, April 2018. https://www.iaeme.com/MasterAdmin/uploadfolder/IJCIET_09_04_136/IJCIET_09_04_136.pdf (accessed Jan. 1, 2020)

- Different materials mixed in with the cement and aggregate can give the concrete different properties. Reinforcing rods (rebar) set into reinforced concrete increase the concrete's strength in tension (especially to protect against earthquakes). A 2% addition of biochar (sawdust baked into a flaky charcoal) causes reduced setting time, reduced water penetration, increased mechanical strength, and it gives a permanent home to some carbon.[603] Including particles of graphene can makes concrete that is four times more water resistant and twice as strong as conventional concrete—allowing use of less concrete—thus, less greenhouse carbon dioxide emitted.[604] Recycled plastic, with some gamma radiation for stiffening, also gives concrete greater strength and flexibility.[605,606] Saturating any concrete with carbon dioxide also increases the strength—and finds a home for some more of that excess carbon dioxide we do not want to put in the sky.

- Builders began including steel bars inside concrete during the 1800s. Rebar provides greater strength and lighter weight than Roman concrete used in many great buildings and aqueducts still standing today. However, after some decades, the builders realized that water and minerals slowly moving through the porous concrete causes steel rebar to rust, eventually destroying the structure. If the ancient Romans had used rebar, maintaining their structures would have needed rebuilding 16 times by now.[607] Corrosion-resistant rebar greatly reduces the need for major rebuilding and the resulting carbon footprint. One of the first responses was galvanizing the steel with zinc, which sacrificially protects the steel until it corrodes, thus delaying the steel corrosion. Galvanized rebar costs only 10% more versus other rebar systems, and it increases the structural lifetime two times to four times according to some studies.[608] Other studies suggest that galvanizing only

---

[603] Souradeep Gupta, Harn Wei Kua, Chin Yang Low, "Use of Biochar as Carbon Sequestering Additive in Cement Mortar," *Cement and Concrete Composites*, vol. 87, pp. 110–129, March 2018.

[604] Dimitar Dimov, Iddo Amit, Olivier Gorrie, et al., "Ultrahigh Performance Nanoengineered Graphene–Concrete Composites for Multifunctional Applications," *Advanced Functional Materials*, vol. 28, 1705183, 2018. https://onlinelibrary.wiley.com/doi/epdf/10.1002/adfm.201705183 (accessed Dec. 29, 2019)

[605] Jennifer Chu, 'MIT Students Fortify Concrete by Adding Recycled Plastic," *MIT News*, Oct. 25, 2017. http://news.mit.edu/2017/fortify-concrete-adding-recycled-plastic-1025 (accessed Dec. 3, 2019)

[606] Carolyn E. Schaefer, Kunal Kupwade-Patil, Michael Ortega, et al., "Irradiated Recycled Plastic as a Concrete Additive for Improved Chemo-Mechanical Properties and Lower Carbon Footprint," *Waste Management*, vol. 71, pp. 426–439, Jan. 2018. https://www.sciencedirect.com/science/article/abs/pii/S0956053X17306992 (accessed Nov. 30, 2020)

[607] Robert Courtland, *Concrete Planet: The Strange and Fascinating Story of the World's Most Common Man-Made Material*, Prometheus Books; Nov. 22, 2011.

[608] Tyler Ley, "What Is Epoxy Coated Rebar and Why Is It Being Banned?" *YouTube*, Sept. 28, 2018. https://www.youtube.com/watch?v=xVDy84rR5Z8 (Oct. 26, 2020)

adds a few more years to the structural life.[609] Because of the corrosion issue, in the 1970s, the National Institute of Standards and Technology (NIST) concluded in 1971 that epoxy-coated reinforcing materials would perform better than unprotected steel. Since then, builders have developed additional corrosion-resistant rebar materials including stainless steel, glass fiber–reinforced polymer, and aluminum–bronze. However, new research offers two compelling non-corrosive alternatives: continuous basalt fiber (CBF) made from the dense and abrasion-resistant igneous basalt rock; and woven fiber made from densified bamboo.[610] These two improved types of rebars, and other rebar alternatives, are more expensive and are only being used in small quantities. The bottom line is that more affordable corrosion-resistant rebar must be developed before any major rebuilding of concrete infrastructure. Otherwise, there would be a need to rebuild it all again within several decades.

- Another experimental solution is to go back to nature with biomineralization. Microbes in a nutrient mix have been used to precipitate the mineral of limestone, calcium carbonate ($CaCO_3$). It has been used for soil stabilization, concrete crack repair, and sealing fractures of oil and gas deposits. Processed around sand, it makes a concrete. High strength for buildings has not been achieved; however, where it can be used, such bioconcrete would not just be carbon neutral—it would be carbon negative.[611]

- The best solution would be if we could find some other binder that is cheaper than baked limestone and has a smaller carbon footprint. One possibility is carbonated calcium silicate ($CaSiO_3$). This cement uses about 30% less energy and emits about 30% less carbon dioxide than the conventional cement used in concretes.[612]

8.7.8.4 Nanotechnology

Nanotechnology is often thought of as just tiny electronics. Although fabrication of electronics down to nanometer sizes is a major use, nanotechnology is so much more. It includes controlled processing of tiny amounts of materials, and at those tiny sizes, some materials have

---

[609] Fujian Tang, "Steel Rebar Coatings for Concrete Structures," *Structure Magazine*, pp. 20–22, 2016. https://www.structuremag.org/wp-content/uploads/2016/05/C-BuildingBlocks-Tang-Jun161.pdf (accessed Sept. 1, 2020)

[610] Blaine Brownell, "Two Natural Rebar Alternatives for Concrete," *Architect*, March 18, 2015. https://www.architectmagazine.com/technology/two-natural-rebar-alternatives-for-concrete_o (accessed Jan. 1, 2019)

[611] Chelsea M. Heveran, Sarah L. Williams, Jishen Qiu, et al. "Biomineralization and Successive Regeneration of Engineered Living Building Materials," *Matter*, vol. 2, pp. 1–14, Feb. 5, 2020. https://www.cell.com/matter/pdfExtended/S2590-2385(19)30391-1 (accessed in press, Jan. 16, 2020)

[612] Sada Sahu, Mengesha Beyene, and Richard C. Meininger, "Characterization of carbonated calcium silicate cement-based concrete," *17th University of Toronto Executive MBA Program (EMBA)*, University of Toronto, Toronto, Canada, May 20–23, 2019. http://civmin.utoronto.ca/wp-content/uploads/2019/05/31.pdf (accessed Jan. 3, 2020)

entirely different properties. For instance, tiny grains of tungsten carbide are much more reactive than larger grains; hence, they can replace expensive platinum as a catalyst in many processes—better hydrogen fuel cells are a possibility.[613] This is important because (as noted earlier) the cost of platinum catalysts has been one of the greatest barriers to widespread use of fuel cells.

Another nanotech application goes back to the previous subsection, Greener Concrete. Synthesized 10- to 25-nanometer (0.0000004 to 0.000001-inch) particles of silicon dioxide (SiO_2, often called nano-silica) fit in the voids between concrete particles better than previously available silicas. The tiny particles also have more surface area per unit mass. Because of these two factors, nano-silica can increase concrete compressive strength by 20%, shorten curing time, and decrease the resulting carbon dioxide emission by 3%.[614]

Small improvements add up, and this is another area where capitalism is better than government. Government romantics want the big change, the revolutionary change. Business, particularly the entrepreneurs of small business, look for the one-percents and the two-percents to make a profit on the bottom line. The Japanese describe it is the *kaizen* of small improvements that can build up into a revolutionary large change.[615]

8.7.8.5 Light-Emitting Diodes (LEDs)

Beginning in the early 2000s, many governments enacted regulations banning various sizes of the incandescent light bulb that Thomas Edison invented. The bans were to encourage their citizens to switch to more efficient alternatives including halogen bulbs and especially compact fluorescent bulbs (that often came in curlicue shapes), and light-emitting diodes (LEDs).

Unfortunately, the alternative lighting sources were not ready for widespread use. Their initial costs were all significantly more expensive than the incandescent bulbs they replaced. Furthermore, they had other disadvantages. The halogen lights were roughly 40 percent more efficient than incandescent bulbs, but they operated at higher temperatures, which often caused a fire danger. The compact fluorescent bulbs were even more efficient but their color ranges were often undesirable to users. Worse, they often began blinking or failed entirely within a few weeks, and they contained mercury, so a broken or discarded bulb really needed a hazardous materials team to clean up.

Finally, years later, the LEDs dropped in price and became more efficient, more color flexible, and more durable than the incandescent and fluorescent lights that they are replacing, and

---

[613] Julie Stewart, "New Material Could be the Catalyst to an Eco-Friendly Fuel Cell, *Electronic Design*, Oct. 9, 2017. https://www.electronicdesign.com/power-management/article/21805684/new-material-could-be-the-catalyst-to-an-ecofriendly-fuel-cell (accessed Dec. 25, 2019)

[614] A. Lazaro, G. Quercia, H. J. H. Brouwers, and J. W. Geus, "Synthesis of a Green Nano-Silica Material Using Beneficiated Waste Dunites and Its Application in Concrete," *World Journal of Nano Science and Engineering*, vol. 3, pp. 41-51, 2013. https://file.scirp.org/pdf/WJNSE_2013090517174064.pdf (accessed Jan. 4, 2020)

[615] Masaaki Imai, *Kaizen: The Key to Japan's Competitive Success*, Random House, New York, 1986.

their improvements are continuing.[616] Initially, they had the same problems as the other alternative bulbs, but their color ranges, durability, and affordability have all steadily improved.

LEDs are a largely unsung success story, and they are a cautionary tale warning against using government regulations to choose technology winners and losers. LEDs probably came into the market faster because of the government regulations, but they would probably have achieved market penetration just a few years later with less hostility if governments had not ordered the preferred technology.

8.7.8.6 Superconductors, Sorry, No—But Okay, High-Voltage Electrical Transmission

For more than a century, scientists have known that near absolute zero –459 degrees Fahrenheit or –273 degrees Celsius or zero in the Kelvin scale) certain materials lose all resistance to electricity. This means that a tiny wire could carry huge amounts of electrical current so motors and generators could be much smaller, much lighter, and more efficient. Also, electrical current could be transmitted in cables without the usual 5% to10% losses, which would save billions of dollars each year. It would make levitating trains practical. Theoretically, it might make it practical to have magnetic cages powerful enough to allow fusion reactors to actually achieve that long-promised unlimited energy of hydrogen atoms fusing into helium. These definite and theoretical possibilities would all reduce global warming emissions by decreasing energy losing, or produce virtually no emissions in the case of fusion power.

Unfortunately, that near-absolute-zero temperature requires fabulously large amounts of expensive refrigeration to bath the superconducting wires in liquid helium. Consequently, the modest proposed uses have not been widely used, and the theoretical uses are still just a dream.

In the 1980s, scientists found a new family of "high-temperature superconductors" that superconduct at the much warmer—hence cheaper—temperature range of liquid nitrogen rather than liquid helium. There was a great burst of hope, but then the scientists discovered that the magnetic field produced by a flowing electric current caused this family of high-temperature superconductors to lose their superconducting condition. Thus, a much more modest quest continues worked on by a few researchers. and it is fervently hoped that scientists will be able to develop a family of "room-temperature" superconductors that would operate without any refrigeration.

The quest for high-temperature superconductors gives us another cautionary note. As with fuel cells in the 1970s, predictions were made about the vast potential of the new technologies., The proposed technological revolution was theoretically good, but engineers have not yet succeeded building affordable and working systems.[617] From 2003 until 2010, the U.S.

---

[616] Philipp Pust, Peter J. Schmidt, and Wolfgang Schnick, "A Revolution in Lighting," *Nature Materials*, vol. 14, pp. 454–458, May 2015.

[617] "Superconductivity Program overview," United States Department of Energy web page, revised July 14, 2009.
https://www.energy.gov/sites/prod/files/oeprod/DocumentsandMedia/Supercon_Overview_Fact_Sheet_7_14_09.pdf (accessed Sept. 5, 2020)

Department of Energy High Temperature Superconductivity (HTS) Program worked in partnership with industry attempting to commercialize high-temperature superconducting cables and other uses.[618] When the program failed, it quietly faded away.

If the government had attempted a revamping of the electrical utility industry based on the promising—but ultimately impractical superconductors, such a program would have failed miserably and expensively. Government (or corporate) crash programs to develop entirely new technologies on an industrial scale can easily fail. This maxim has applied from Thomas Edison's attempt to develop magnetic mining of magnetic iron-ore separation[619,620] to the U.S. government's attempt to develop fusion power reactors.[621] (Edison's ore separation was harder than he thought, and a giant new iron-ore mining area opened up in the Mesabi Range of Minnesota rendering his iron production uncompetitive. Fusion reactors have produced additional energy but not enough to be profitable. Worse, there are many engineering issues that make profitable systems hard to achieve.

Meanwhile, the more prosaic development of long-distance power lines shows the advantages of a technology that can be developed in stages. It started with Nikola Tesla building an alternating current generating plant at Niagara Falls in 1895, and within a few years, transmitting electrical power to New York City 700 kilometers (430 miles) away. Power lines have been evolving for more than a century. In the 1970s, transformable direct current arrived. High-voltage direct current carries electricity much more affordably than alternating current lines because has lower transmission losses and it only needs one cable rather than the three for alternating current.

Regional-sized and even continental-sized grids have advantages of being able to "wheel power" from areas with extra power to areas willing to buy. The European Union's electricity grid is the largest most interconnected continental power network in the world,[622] and its operation suggests that regional grids will continue to grow in size and capabilities. The Europeans also have the longest submarine power line to date, the 580-kilometer (360-mile) NorNed high-voltage direct-current line carrying power from Norway to Holland and the deepest cable that runs between

---

[618] "High Temperature Superconductivity Program," United States Department of Energy web page, undated. https://www.energy.gov/oe/high-temperature-superconductivity-program (accessed Sept. 5, 2020)

[619] "The Edison Mines," *Geocaching*, Sept. 6, 2007. https://www.geocaching.com/geocache/GC15QP1_the-edison-mines (accessed Sept. 5, 2020)

[620] Martin Woodside, *Thomas A. Edison: The Man Who Lit Up the World.* Sterling Publishing Company, Inc. pp. 73–74, 2007.

[621] Steve Minsky, "Where's My Fusion Reactor?" *Scientific American*, pp. 51–57, March 2010.

[622] Bruno Lajoie, "Europe's Interconnected Electricity System: an in-Depth Analysis," *Medium.com*, June 8, 2018. https://medium.com/electricitymap/what-does-it-take-to-decarbonize-europe-d94cbed80878 (accessed Sept. 1, 2020)

8 Fixes from Energy and Material

Italy and Sardinia.[623] Such submarine power transmission will be a key component in the ocean development described in Chapter 9.

North America has several grids that are loosely connected. The Climate Energy Institute (a group supported by a number of rich business people) has proposed the North American Supergrid (NAS). The NAS was inspired by a 2016 paper proposing an advanced electrical grid the size of North America to wheel large amounts of solar and wind that would be wasted because of oversupply in some local areas and thus replace a much greater fraction of fossil-fuel-generated electricity with solar and wind.[624] The NAS proposal[625] would have several details that apply to GND:

- It would provide greater efficiency and greater reliability, which are key items for applying new energy technologies. Indeed, a report from the American Society of Civil Engineers is entitled, *Failure to Act: Electric Infrastructure Investment Gaps in a Rapidly Changing Environment*,[626] The main theme of that report was that much of the American electrical grid was installed in the 1950s and 1960s. That equipment is obsolete and has already exceeded its design lifetime. Even if the NAS does not wheel as much solar and wind as proposed, just upgrading the grid will provide more power and more dependable power per energy input from whatever source.

---

[623] Mircea Ardelean and Philip Minnebo, "6, Long Cable Examples," *HVDC Submarine Power Cables in the World*, Report EUR 27527 EN, European Union, 2015.
https://publications.jrc.ec.europa.eu/repository/bitstream/JRC97720/ld-na-27527-en-n.pdf (accessed Sept. 3, 2020)

[624] Alexander E. MacDonald, Christopher T. M. Clack, Anneliese Alexander, et al., "Future Cost-Competitive Electricity Systems and Their Impact on US CO_2 Emissions," *Nature Climate Change*, vol. 6, pp. 526–531, May 2016.
http://denning.atmos.colostate.edu/readings/Solutions/Macdonald.HVDC.pdf *accessed Sept. 2, 2020)

[625] *North American Supergrid: Transforming Electrical Transmission*, Climate Institute, Washington, D.C., (original publication 2017) updated Sept. 2019.
http://northamericansupergrid.org/ (accessed Sept. 1, 2020)

[626] *Failure to Act: Electric Infrastructure Investment Gaps in a Rapidly Changing Environment*, American Society of Civil Engineers, 2017.
https://www.asce.org/uploadedFiles/Issues_and_Advocacy/Infrastructure/Content_Pieces/failure-to-act-electricity-report.pdf (accessed Sept. 2, 2020)

- It would provide greater protection against solar flares and artificial electromagnetic pulses. This is something that has been urgently suggested for decades.[627] [628]

- Constructing the NAS would be roughly a 30-year project (NAS, Introduction, p. 8). This is another instance in which investments in advanced technologies can greatly reduce global warming, but these investments may not begin seriously providing those returns for years or even decades. Once again, we will probably need the "all of the above" approach, and it may get warmer before it gets cooler.

- The NAS plan proposes a cost of about $500 billion that would be largely privately financed and paid for through consumer bills (NAS, Introduction, p. 8).

- The NAS plan includes an underwater cable shortcut between the American Northeast with the American Southeast (NAS, Case Study 2 – Atlantic Coast Submarine Project).[629] This is another application, that could provide experience and technology development for more ambitious projects. Such submarine power transmission will be a key component in the ocean development described in Chapter 9.

Wild-eyed proposals for various regional and worldwide grid have been made for decades because such a grid could theoretically balance all the energy low-use and high-use periods throughout the world. This would be particularly so for solar peak power that rolls inexorably westward with the flow of sunlight on the Earth. Some of them include:

Futurist Buckminster argued for a world grid as one of his most important proposals for decades.[630] Furthermore, there have been ambitious proposals for regional access to power, such as solar power from the Sahara Desert to Europe, wind power from the Gobi Desert to more populous parts of China, and wind energy from Greenland across the Atlantic to Europe. The ultimate result would be the proposal, "Let's Build a Global Power Grid "[631]

---

[627] Richard A. Lovett, "What If the Biggest Solar Storm on Record Happened Today? *National Geographic*, March 4, 2011. https://www.nationalgeographic.com/news/2011/3/110302-solar-flares-sun-storms-earth-danger-carrington-event-science/ (accessed Sept. 2, 2020)

[628] Samuel Feinburg, "A Catastrophic Blackout is Coming - Here's How We Can Stop It," *TEDxBaylorSchool*, April 10, 2020. https://www.youtube.com/watch?v=ZGan5NwJ-LM (Dec. 7, 2020)

[629] Rachel Levine, José Alfredo Durand Cárdenas, Xie He, et al., Technical Feasibility and Environmental Challenges," *North American Supergrid: Transforming Electrical Transmission*, Climate Institute, Washington, D.C., (original publication 2017) updated Sept. 2019. http://northamericansupergrid.org/ (accessed Sept. 1, 2020)

[630] "Buckminster Fuller on the Global Energy Grid," *Global Energy Network Institute Newsletters*, second quarter 1995. http://www.geni.org/globalenergy/library/newsletters/1995/buckminster-fuller-on-the-global-energy-grid.shtml (accessed Sept. 1, 2020)

[631] Clark W. Gellings, "Let's Build a Global Power Grid," *IEEE Spectrum*, July 28, 2015. https://spectrum.ieee.org/energy/the-smarter-grid/lets-build-a-global-power-grid (accessed Sept. 1, 2020)

8.7.9 Putting It All Together: Hybrid Systems of Combustion, Ultracapacitors, ... and Batteries

A combination of things often works better than just one. That is often the case with power and transportation systems. For solar-heat power systems, burning natural gas just during times of darkness or cloudy weather would allow these systems to provide dependable (base load) power while still burning much less fuel and emitting much less greenhouse-warming carbon dioxide than conventional thermal generating systems.

For trains, the combination of powers in diesel–electric locomotives that largely replaced steam engines from 1935 through 1960 for moving railroad trains. [632] A diesel–electric locomotive carries its own generator that powers motors at the wheels, rather than having drive shafts. Similarly, ships are often diesel–electric or turbine–electric for a number of reasons. Some of the most important reasons are:[633]

- Eliminating the long propeller shaft (or shafts) eliminates weight and stress on the ship
- Eliminating the long propeller shaft (or shafts) makes more space for cargo or other equipment
- Using electric propeller pods, rather than shaftline propulsion allows much greater maneuvering control, especially for docking
- Having additional electrical power can support auxiliary functions such as passenger amenities on cruise ships and lasers on warships
- Achieving greater efficiency to reduce fuel use and exhaust emissions.

The advantages for cars and trucks are even greater because these vehicles have major loads for accelerating and a major energy source in slowing down that can be tapped for regenerative braking when the wheel motors function as generators.[634]

Turbines could get a similar benefit. Turbines are more efficient than Otto or Diesel cycles, and they last longer than those engines. But their better efficiency only works at full throttle, and their acceleration is even worse than a diesel. However, a hybrid with battery-powered wheels can

---

[632] "Diesel Traction," *Encyclopedia Britannica*. (accessed Dec. 24, 2019)

[633] Raunek Kantharia, "Electric Propulsion System for Ship: Does it have a Future in the Shipping? *Marine Insight*, Oct. 11, 2019. https://www.marineinsight.com/marine-electrical/electric-propulsion-system-for-ship-does-it-have-a-future-in-the-shipping/ (accessed Dec. 24, 2019)

[634] CSIRO Australia (2008, January 18). "UltraBattery Sets New Standard for Hybrid Electric Vehicles," *ScienceDaily*, Jan. 18, 2008. http://www.sciencedaily.com/releases/2008/01/080118093341.htm (accessed Jan. 20, 2008)

combine the best of both worlds just as diesel-electric locomotives dominate railroad transportation.[635]

Major Things for Governments and Businesses to Do

- Government: Run more programs for research and subsidizing use of radiation-hardened electronics for key uses such as construction and maintenance of nuclear reactors, outer space facilities, and Internet communications, and utilities.

- Government: Run a major development program for development of additional high-temperature metal alloys and composites.

- Business: Invest in major electrical grid upgrades such as the North American Supergrid.

- Government: Continue programs, such as the NASA Solar Shield,[636] for observing the Sun with the goal of providing longer waning times before particle impacts from solar flares so that electrical equipment can be better protected.

- Business: Adopt large power units to supercritical carbon dioxide turbines.

- Business: Develop and sell radiation-resistant electronics.

What You Can Do

- Increase your connections for and attend remote (also called virtual) meetings and events. Much like replacing old Dobbin in the pasture with an automobile, it's the new age. Remote meetings will only become more common in all types of social connections.

- If you have the money, make your next car a hybrid for greater fuel economy. If the battery range doubles again (maybe 700 miles or 1100 kilometers), and the local climate does not have battery-numbing cold, consider going full battery.

- Add electrical protections against electrical surges/pulses. This starts with surge protectors coming from wall plugs to more expensive protective hardware in constructing new or upgrading your building's electrical systems.

---

[635] Ian Wright, "How Jet-Powered Garbage Trucks Can Save the World, *TEDx Talks Christ Church*, Nov. 9, 2015. https://www.youtube.com/watch?v=F4H3FE0Z4QQ (accessed July 14, 2020)

[636] Tony Phillips, "Solar Shield--Protecting the North American Power Grid," National Aeronautics and Space Administration web page, Oct. 28, 2010. https://science.nasa.gov/science-news/science-at-nasa/2010/26oct_solarshield/ (accessed Sept. 2, 2020)

8.8 Electric, Autonomous, Transportation Utilities—Elon Musk Strikes Again

Then, out of the blue, there are new concepts that were never considered before, technological disruptions that could massively change everything. One such disruption is the confluence of dial-up car rentals (such as Uber and Lyft), electric cars, and autonomously driving cars.[637]

Combining these three trends could lead to peak automobiles as people switch largely away from personally owned and company-owned vehicles to transportation rental services provided by fleets of autonomous vehicles—probably electric. Advantages of the fleet renal business model are that:

1. Considering that personal vehicles are generally only used 4 percent of the time, the cost of just occasionally renting vehicles could spread over many households, thus lowering costs and increasing disposable income for those households.

2. Because electric vehicles have fewer complex parts, they might generally last three times as long as combustion-engine vehicles. This would further reduce costs for the users.

3. If the additional electricity for powering this fleet were to come from nonfossil-fuel sources, there would be a major decline in automobile emissions that contribute to global warming and to smog.

4. The efficiencies in vehicle use would allow a reduced number of vehicles in the driving fleet. There would be a time of "peak autos" followed by some years of decline. This, in turn, would cause an additional decline in energy use and emissions.

5. City centers could also reduce the significant percentage of prime land area used as parking lots because the autonomous vehicles could be ordered to stage a short distance away in less urbanized areas. That would free up high-cost land presently in those parcels for higher value activities.

6. Traffic safety would be vastly improved so the costs, injuries, and fatalities of accidents would decline.

7. People would have more personal time available when riding rather than driving in their vehicles.

8. The cumulative efficiencies should provide a major increase in living standards.

one major cost would be installing the additional electrical power generation capability needed to power the electric vehicle rental fleets. Furthermore, those same issues of transporting

---

[637] Tony Seba, *Clean Disruption of Energy and Transportation: How Silicon Valley Will Make Oil, Nuclear, Natural Gas, Coal, Electric Utilities and Conventional Cars Obsolete by 2030*, publisher: Tony Seba, May 20, 2014.

energy in without overloading the grid would apply. Might there be a number of relatively small nuclear reactors or generating stations burning ammonia, natural gas, or hydrogen? Stay tuned.

8.9 Electric Airplanes … Really!

Yes, there really are practical arguments for electric airplanes … or at least engine–electric hybrids.[638] The electric hybrid airplane can be a case study for the power of innovation to decrease fuel use and decrease global-warming emissions.

Let's take a look. For the GND and the environment in general, jet engines for aviation have major drawbacks.

- They are noisy when near the ground for takeoffs and landings.
- They increase the carbon dioxide load in the atmosphere.
- They emit oxides of nitrogen that contribute to smog at low altitudes, attack ozone in the stratosphere, and are greenhouse gases themselves.
- Most of all, they are essentially giant flying buses that (like diesel buses) spew soot. High in the atmosphere (in the lower stratosphere) where commercial jets spend the majority of their flight time, soot particles have a major greenhouse warming effect. Soot particles absorbing many wavelengths of light and then radiate the resulting heat to the air nearby.[639] This not only warms the surrounding air, but the warmth causes nearby clouds to evaporate, and fewer cloud results in more sunlight passing through and further warming the ground. At stratospheric altitudes, the soot provides nucleation sites for thin white cirrus clouds that are too thin to reflect much sunlight, but ice crystals inside them traps heat for a significant net warming effect.[640] Unlike low-level clouds that have a net cooling effect, these stratospheric soot-generated clouds warm the planet.

---

[638] For a quick summary, see Susan Fourtané, "The Future of Aviation: Electric Airplanes Will Decarbonize the Aviation Industry," *Interesting Engineering*," Dec. 24, 2019. https://interestingengineering.com/the-future-of-aviation-electric-airplanes-will-decarbonize-the-aviation-industry?_source=newsletter&_campaign=XqMva21Llo9QY&_uid=LDdwpX6Re1&_h=9616d98f82a7a631519bae2822a86b243f923bbc&utm_source=newsletter&utm_medium=mailing&utm_campaign=Newsletter-24-12-2019 (accessed Dec. 26, 2019)

[639] V. Ramanatha and G. Carmichael, "Global and Regional Climate Changes Due to Black Carbon," *Nature Geoscience*, vol. 1, pp. 221–227, 2008.

[640] Katie Camero, "Aviation's Dirty Secret: Airplane Contrails Are a Surprisingly Potent Cause of Global Warming," *Science*, June 28, 2019. https://www.sciencemag.org/news/2019/06/aviation-s-dirty-secret-airplane-contrails-are-surprisingly-potent-cause-global-warming (accessed Nov. 29, 2019)

Batteries are not a serious competitor yet because jet fuel has 43 times the energy density of batteries. However, a hybrid of a jet turbine or fuel cell with a modest battery running electric propellers has competitive advantages.[641]

- The power source could be more fully enclosed so noise would be limited to the noise of propellers rather than the full noise of jet engines.
- As with hybrid automobiles and diesel–electric railroad locomotives, the emissions can be better controlled than with jet engines.
- Electric cabling to the propellers can eliminate some weight of the jet engines on the wings to allow a lighter structure and hence increased efficiency.

An analysis by a propeller-driven aircraft manufacturer (ATR, a Franco-Italian joint venture) concluded that one of the biggest advantages for propellers vs. jets is the higher bypass ratio, the comparison between the quantity of air drawn into the engine for combustion, and the volume of air going around the engine. A higher ratio means that less fuel is consumed by the aircraft (ka-ching!) and that correspondingly less CO_2 is emitted. For the largest jet engines currently on the market, manufacturers aim for a bypass ratio of 15 or 16. By comparison, an ATR, with its uncovered propellers, achieves a bypass ratio of more than 50.[642]

There are experimental propellers that can attain nearly the speed of present commercial passenger liners, and the slower speed would be less of a handicap for smaller commercial planes serving regional short hops. Engine–electric hybrids would be even more competitive for private aviation that has remained largely propeller driven.

Furthermore, the additive manufacturing (3D printing) techniques mentioned earlier allow novel shapes and unusual material properties—such as propeller blades optimized for strength at one end and flexibility at the other—could radically increase propeller capabilities.

One possibility being explored by jet manufacturer Pratt & Whitney is engines with fewer parts, which would need less assembly and be cheaper to make. There could be radically new aircraft designs "like many engines embedded in a wing for ultra-aerodynamic efficiency.[643] (Such multiple engines would improve both jets and propeller-driven aircraft.)

Back at the electric airplane, totally electric airplanes would still be very limited in range even with higher capacity batteries. However, the hybrid electric airplanes would benefit from more battery power (and/or ultracapacitor power) per unit mass. Just more electrical power storage

---

[641] Evan Gaj, "The Electric Aircraft Is Taking off," *Techcrunch*, July 8, 2018. https://techcrunch.com/2018/07/08/the-electric-aircraft-is-taking-off/ (accessed Nov. 29, 2019)

[642] "The Future of Aviation: It All Revolves around Propellers!" ATR, Oct. 4, 2019. https://www.atr-intolife.com/the-future-of-aviation-it-all-revolves-around-propellers/ (accessed Dec, 1, 2019)

[643] Kevin Bullis, "Additive Manufacturing Is Reshaping Aviation," *MIT Technology Review*, Feb. 6, 2015. https://www.technologyreview.com/s/534726/additive-manufacturing-is-reshaping-aviation/ (accessed Dec. 2, 2015)

would be a weight advantage. Even better, the greatest propulsive thrust is needed during very limited time periods, takeoffs and landings. Increased electrical storage could provide that occasional power rather than paying for the expense and the weight penalty of a more powerful motor.

These advances are another area that will require more than a decade just for development of hybrid–electrics that will require lighter electronic controls, better electronic power management, and development of controls to better use the advantages of electric propulsion.

As with the automotive industry, another revolution in batteries could be a game-changer in which pure electric planes might become a major part of the aviation fleet, particularly for light planes. However, that would be even further into the future.

Major Things for Governments and Businesses to Do

- Government: Research the key technologies and develop regulations for hybrid-electric and all-electric airplanes.

- Business: Build those hybrid–electric aircraft.

- Business: Build and market high-altitude drones for virtual communication satellites, weather observation, and site monitoring.

- Business: build and market drones and drone swarms for use as cranes for loading ships, trucks, and trains.

What You Can Do

- Buy and fly your own hobby mini-drone.

8 Fixes from Energy and Material

9 Fears and Fixes from Sea and Sky

Many of the issues of the Green New Deal (GND) revolve about the oceans. The oceans cover three times the area of the land. The majority of heat storage is not in the land or air; it is in the oceans. The largest carbon sink drawing down global-warming carbon dioxide from the air, and buffering against rapid global warming, is the oceans. Environmental damage to the oceans could rebound and hurt humanity in many, many ways.

Two of the next technological revolutions will be extension of farming into those oceans—onto the relative shallows of the continental shelf and then onward into deep waters.

It was said that there was a bluewater revolution of ocean farming, comparable to the Green Revolution of high-producing wheat, high-producing rice, increased fertilizer use, and increased irrigation. However, this is incorrect. The large increase in farmed aquatic life was largely in vastly increasing the freshwater pond culture (or polyculture) that has been practiced for centuries in the Orient: For instance, runoff from fields and livestock wastes goes to duck ponds (supplying more nutrients) so there are abundant algae to feed various kinds of fish. This same principle, under the name of integrated multi-trophic aquaculture, will extend into the oceans.[644,645]

The oceans will supply much of the food, the energy, and the cooling to ward off the warming apocalypse that many fear.

9.1 Ocean Worries—The Seven Deadly Fins

We came from the sea, and the saltiness of the sea still flows in our blood. It is an entirely different world two-and-a-half times as big (71 percent of the Earth's surface) as all the land that people live on with more minerals than people could use in thousands of years and the capacity to feed the humanity many times over. The ocean world is the dog that wags the climate tail of warm cold, wind, snow, and rain.

Yet that world is rarely visited ... and when visited, it is often used badly: ketchup bottles over the side, nerve gas canisters dumped into the Atlantic after World War II, trash and sewage excreted out to sea via garbage scows and giant sewer pipes, respectively.

---

[644] Paul Southgate and John S. Lucas, Chapter 2, "Principles of Aquaculture," *Aquaculture: Farming Aquatic Animals and Plants*, Third Edition, edited by John S. Lucas, Paul C. Southgate, and Craig S. Tucker, John Wiley & Sons Ltd, Hoboken, New Jersey, pp. 19–39. 2019.

[645] John Hargreaves, Randal Brummet, and Craig S. Tucker, Chapter 27, "The Future of Aquaculture," *Aquaculture: Farming Aquatic Animals and Plants*, Third Edition, edited by John S. Lucas, Paul C. Southgate, and Craig S. Tucker, John Wiley & Sons Ltd, Hoboken, New Jersey, pp. 617–636. 2019.

For centuries, we could wash of our problems out to sea and be done with them. Now, they are sloshing back, sometimes multiplied by the awesome power of the ocean and joined by effects from the ocean of air, the sky.

Of all the future climate warnings, none are worse than those about the oceans. Fortunately, some old and new technologies are finally coming together to provide practical solutions. I call these solutions "The Green Water Revolution."

To start, we need to introduce the bad guys. The greatest real and potential dooms can be thought of as "The seven deadly fins … in the water."

1. Overfishing and bad fishing
2. Global warming—including sea-level level rise
3. Reduced ocean productivity due to increased stratification caused by warmer temperatures
4. Decreased carbon drawdown from the atmosphere by sea life due to warming and ocean acidification
5. Pollution and trash building up in the ocean
6. People crowding infrastructure into beaches and wetlands—greater vulnerability and more of an increase in global warming
7. Earthquakes, volcanoes, giant waves, rocks from the sky, and nuclear bombs

9.1.1 Over Fishing and Bad Fishing

Fishing helped build civilizations for centuries,[646] But today, oceanic fishing applies the primitive hunter-gatherer modality with modern diesel power, sonar, factory ships, and super plastic netting. It is a system continuing a steady series of sea-catch-population crashes; only the high-tech has made it a worldwide crisis.[647] It starts with destruction of top predators, which destabilizes the ecosystem. Each fishery is then emptied of progressively smaller fish and other sea life while damaging the bottom as a byproduct. This is a progression of sea-life crashes that has continued for centuries, but which has increased ominously as growing industrial capabilities have increased attack capabilities.

If space aliens started fishing Earth the same way as people fished the oceans, the flying saucers might start with baited hooks to catch eagles and hawks. Then they would sweep the sky with giant seine nets to collect flocks of ducks and geese, followed by progressively finer nets for pigeons, robins, and finally moths and flies. Then they would have trawl lines dragging on the ground for deer, buffalo, moose and elk. Next, deeper drag lines would take gophers, snakes, and angle worms. Ultimately, they would abandon the ruined planet.

---

[646] Brian Fagan, Fishing: *How the Sea Fed Civilization*, Yale University Press, Sept. 2017.
[647] Callum Roberts, *The Unnatural History of the Sea*, (originally Shearwater Books, 2007) Island Press, 2008.

That is what is happening in the ocean … we just can't see the havoc wreaked below. Likewise, most people did not notice that the catch of wild fish peaked about 1988. Fish farming is already replacing wild catch, but that is another story in 9.2.

The driver of overfishing and bad fishing is something Garrett Hardin called "the tragedy of the commons."[648,649] In colonial New England, the little towns each had a common grazing area where everybody could graze their sheep. If the commons were to be overgrazed, one herder might take his sheep away to let the commons recover, but then another herder could bring his herd in to eat every last blade of grass while the conscientious herder lost out.

The ocean fisheries are the same, and it could easily be fished to the last useful fish.

But, as the late-night sales infomercials say, there's more. Secondary affects add to the pain. These include bycatch where unwanted or illegal-to-fish species are thrown back overboard. Unfortunately, these bycatch animals are already dead or dying from the trauma of netting.

Intensive fishing often upsets ecological balances by removing large predators at the top of the food chain (apex predators). As the wise old Indian says in *Never Cry Wolf*, once the people killed off the predators like the wolves and the bears, but then they found that the caribou became sick.[650] Similarly, in the ocean, overfishing certain species can have immense impacts on others.

In the ocean, the apex predators are often the most prized catches, or the most feared, so they have been fished most intensively. This includes sharks, spearfish, tuna, and cod. What remains are the smaller predators that may cause a range of new larger problems.[651,652,653] For example, surging jellyfish populations may be the result of an ocean with far fewer apex predators than before modern fishing started.

---

[648] Garrett Hardin, "The Tragedy of the Commons," *Science*, vol. 162, no. 3859, pp. 1243–1248, Dec. 13, 1968. DOI: 10.1126/science.162.3859.1243

[649] William Forster Lloyd first described the tragedy as an economic theory whereby individuals use up a resource to benefit themselves if it is theoretically shared by many. Garret Hardin (previous reference) applied the metaphor to environmental concerns, especially fishery resources.

[650] Farley Mowat, *Never Cry Wolf: Amazing True Story of Life Among Arctic Wolves*, McClelland and Stewart Limited, Toronto, Ontario, Canada, 1963.

[651] "Loss of Top Predators Causing Ecosystems to Collapse," *Live Science*, Oct. 1, 2009. https://www.livescience.com/9716-loss-top-predators-causing-ecosystems-collapse.html (accessed Nov. 16, 2018)

[652] Caroline Fraser, "The Crucial Role of Predators: A New Perspective on Ecology," *YaleEnvironment360*, Sept. 15, 2011. https://e360.yale.edu/features/the_crucial_role_of_predators_a_new_perspective_on_ecology (accessed Nov. 16, 2018)

[653] John Terborgh and James A. Estes (editors), *Trophic Cascades: Predators, Prey, and the Changing Dynamics of Nature*, 1st Edition, Island Press, April 20, 2010.
James A. Estes, John Terborg, Justin S. Brashares, et al., "Trophic Downgrading of Planet Earth," *Science*, vol. 333, issue 6040, pp. 301–306.

Another ecology deranged by predator loss is the set of species including bull kelp, sheephead fish, sea otters, and sea urchins. Bull kelp areas have often been called the redwood forests of the Pacific Coast of North America. These magnificent plants (often called the redwoods of the sea) grow from their "holdfasts" gripping rock on the seabed, but that is their weak point when voracious sea urchins eat the holdfasts and cause the kelp stalks to drift onto shore. Sea urchins have become a greater menace to bull kelp because its two main predators attacking sea urchins have severely declined in population. First, the sea otters were nearly hunted to extinction for their pelts in the 1800s. Then, in the 1900s to 2000s, the sheephead were overfished to the point that their population is mostly small specimens that are often too small to prey on the sea urchins.[654]

9.1.2 Global Warming—Including Sea-Level Rise

The second deadly fin of global warming, as noted elsewhere, is where certain gases and aerosols let visible sunlight in but block infrared heat waves from getting out. That's a good thing because, without the greenhouse effect, Earth would be below freezing and be an icy desert like Mars. The worry is that massive amounts of global-warming materials from our industrial society could make things ... too warm for our comfort. This is true in the oceans as well as on land.

However, there are more warming effects for the ocean. Historically, the ocean has been rising roughly a foot (a third of a meter) per century, and warming can be expected to increase that speed of rise. Also, historically, people built **and continue to build** as close to the water as possible and then rebuild after storm damage. Moreover, they often cause land to sink (subsidence) by pumping out water and oil from underground for use at the surface. Consequently, an increasing percentage of humanity's population and wealth are threatened by rising sea levels.

9.1.3 Reduced Ocean Productivity Due to Increased Stratification

The third fin, reduced ocean productivity due to increased stratification, could multiply the impact of the other two. Much of the deep ocean is blue water desert because mineral nutrients from the ocean bottom are far below. However, the interplay of winds and current causes water to sink in some areas and rise in others. The sinking carries carbon dioxide down into deeper water reducing global warming. Even better, the sinking water swirls around with other currents ... and sometimes causes deep water to upwell to the surface elsewhere. These upwelling areas are just 1 percent of the ocean surface but half of the fish catch today. That suggests these areas are tremendously fertile—more sea life and less carbon dioxide on the loose—less global warming.

However, increasing global temperatures reduce the upwelling.[655] Increased temperatures warm the upper layers (or strata) of water first. Those warmer water strata become less dense, so they are more likely to maintain stable warm strata at the top. This decreases the amount of cool

---

[654] "Sheephead, Sea Urchins & Kelp: A Dynamic Food Chain," FISHBIO, May 28, 2018. https://fishbio.com/field-notes/the-fish-report/sheephead-sea-urchins-kelp-dynamic-food-chain (accessed Oct. 16, 2020).

[655] Guancheng Li, Jiang Zhu, Kevin Trenberth, et al., "Increasing Ocean Stratification over the Past Half-Century," *Nature Climate Change*, Sept. 28, 2020. https://doi.org/10.1038/s41558-020-00918-2 (accessed Oct. 16, 2020)

denser fertile waters upwelling to the surface. Lesser amounts of fertile water reaching the surface mean less ocean productivity. That means less carbon dioxide converted into food and less byproduct oxygen produced. Thus, there would be a self-reinforcing cycle with less food produced and more warming.

Marine biologists refer to this effect as eutrophic amplification where a change in the plankton population can cause a multiplied increase (or decrease) in other things.[656] In the case of warmer surface waters reducing nutrient flow to the sunlit (euphotic) zone, the result is decreased plankton and less draw down of carbon dioxide, which results in more net increase of carbon dioxide in the atmosphere and more warming. As noted in In Subsection 9.4, A Step Even Further: The Blue-to-Green Revolution, we shall look at ways to increase nutrient flow to the surface, increase plankton productivity, and create a negative temperature amplification.

9.1.4 Decreased Carbon Drawdown from the Atmosphere

Regarding deadly fin number four, many of the ocean's limestone-producing sea life forms are under attack, and this may significantly reduce the ocean's drawdown of carbon dioxide from the atmosphere. This produces another positive feedback of more carbon dioxide in the atmosphere causing even more warming.

The most noticeable oceanic ecosystems drawing down carbon dioxide are the coral reefs. Corals are a species of animal that live symbiotically with internalized algae to build large reefs of limestone (calcium carbonate, $CaCO_3$). Coral reef building is a major ocean mechanism that removes carbon dioxide from the air and reduces the potential dangers of global warming. Unfortunately, corals are sensitive to several damaging trends, including:

1. Warming water temperatures (although new coral reefs would develop farther poleward, this probably would not happen as fast of the die offs in warmer areas closer to the Equator),

2. Increased levels of carbon dioxide,

3. Changes in the species balance caused by overfishing,

4. Fishing attacks on coral reefs to sell exotic aquarium fish, and

5. Pollution.

Warming water temperatures and increased levels of atmospheric carbon dioxide are the best-known threats to corals. Periods of particularly warm water are associated with coral die offs. However, it is the combination with the other threats that most endangers the coral reefs … and their carbon-drawdown capability.

Increased levels of carbon dioxide in the atmosphere translate into slightly more acidic surface waters where most corals live. This higher acid level results in the corals needing to work

---

[656] Guillem Chust, J. Icarus Allen, Laurent Bopp, et al., "Biomass Changes and Trophic Amplification of Plankton in a Warmer Ocean," *Global Change Biology*, March 7, 2014. https://doi.org/10.1111/gcb.12562 (accessed Aug. 10, 2020)

harder to draw carbon dioxide out of the atmosphere. This extra work is another stress on the corals that makes them more susceptible to the other damaging mechanisms.

As a side note, humanity caught a break with the coccolithophores, a type of plankton that accrete tiny calcium carbonate shells and consequently draw down major amounts of global-warming carbon dioxide from the atmosphere, often turning areas of the surface milky white. Some varieties of coccolithophores grow better with higher levels of carbon dioxide and warmer temperatures. Count your blessings.

Changes in species balance was mentioned before, and it is causing major damage to coral reefs throughout the world. In all oceans except the Atlantic, the most dangerous enemy of corals reef ecology is the *Acanthaster planci*, or crown of thorns seastar (COTS) that directly attacks corals. A single COTS can eat its body weight in coral during a single night, leaving dead bleached-white remains. One individual can thus kill roughly 13 square meters (140 square feet) of reef per year. In the past, the trumpet triton, helmet snail, wrasse, triggerfish, and puffer fish all helped keep COTS populations in check, but these species have been overfished, and the corals are suffering.[657] Major declines in corals start a cascading effect damaging other related reef communities. The result is often an increase in the seaweed competitors to corals.[658]

In the Atlantic, there are no crown of thorns seastars, but Atlantic seaweed species compete with the corals that build coral reefs. Parrotfish are the predator that kept the seaweeds in check to maintain coral reef health.[659] Again, parrotfish have been heavily fished, and the corals suffer as a result.[660,661]

A related cause of damage is direct attacks by people on the coral reefs. Exotic multi-colored coral-reef fish are highly prized for aquariums. Hence, they command high prices, and many impoverished islanders attack them with dynamite or poisons to stun the reef fish for

---

[657] James Barrat, "COTS: Aitutaki," Khaled bin Sultan Living Oceans Foundation, undated. https://www.livingoceansfoundation.org/science/crown-of-thorns-starfish/coral-killers/ (accessed Nov. 2, 2020)

[658] Christopher Fulton, editor; Mohsen Kayal, Julie Vercelloni, et al., "Predator Crown-of-Thorns Starfish (*Acanthaster planci*) Outbreak, Mass Mortality of Corals, and Cascading Effects on Reef Fish and Benthic Communities," *Journal of PLoS One*, vol. 7, no. 10, e47363, 2012.

[659] Michael Slezak, "Limiting Catch of One Type of Fish Could Help Save Coral Reefs, Research Finds," *The Guardian*, April 4, 2016. https://www.theguardian.com/environment/2016/apr/05/limiting-catch-one-type-fish-help-save-coral-reefs (accessed Aug. 8, 2020)

[660] "Parrotfish Are Critical to Coral Reef Health, Study Finds," Science Daily, Jan. 23, 2017. https://www.sciencedaily.com/releases/2017/01/170123094552.htm (accessed Nov. 8, 2018) "Parrotfish Critical to Coral Reefs: Permanent Damage Likely Unless Urgent Action Taken," *Science Daily*, Nov. 5, 2007. https://www.sciencedaily.com/releases/2007/10/071031112907.htm (accessed July 8, 2020)

[661] Mario Aguilera, "Scripps Leads First Global Snapshot of Key Coral Reef Fishes: Fishing Has Reduced Vital Seaweed Eaters by More Than 50 Percent, Report Reveals," Scripps Institute of Oceanography, La Jolla, CA, Dec. 3, 2013. https://scripps.ucsd.edu/news/14014 (accessed Nov. 8, 2018)

collection. The collected fish are sold to wholesalers who eventually pass them on to aquariums in the developed countries. Vastly more fish die in this fishing process and in the shipping than reach the aquariums. The damage to the reefs is dire and long lasting.[662]

Any one of the three deadly fins of overfishing, warming and water stratification can damage corals and other seal life. In combination, their damage is multiplied. Still, there is one more deadly fin that is worse than the others.

9.1.5 Pollution and Trash

The fifth deadly fin, pollution and trash damages ocean life, particularly corals, the most. Corals have evolved to be supremely able to thrive in a nutrient-constrained environment. The 1900s and 2000s have brought unprecedented levels of nutrients to the ocean in the form of wastes from livestock operations. In Australia wastes from cattle ranching have impacted the Great Barrier Reef off the coast. In Hawaii the cowboys are called paniolos, but the steer manure has the same nitrates and phosphates as downstream from other cattle and pig operations.

Elsewhere, increasing human populations send their own manure—much like steer manure out to sea. Raw sewage is the worst, but even advanced sewage treatment that removes solids and kills bacteria leaves nitrates and phosphates in the water. Runoff of farm fertilizer carries the same nutrients of nitrates, phosphates, and trace amounts of many others that can be too much of a good thing for corals. In such highly fertilized waters, various algae and plants overgrow the corals and shade them from lifegiving sunlight.

Just dirt from road construction, dredging of harbors and channels, coastal construction, agriculture (including those cattle again), and urban storm runoff all throw a silt blanket over the coral reefs and kelp beds, slowing their growth and making them more vulnerable to competing and predator species.

Toxins from pesticides, industrial waste, and even tanning lotion can damage certain types of sea life and further skew ecological balances. Similarly, plastic wastes have been associated with increased disease among corals by transporting bacteria.

But sadly, there is more. Areas with intense pollution have more severe damage than loss of corals. The fertilizer- and sewage-enriched waters of the mighty Mississippi River have produced a mighty super bloom after each spring runoff. By late summer and fall, much of this bloom has died, and organic matter floats toward the bottom. As this bloom material decays, it sucks nearly all the oxygen from the water and kills all life. This is one of the world ocean's largest "dead zones," and at its largest extent it has been the size of New Jersey. Other such dead zones occur at other river mouths, in the Baltic Sea, in the Black Sea, in Chesapeake Bay, and in Lake Erie.

Yet, it can be worse. Some algae in overly fertile waters are poisonous. These harmful algae blooms are often red; they are sometimes called red tide. They defend themselves against

---

[662] Alastair Bland, *Pacific Standard*, "Ending the Ecologically Harmful Capture of Tropical Fish," *Pacific Standard*, Jan. 9, 2018. https://psmag.com/environment/tropical-fish-collection-kills-coral-reefs (accessed June 30, 2020)

predators by producing large amounts of neurotoxins that kill fish, mammals, and birds. These die-offs have been associated with runoff from hog operations in North Carolina and runoff from chicken operations in Maryland.[663]

The toxins accumulate as they go up the food chain. Again, large die-offs of apex predators are another injury to the marine food chain. Too much can be worse than too little.

Most noticeable to us on the surface is plastic trash floating near the ocean surface. This was not an observed problem in the past for two reasons. First, everybody did it. Garbage scows made regular runs from the cities of the developed world a few miles offshore and dumped their trash—no muss, no fuss. Second, the dumped garbage had very little long-term floating material; most material soon became waterlogged and sank.

The change came from two factors. First, the developed world grew prosperous enough to become more environmentally virtuous. The richer countries have stopped dumping garbage at sea by the multi-ton load; they now send it to landfills and condemn others who still dump at sea. Second, beginning in the 1950s, plastics started becoming a growing percentage of the garbage flow. Today, the nearly indestructible plastic items of plastic bottles, Styrofoam cups, lost fishing nets (ghost nets), and many other items increasingly drift at or near the ocean surface. These stray items are often eaten by wildlife sometimes with fatal consequences. Speculations about tiny plastic particles entering and catastrophically poisoning the food chain are unproven but worrisome.

Suggested solutions of banning plastic bags and straws in developed countries would have little effect because these countries are already putting trash in landfills. An international treaty banning ships dumping trash at sea would be useful, but ships are still only a small part of the plastic trash flow. The preponderance of trash today comes from developing countries that are producing much greater tonnages of goods (and the resulting trash) while still not being prosperous enough to find better ways of trash disposal than garbage scows and throwing trash in the street when a hard rain comes.

9.1.6 People Crowding Infrastructure into Beaches, Wetlands, and reefs

The sixth deadly fin is people crowding expensive infrastructure into places where they should not be. People should minimize infrastructure construction directly on the shore for two reasons. First, beaches, wetlands, and reefs protect houses and other infrastructure from storms and tsunamis.[664] Second, wetlands provide vital support for ocean health by buffering nutrient

---

[663] "Why Are Outbreaks of Pfiesteria and Red Tides Suddenly Threatening Our Oceans?" *Scientific American*, April 20, 1998. https://www.scientificamerican.com/article/why-are-outbreaks-of-pfie/ (accessed Nov. 2, 2020) (accessed Nov. 2, 2020)

[664] Fanglin Sun and Richard T. Carson, "Coastal Wetlands Reduce Property Damage during Tropical Cyclones," *Proceedings of the National Academies of Sciences of the United States of America*, vol. 117, no. 11, pp. 5719–5725, March 17, 2020 (first published March 2, 2020). https://www.pnas.org/content/117/11/5719 (accessed June 28, 2020)

flows into the ocean and providing habitat for young sea life, thus protecting the ocean, providing more seafood for people, and reducing global warming.

Let's be clear about damage to people's infrastructure. The ocean does not deliberately destroy houses, hotels, apartments, and other structures near the shoreline. People are not content to picnic, canoe, or otherwise enjoy beach or wetland areas near the water. They crave life near the water, and they bring trillions of dollars of breakable infrastructure with them to the water's edge.[665]

- People build on flood plains near rivers and the ocean where there are already periodic floods with today's sea level and climate. Superstorm Sandy in 2012 that did $70 billion of damage around New York City was said to "show the fingerprints of global warming."[666] The Long Island Express of 1938 (also called the Yankee Clipper) had a higher storm surge then Sandy; it is just that Sandy hit at high tide.[667] Adding a foot-and-a-half (a half meter) of high tide did a lot more damage than couple inches of sea-level rise. Meanwhile, extensive developed areas around New York are in flood plains. Those areas damaged by Superstorm Sandy have been largely rebuilt, and new flood plain developments are still being built.[668] In the United States, more than a trillion dollars of assets are within 700 feet (215 meters) of the ocean.[669] Most of these assets are already in danger WITHOUT ANY GLOBAL WARMING.

- People accumulate progressively larger levees beside rivers and then build structures just on the other side in what was flood plain so that when a long-delayed flood eventually breaks the levee, it becomes a catastrophe instead of an expense of repairing fences and roads in the farm areas that should have remained farm

---

[665] Nearly forty percent of Americans lived in counties directly on the ocean shoreline per a United States National Oceanic and Atmospheric Administration report entitled "What Percentage of the American Population Lives Near the Coast? (last update June 25, 2018) https://oceanservice.noaa.gov/facts/population.html (accessed June 30, 2020)

[666] Matt Sledge, "Hurricane Sandy Shows We Need to Prepare for Climate Change, Cuomo and Bloomberg Say," *The Huffington Post*, Oct. 30, 2012. http://www.huffingtonpost.com/2012/10/30/hurricane-sandy-cuomo-bloomberg-climate-change_n_2043982.html

[667] Cherie Burns, *The Great Hurricane: 1938*, Grove Press; Reprint edition, June 5, 2006.

[668] Ted Steinberg, "*Development of a Disaster*," Discover, pp. 50–54, May 2014.

[669] Patrik Jonsson, "Sea Levels Are Rising, So Why Is Coastal Construction? *Christian Science Monitor*, Oct. 11, 2019. https://www.csmonitor.com/Environment/2019/1011/Sea-levels-are-rising-so-why-is-coastal-construction (accessed Oct. 15, 2019)

land.[670,671] One example is the Great Mississippi River Flood of 1927 that started from floods on Illinois rivers, swelled to great levee breaks on the Mississippi River in Arkansas and Mississippi, and continued down through New Orleans where the authorities dynamited the levees (in two poor parishes) to relieve pressure. Damage for the flood was estimated to be about a third of the federal budget at that time.[672] Katrina in 2005 was a much milder echo of the Great Mississippi Flood.

- People build in flood plains in areas prone to hurricanes. The 1928 Okeechobee hurricane in Florida killed more people than Hurricane Katrina at a time when the Florida population was much less. People have continued building in those flood-plain areas in the Florida construction boom that has continued to this day.

- Worse, American houses already in flood plains facing the ocean are collectively a financial time bomb.[673] Large areas of concentrated wealth (such as New York and Miami) may be able to afford to dike up against rising sea level. More diffusely settled residential areas may simply be abandoned with the federal government stuck for the unpaid mortgages.

- People build structures on barrier islands that are drifting sand bars that could be displaced by periods of extreme wave action. For example, much of Miami Beach in Florida is built on what was a barrier island with an adjoing mangrove swamp. Miami residents now have a beautiful unobstructed view of oncoming hurricanes. Real estate developers call it permanent; nature has other plans—count on it! Don't be surprised when a barrier island goes. Don't be surprised when the rent-seeking owners demand compensation from the federal government for the fruits of their folly.

- People dredge sand up from the near-shore sea bottom or transport in material from land to make artificial land for neighborhood developments. However, such land "reclaimed from the sea" may sink a bit over time, and any major currents might cut into the reclaimed land. This could present problems for a number of urban areas such as (there it is again) Miami. Also, landfilled areas (such as the Boston

---

[670] John Flesher and Cain Burdeau, "AP IMPACT: Deficient Levees Found across America," Associated Press, Jan. 17, 2013. http://news.yahoo.com/ap-impact-deficient-levees-found-across-america-080619976.html (accessed Jan. 17, 2013)

[671] Jessica Ludy and G. Matt Kondolf, "Flood Risk Perception on Lands 'Protected' by 100-YearLevees," *Natural Hazards*, vol. 61, pp. 829–842, 2012. https://link.springer.com/article/10.1007/s11069-011-0072-6 (accessed Nov. 2, 2020)

[672] John M. Barry, *Rising Tide: The Great Mississippi Flood of 1927 and How it Changed America*, Touchstone, New York, NY, p. 40, 1997 (Touchstone edition 1998).

[673] Zack Colman, "How Climate Change Could Spark the Next Home Mortgage Disaster," *Politico*, Nov. 30, 2020. https://www.politico.com/news/2020/11/30/climate-change-mortgage-housing-environment-433721?utm_source=Daily%20on%20Energy%20120120_12/01/2020&utm_medium=email&utm_campaign=WEX_Daily%20on%20Energy&rid=9580 (accessed Dec. 1, 2020)

Back Bay and comparable areas around San Francisco can also liquify during earthquakes causing even well-built structures to topple. This was a major part of the damage in the 1906 San Francisco earthquake.

The beaches, wetlands, and reefs also protect people and infrastructure from the force of ocean waves. But more importantly, they are some of the most powerful areas acre-for-acre that draw down carbon dioxide from the air to reduce global warming. They do that by:

- Being the lungs that breath in the carbon dioxide so that the assortment of plants living on them can make organic matter and breath out oxygen.

- Acting as kidneys to filter nutrients so the nearby ocean is moderate and steady, less likely to generate overfertilized dead zones.

- Yet providing a storehouse that that buffers the flow of nutrients from major rain-driven surges of nutrients to the nearby ocean at a steady rate so that there will be less blue desert from the opposite—underfertilization.

- And finally, being a nursery where young of many oceanic species are protected by the extensive maze of stems and roots holding leaves and nutrients from shore, his maze provides habitat and food for sea life. In the tropics, the young from as many as three-quarters of the tropical deepwater fish species live in such wetlands.

Despite all their benefits for ocean ecology, ocean productivity, and climate stability, the beaches, wetlands, and reefs are under attack for a number of reasons.

- Throughout history, farming has taken progressively more wetlands and changed them into farmland. The Dutch built much of their country by diking off marshes and areas of shallow sea. The new state of Florida immediately set about draining much of the wetlands inland and along the coasts to increase available farm land. In rural tropical areas, as we commence the Twenty-First Century, the broad shallow areas favoring mangrove swamps can become highly profitable fish farms, especially shrimp/prawn operations. (Remember that fish farms now produce more sea food than catches of wild fish.)

- Also, throughout history, cities have encroached on the natural wetlands. People in northern Italy began colonizing an area of swamps and low islands as a refuge from Attila and his rampaging Huns in 453 A.D. That was the start of Venice, a major Mediterranean power until discovery of the Americas carried the axis of trade westward into the Atlantic. In the 1900s, some New Jersey swamp land became a giant waste dump, "Mount Trashmore," the highest elevation east of the Appalachian Mountains. In the 2000s, Singapore changed wetlands and several nearby islands to city land. As noted earlier, parts of the Boston back bay and the San Francisco Bay became housing tracts. The mouth of the Los Angeles (California) River is in the port city of Long Beach, and its previous mouth of

Ballona Creek became the upscale Marina Del Rey area of Los Angeles. This pattern is repeating, but more strongly, in the developing world where many coastal cities are growing into giant megapolitan sprawls, and they often sprawl foolishly onto land "reclaimed" (probably temporarily) from the sea.

- Meanwhile, sprawl is also attacking reefs near the coast. Sandy reefs are the cheapest source of dredged sand to build new land and replenish beaches that are eroding. Reefs are often cleared away to help ship navigation. Last but not least, pollution attacks many of the life forms on the reefs.

- The historic rate of sea level rise, plus that ill-defined but hotly debated increase in the rate of sea level rise, is eating into coastal wetlands. In the past, rising sea levels simply caused coastal wetlands to retreat from the sea and correspondingly advance inland. However, all the populations of existing and developing urban areas can be expected to defend themselves with dikes.

- Channels dug to support harbors, oil and gas drilling, and logging allow increasing erosion that contributes to retreat of coastal wetlands.

- Finally, many areas around coastal wetlands and urban areas are also sinking in addition to any sea levels rise. Some of this is from natural causes as along the Louisiana and Texas coast near the Mississippi Delta where thousands of years of sediment are teeter-tottering the coast gradually downward while hills slowly rise some distance inland. More often, farmers and city water utilities have pumped out water from wells, causing compaction of the sediment below showing as subsidence at the surface.[674] Severe subsidence has been observed in the areas around Jakarta, Indonesia; Venice, Italy; and New Orleans, Louisiana. (There's New Orleans again!)

The result is that half of the world's coastal wetlands were lost in the 1900s, and this rate has continued in the 2000s. A report from China, one of the most recent countries to experience major industrialization, stated that from the 1950s until 1912 China's mangrove swamps decreased by 73% in area, coral reefs decreased by 80%, and coastal wetlands by 57%.[675] Loss of nearshore reefs, though harder to gauge, has also been severe. As can be imagined from the previous text, this loss means that the planet is liable for more global warming because of the reduced drawdown of carbon dioxide. Meanwhile, the growing urban areas on the coast are even more vulnerable because they are losing their protective coastal beaches, wetlands, and reefs.

---

[674] Joel K. Bourne, Jr., "Louisiana Wetlands: Gone with the Water," *National Geographic*, Oct. 2004. http://ngm.nationalgeographic.com/2004/10/louisiana-wetlands/bourne-text (accessed Jan. 13, 2013)

[675] Jane Qiu, "Chinese Survey Reveals Widespread Coastal Pollution," *Nature*, Nov. 6, 2012. https://www.nature.com/news/chinese-survey-reveals-widespread-coastal-pollution-1.11743 (accessed March 22, 2020)

This is all important because increased global greenhouse warming, sea-level rise, offset climate bands, and other associated changes have happened on Earth many times.[676] The changes during the last three thousand years have been relatively mild compared to some other times. Cramming so much infrastructure onto the shoreline is a recipe for disaster even if there were no heating from human-caused climate effects.

9.1.7 Earthquakes, Volcanoes, Giant Waves, Large Objects from the Sky, and Nuclear-Bombs

The seventh deadly fin in the water is a hodge-podge of rare events that start elsewhere but impact the ocean … and the land. They include earthquakes, volcanoes, giant waves, large meteorite strikes (which only happen at intervals of decades, centuries, millennia, or even millions of years), and nuclear explosions that might become more common.

Nevertheless, these dangers and possible profitable mitigations should be considered. There must be some profit involved because voters, and hence their legislators, do not like to pay for policies that may protect against policies hundreds of years in the future with no return at all for them. Before discussing possible mitigations, we shall first examine the dangers.

Earthquakes happen every day, but they are usually minor. Even severe earthquakes that can harm a local area happen every few years, but any particular area may not see such a quake for decades or even centuries. A severe earthquake at sea or near the shore that causes a giant wave or tsunami (Japanese for harbor wave) capable of killing thousands or more throughout a wide region only occurs (on average) once in decades.

On a scale of thousands of years, there have been greater disasters associated with various combinations of earthquakes, volcanoes, large objects from the sky, and triggered landslides. Some of the past and potential future giant wave incidents follow.

- About 66 million years ago, a comet or asteroid[677] 10 to 15 kilometers (6 to 9 miles) wide impacted the sea in an area that is now part of the Yucatan Peninsula in Mexico. (The resulting Chicxulub crater of 150 kilometers (93 miles) in diameter has been buried by sediment, but it shows in seismic surveys.) Besides the tsunami around the Caribbean Sea and Atlantic Ocean, the shock induced earthquakes and more tsunamis all over the world. Splattered hot rocks landed over much of the world, much of the sky was heated like an oven, and then lofted dust and sulfur dioxide shaded the Earth, cooling from inferno into what has been called a volcanic winter for several years. The Chicxulub event caused or was a major contributor

---

[676] Vivien Gornitz, *Rising Seas: Past, Present, Future*, Columbia University Press, New York, 2013.

[677] Comet or asteroid is an issue with astrogeologists. A comet is a heavily ice body that comes in from the edge of the Solar System in a steep orbit. An asteroid is a body with less water and in a more circular orbit. Approaches to the Sun could evaporate much of the water, and glances off the gravity fields of different planets could make a comet's orbit more circular, thus transforming a comet into an asteroid.

to three-quarters of all plant and animal species going extinct. The alternate theory is that more life-killing effects came from poisons emitted from a giant volcanic area in India called the Deccan Traps, and in that case, the space impact was just the final blow.

- About 35 million years ago, an asteroid hit the ocean off the East Coast of North America, forming a 25-mile (40-kilometer) diameter crater now buried beneath the Chesapeake Bay in Virginia and Maryland. The nearby area experienced fires, earthquakes, an air blast, and a devastating tsunami. (Western Europe also got a major tsunami but not as large.)

- About 1.2 million years ago, the 175-square-mile Valles caldera in the middle of northern New Mexico, west of Santa Fe exploded. piling up 150 cubic miles (625 cubic kilometers) of rock and blasting ash as far away as Iowa. As with other calderas, there are still signs of heat below; hot springs are still active around the Valles caldera.

- About 760,000 years ago, the Long Valley volcano erupted in east-central California near the present Nevada state line. After erupting, it sank down to become a caldera. Technically, that eruption was not a supervolcano eruption—it was only 140 cubic miles (584 cubic kilometers). Still, this Long Valley eruption unleashed 2,000 to 3,000 times as much lava and ash as Mount St. Helens, in Washington in 1980. Some of the ash reached as far east as Nebraska. What is worrisome is that the area has been active intermittently until the present. There was a swarm of strong earthquakes in 1980 and the 10-inch (25-centimeter) rise in one area of the caldera floor. Then, in the early 1990s, large amounts of carbon dioxide gas from magma below began seeping up through the ground and killing trees in the Mammoth Mountain part of the caldera. Improved sensing techniques now indicate that 240 cubic miles (1000 cubic kilometers) of molten rock are beneath the Caldera[678]. If that were to erupt, it could be a supervolcano. It probably will not erupt for thousands of years, … or it could erupt tomorrow.

Note: When a very large volcano erupts explosively, the remaining rock collapses, leaving a giant bowl-shaped hole or caldera behind. Between eruptions, much of the remaining rim may erode and various land forms may complicate the area inside the caldera with stream beds and dunes. Consequently, such calderas may be hardly noticeable except by geological survey. Long Valley, and Yellowstone to be mentioned next, are such calderas. They do not fit the common image of a volcano,

---

[678] Trevor Nace, "240 Cubic Miles of Magma Was Just Discovered Beneath California's Supervolcano," *Forbes*, Aug. 16, 2018.
https://www.forbes.com/sites/trevornace/2018/08/16/a-massive-240-cubic-miles-of-magma-was-just-discovered-beneath-californias-supervolcano/#5d1f901e6cb9 (accessed Oct. 28, 2019)

but they are active volcanoes that could have a minor eruption or a super volcanic eruption at any time.

- About 630,000 years ago, the Yellowstone supervolcano in present-day Wyoming erupted and deposited 1,000 km³ (240 cubic miles) of volcanic material (this size is the smallest super volcano). Much of western North America was covered with abrasive volcanic dust, and there were several years of worldwide volcanic winter. However, being inland, there would have been no tsunami. The three recorded mega-eruptions from the Yellowstone caldera have been at roughly 600,000-year intervals, so Yellowstone is due (plus or minus a few thousand years).

- About 74,000 years ago, the super volcano, Mount Toba on Sumatra in Indonesia, exploded and emitted more than 2,800 cubic kilometers (670 cubic miles) of igneous material in possibly the biggest eruption in the last 25 million years. The blast is thought to have ejected about 800 cubic kilometers of ash into the atmosphere, producing the world's largest crater lake (yes, another caldera) that is 100 kilometers long and 35 kilometers wide (62 by 22 miles). The Toba eruption caused a volcanic winter with a worldwide decrease in temperature between 3 to 5 °C (5.4 to 9.0 °F), and up to 15 °C (27 °F) in higher latitudes. One theory is that the extreme cold and resulting low food supplies nearly caused humanity to go extinct.[679]

- About 73,000 years ago, the eastern side (flank) of the Pico do Fogo volcano on the island in the Cape Verde Islands west of Mauretania, collapsed into the ocean, creating a tsunami 170 meters (560 feet) high which struck the nearby island of Santiago. This demonstrated the power of nearby tsunami events; and a recent discovery on the western side of the volcano stirred fears of megatsunamis that might go far inland on the east coasts of the United States, the easternmost Caribbean islands, and Brazil. However, the good news is that the waves widen over distance so that the diluted height would only (?) be about 10 meters (33 feet) after travelling a few hundred miles (as was demonstrated by the 2004 Indonesian earthquake wave described later). That wave height would still be a mega disaster especially for coastline areas that have never had tsunamis in recorded history and have made no preparations.

- Similarly, any large active island volcano can over-steepen and build up a large unstable slope as molten lava solidifies on contact with seawater. Occasionally, at intervals of tens or hundreds of thousands of years, such volcanoes can have a major landslide and produce a mega-tsunami.[680] The big island of Hawaii in the Hawaiian

---

[679] Donald R. Prothero, *When Humans Nearly Vanished: The Catastrophic Explosion of the Toba Volcano*, Smithsonian Books, Oct. 16, 2018.

[680] Raphaël Paris, Ricardo S. Ramalho, José Madeira, et al., "Mega-Tsunami Conglomerates and Flank Collapses of Ocean Island Volcanoes," *Marine Geology*, vol. 395, pp. 168187, Jan. 1, 2018.

Islands comes to mind, and there is evidence of tsunami action from this island roughly every two-hundred-thousand years.

- About 39,000 years ago, there was an eruption near Mount Vesuvius (and perhaps linked to it). The location is a caldera called Campi Flegrei (Burning Fields in Italian). It is a large complex, much of it underground or under the Mediterranean Sea, that includes 24 craters, as well as geysers and vents that release hot gas. The 39,000-years-before-present eruption sent 300 cubic kilometers (72 cubic miles) of molten rock high into the stratosphere, along with 450,000 tons of sulfur dioxide and poisonous fluorine compounds. Together these effects led to a several-year volcanic winter and may have helped kill off the Neandertals.[681,682]

- The Earth is now in an ice age. It has been in an ice age for roughly the last two-and-a-half million years with continent-spanning ice sheets covering places such as Minneapolis, Chicago, New York, and almost to London followed by short warmer intermissions (called interstadials) that still have continental ice sheets in Greenland and Antarctica. (Note: glaciologists consider our present day to be part of an interstadial with the thought that the ice sheets might start growing back—that might be worse than global warming.) The latest glacial period started about 115,000 years before present and had its coldest point about 22,000 years ago. Then, the warming of our present interstadial began. From then until now, there has been a general pattern of warming (with periods of relatively slight periods of warming and cooling such as the Medieval Warm Period (950 to. 1250 A.D.) and the Little Ice Age (late 1200s to 1850).

- During the period of roughly 12,900 to 11,700 years before present. The retreating ice-age and warming temperatures went into reverse with a period of severe chilling. Glaciologists call it the Younger Dryas because the dryas flower common in tundra and the high Alps became more common during this cold time. The change was relatively sudden, happening in decades, and it resulted in a decline of 2 to 6°C (3.6 to 10.8°F) and advances of glaciers and drier conditions until warming resumed 1200 years later. There are two competing theorized causes. One is that great ice dams holding back massive amounts of water broke and released a cascade of cold fresh water that spilled out into the North Atlantic and changed Atlantic Ocean circulation patterns.[683]

---

[681] Erik Plammeti, "Campi Flegrei in Italy Adopts an Orphan Massive Eruption," Discover, April 25, 2019. https://www.discovermagazine.com/the-sciences/campi-flegrei-in-italy-adopts-an-orphan-massive-eruption (accessed Feb. 21, 2020)

[682] Charles Q. Choi, "Ancient Super-Eruption Larger Than Thought," LiveScience, June 21, 2012. https://www.livescience.com/31560-ancient-super-eruption-larger.html (accessed Feb. 21, 2020)

[683] Wallace S. Broecker, "Was the Younger Dryas Triggered by a Flood?" Science, vol. 312, no. 5777, May 26, 2006, pp. 1146–1148.

A newer theory is that a passing comet torched much of North America and dropped one or more large meteorites into Greenland.[684] The fire and impact caused the breaking of ice dams, and aerosols ejected from the impacts contributed to a major cooling. This second theory is disquieting because smaller impacts are much more common than impacts the size of the one associated with the end of the dinosaurs. If this theory were to be correct, there might be more cosmic-impact winters.

- During the period of roughly 6225–6170 B.C., three giant landslides occurred under water on the continental slope west of Norway. These *Storegga* (Norwegian for "the Great Edge") landslides are the greatest underwater landslides yet discovered. The first direction was westerly, but Scotland to the southwest has sediment deposits as far as 80 kilometers (50 miles) inland along river valleys and 4 meters (13 feet) above current normal tide levels.[685]

- About 1650 B.C., the volcanic island of Thera exploded in the Aegean Sea, leaving the remaining crescent-shaped island of Santorini. A major tsunami wave and the resulting ash fall on nearby Crete may have led to the collapse of the Minoan civilization. Chinese records about this time describe a very cold year.

- In 535 A.D. (and possibly continuing into 536 A.D.), one or more volcanoes erupted and caused a dust veil and fog mentioned in accounts from the Mediterranean to China. There was one particularly bad year followed by a slow recovery of several years. These years had the Sun shining weakly, producing cold temperatures and resulting famines. Those all contributed to a major bubonic plague, referred to as Justinian's Plague in the Mediterranean world.[686] The end result was social catastrophes across the world and the beginning of the Dark Ages in Europe. The most likely suspect volcanoes are the. Ilopango volcano (now a crater lake) in El Salvador and an unidentified volcano in the tropics (possibly Rabaul volcano in New Guinea or an earlier version of the Krakatoa volcano in

---

[684] Wendy S. Wolbach, Joanne P. Ballard, Paul A. Mayewski, et al., "Extraordinary Biomass-Burning Episode and Impact Winter Triggered by the Younger Dryas Cosmic Impact ~12,800 Years Ago. 1. Ice Cores and Glaciers," *The Journal of Geology*, vol. 126, no. 2, pp. 165–184, March 2018.

[685] S. Bondevik, F. Lovholt, Carl Harbitz, et al., "The Storegga Slide Tsunami—Comparing Field Observations with Numerical Simulations, *Marine and Petroleum Geology*, vol. 22, Nos. 1–2, pp. 195–208, 2005.

[686] William Rosen, *Justinian's Flea: The First Great Plague and the End of the Roman Empire*, Penguin Books, 2007.

Indonesia) or an unidentified volcano in Iceland. The tropical eruption locations would probably have also caused major tsunamis.[687,688]

- In 1257 A.D., The Samalas volcano erupted on the Indonesian island of Lombok. The event only emitted 10 cubic kilometers (2.4 cu mi) of rocks and ash, but it had a high volcanic explosivity index making it one of the largest volcanic eruptions during the current Holocene epoch and it had a high sulfur content. It created eruption columns reaching into the stratosphere. The aerosols injected into the atmosphere reduced the solar radiation reaching the Earth's surface, cooling the atmosphere for several years and leading to recorded crop failures and famines from Europe to Korea. The Samalas volcano, plus several other eruptions may have started the Little Ice Age cooling,[689] that started with spectacular famines, plagues, and wars in the second and third decade of the 1300s.[690]

- On February 19, 1600, the Huaynaputina volcano in southern Peru began two weeks of eruption that included the largest known volcanic explosion in South America. The volume of material ejected was modest at 30 cubic kilometers (7.2 cubic miles), but the sulfur dioxide released, along with a period of low solar activity, shifted the climate into much colder weather that showed starting with the Russian famine of 1601 through 1603 that killed two million people, one third of the Russian population. Feedbacks from that cold weather, several other smaller volcanoes, and low solar activity shifted the world climate into the coldest part of the Little Ice Age.[691] Geoffrey Parker argues that the extreme cold weather led to famines, plagues, and increased wars.[692]

- In April 10, 1815, Mount Tambora on Sumbawa Island in Indonesia exploded taking off the top third of the mountain. Although not a supervolcano by size, the explosion was the greatest in recorded history. More than 100,000 people died in

---

[687] Tim Newfield, The Global Cooling Event of the Sixth Century, Mystery No Longer?" *HistoricalClimatology.com*, May 1, 2016. https://www.historicalclimatology.com/blog/something-cooled-the-world-in-the-sixth-century-what-was-it (accessed Dec. 27, 2018.

[688] Ann Gibbons, "Why 536 was 'the worst year to be alive'" *Science*, Nov. 15, 2018. https://www.sciencemag.org/news/2018/11/why-536-was-worst-year-be-alive (accessed Dec. 27, 2018)

[689] Naomi Schalit, "Indonesia's Samalas Volcano May Have Kickstarted the Little Ice Age," *The Conversation*, Sept. 30, 2013. *https://theconversation.com/indonesias-samalas-volcano-may-have-kickstarted-the-little-ice-age-18772 (accessed July 19, 2020)*

[690] William Rosen, *The Third Horseman: A Story of Weather, War, and the Famine History Forgot*, Viking, May 15, 2014.

[691] Andrea Thompson, "Volcano in 1600 Caused Global Disruption, Study Suggests," *LiveScience*, May 5, 2008. https://www.livescience.com/4912-volcano-1600-caused-global-disruption-study-suggests.html (accessed Dec. 27, 2018)

[692] Geoffrey Parker, *Global Crisis: War, Climate Change & Catastrophe in the Seventeenth Century*, Yale University Press, New Haven, Connecticut, 2013.

the immediate effects of blast, fire, ash, and tsunamis. The amount of ejected material was only roughly 150 cubic kilometers (36 cubic miles) of ash, pumice and other rock, and aerosols—including an estimated 60 million tons of sulfur dioxide that were lofted high into the atmosphere.[693] The dust particles actually had a greenhouse effect because they were large enough to absorb or reflect infrared heat waves from the ground, while the sulfur dioxide formed smaller liquid aerosol particles that acted in a greenhouse fashion, reflecting visible light back into space. After about a year, the larger and heavier dust particles had largely settled out of the atmosphere, leaving the cooling sulfur dioxide. The average global temperature dropped by as much as 3 °C (5.4 °F). New Englanders called it "the year without a summer" or "eighteen hundred and starve to death." Besides inspiring the writing of *Frankenstein*, the cold rainy weather caused famines, and many government collapses throughout the world.

- On August 26, 1883, an eruption began on Mount Krakatoa, which comprised an island by the same name in the Sunda Strait between the islands of Java and Sumatra in Indonesia. The eruption climaxed the next day when more than 70% of the island and its surrounding small islands were destroyed in an explosion and submergence of much of the remainder. Of the estimated 21 cubic kilometers (5 cubic miles) erupted, it is estimated that about 4 cubic kilometers (1 cubic mile) of ash and sulfur dioxide were injected into the stratosphere above 25 kilometers (16 miles). Global temperatures were lowered by as much as 1.2°C (2°F) on the average and did not return to normal for 5 years. The eruption was much smaller than that of Tambora, but it was the first catastrophe reported worldwide due to trans-ocean telegraph lines being in place.[694]

- On July 9, 1958, a 7.8 Richter scale earthquake triggered a rockslide of 40 million cubic yards (30 million cubic meters and about 90 million tons) into the narrow inlet of Lituya Bay, Alaska. that caused a megatsunami that washed out trees to a maximum elevation of 1,720 feet (520 m) in the bay, the largest megatsunami in recorded history. The good news is that this only happens in a confined area.

- In May 18, 1980, Mount Saint Helens in Washington state erupted. It was far from a supervolcano with only 0.05 cubic miles (0.21 cubic kilometers) of ejected material. However, it demonstrated how far volcanic ash particles can be spread by even a small volcano. Yakima, Washington, 90 miles (140 kilometers) away, received four to five inches (100 to 130 mm), and Spokane at 36 miles away (580 kilometers) experienced darkness at noon with visibility reduced to 10 feet

---

[693] Gillen D'Arcy Wood, *Tambora: The Eruption That Changed the World*, Princeton University Press, Sept. 15, 2015.

[694] Simon Winchester, Krakatoa: *The Day the World Exploded: August 27, 1883*, HarperCollins, April 4, 2003.

(3 meters). It was not cataclysmic on a world scale; it only provides a point of comparison against the other larger eruptions mentioned.

- In June 1991, Mount Pinatubo on the Island of Luzon erupted. Its size was nowhere near that of a supervolcano. However, this was the first eruption for which observational tools were available to clearly measure the cooling power of volcanoes. Mount Pinatubo lofted between 15 and 30 million tons of sulfur dioxide gas into high into the atmosphere within two hours of the eruption, attaining an altitude of 34 km (21 miles). In 1992 and 1993, the average temperature of the planet dropped 0.4 to 0.5°C (0.7 to 0.9°F).[695] The United States experienced its third coldest and third wettest summer in 77 years during 1992.[696]

- In 2004, a 9.1 Richter Scale earthquake shook for 10 minutes along an epicenter in the Indian Ocean west of Sumatra in Indonesia because of a movement at a fault. The movement at the fault caused a megatsunami. The earthquake and the tsunami caused about a quarter million deaths. Some nearby coastlines experienced waves as high as 100 feet (30 meters). Waves were still as high as 30 feet (9 meters) in India, Shri Lanka, Thailand, and several other nearby countries that experienced two-fifths of the deaths. Fortunately, the wave front of such tsunamis spreads, and consequently weakens, with distance from the source avalanche just as the light from a candle is dimmer on the far side of the room. The tsunami that killed tens of thousands in India, Shri Lanka, and Thailand was only 1.5 meters (5 feet) when it hit East Africa, and deaths were only in the hundreds.[697]

- In 2000, the British Broadcasting Corporation docudrama called "End Day" included the theory that The Cumbre Vieja volcano on La Palma in the Canary Islands could send a megatsunami 1000 feet (300 meters) high roaring off toward the east coasts of South and North America. It is true that the volcano has had catastrophic collapses in the past.[698] Also, it does have a fault on the west side running roughly north and south so a possible landslide or flank collapse might be expected to send a wave west toward the Americas. More skeptical geologists later

---

[695] Chris Newhall, James W. Hendley II, and Peter H. Stauffer, *The Cataclysmic 1991 Eruption of Mount Pinatubo, Philippines*, U.S. Geological Survey Fact Sheet 113-97, 1997. https://pubs.usgs.gov/fs/1997/fs113-97/ (accessed Nov. 13, 2018)

[696] Matt Rosenberg, "Mount Pinatubo Eruption: The Volcanic Mount Pinatubo Eruption of 1991 that Cooled the Planet," *About.com Geography*, updated Nov. 11, 2010. http://geography.about.com/od/globalproblemsandissues/a/pinatubo.htm (accessed May 30, 2012 (accessed May 30, 2012)

[697] David Oburo, "Impacts of the 26 December 2004 Tsunami in Eastern Africa," *Ocean & Coastal Management*," vol. 49, issue 11, pp. 873–88, 2006.

[698] J. E. Hunt, R. B. Wynn, P. J. Talling, and D. G. Masson, "Multistage Collapse of Eight Western Canary Island Landslides in the Last 1.5 Ma: Sedimentological and Geochemical Evidence from Subunits in Submarine Flow Deposits," *Geochemistry, Geophysics, Geosystems*, vol. 14, issue 7, pp. 2159–2181, July 2013. https://agupubs.onlinelibrary.wiley.com/doi/full/10.1002/ggge.20138 (accessed Nov. 2, 2020)

suggested that the warning was overblown for two reasons. First, previous collapses off La Palma have left avalanche deposits on the sea floor showing a series of collapse subunits as one area removed the support for the area, then after that area a pause that second area began to move, and so on. That would generate a series of smaller waves rather than one large wave. Second, as noted about the 2004 tsunami, the wave front (or fronts) of such waves spreads and rapidly diminishes with distance. Thus, damage from a volcanic flank collapse at sea might be no more severe the 2004 earthquake tsunami. That would still be severe. It would be even more severe because (as superstorm Sandy showed) many areas of the United States East coast are unprepared for high water.[699] Countries along the Pacific Rim have tsunamis every few years, and they have an active warning system. Countries around the Indian Ocean go for decades without a tsunami, and they are sometimes too lax. Countries with eastern Atlantic coasts had only one locally severe tsunami in recent history. That was the 1929 Grand Banks tsunami that only had severe effects on Newfoundland, Canada.[700] The people of the Atlantic shores of the Americas don't have a clue about tsunamis.

- Ever since nuclear weapons were first used in 1945, there has been a type of literary composition (genre) often referred to as the nuclear apocalypse, in which civilization or even all human life might be destroyed by a nuclear war. Life nearly imitated art several times during the Cold War when both the United States and its lead communist adversary, the Union of Soviet Socialist Republics, were poised on the brink of a hot war. Each had tens of thousands of nuclear weapons. In addition to blasts and fire, there would have been poisonous dust similar to the great dust clouds of a supervolcano. That dust would also have included deadly levels of radioactive isotopes, but people were accustomed to that threat. Then, environmentalists warned that the dust cloud would also cause a nuclear winter like the volcanic winters that followed the eruptions of Toba and Tambora.[701] The threat of nuclear winter may have helped cause the Cold War to wind down. Then, when the Soviet Union collapsed into several smaller countries in 1992, the nuclear-armed countries greatly reduced the number of nuclear bombs in their arsenals. The United States reduced its arsenal also. However, over the decades, progressively more countries have built nuclear bombs (nuclear proliferation) and have a strong probability of using them. For instance, India and Pakistan uneasily share the Indian subcontinent. They have fought three wars since independence from Britain in 1947, and they both have nuclear weapons. Similarly, the Jewish state of Israel was recreated in 1948. Israel has fought four wars with its more numerous

---

[699] Douglas Main, "US East Coast Faces Variety of Tsunami Threats," *Live Science*, Nov. 15, 2012. https://www.livescience.com/24813-east-coast-tsunamis.html (accessed July 1, 2020)

[700] Bruce Ricketts, "The Great Tsunami of 1929," *Mysteries of Canada*, Oct. 30, 2014. https://www.mysteriesofcanada.com/newfoundland/tsunami/ (accessed Nov. 2, 2020)

[701] Paul R. Ehrlich, Carl Sagan, Donald Kennedy, et al., *The Cold and the Dark: The World After Nuclear War*, W. W. Norton & Company, 1985.

surrounding Moslem countries and would probably use its nuclear weapons if seriously threatened. Such use of nuclear bombs would be much less damaging than a full-blown exchange during the Cold War might have caused. Still, tens of millions-of-lives could be endangered in such a mini nuclear apocalypse—more from the fallout, cold, and famine than from the initial blasts.

To summarize the potential ocean, geological, cosmic, and nuclear -proliferation dangers:

1. Global warming, the reason for existence for this book, could cause very expensive sea-level rise, made much more painful because people have been concentrating along coastlines and making the coastlines more vulnerable.

2. Global warming and other poor husbanding actions from people could seriously reduce ocean productivity—thus reducing food supply and causing global warming to become more severe. His would be in addition to climate disruption of agriculture on land already discussed. These changes could conceivably become a serious threat to human food supply and damage from rising sea levels.

3. However, warming would have an onset of decades, which would allow more time to develop measures for adapting to change and devising ways to mitigate the change.

4. The most frequent catastrophes with the greatest loss of life have historically been dust and gases from volcanoes that cause famine-making cold for several years at a time with very little warning. Recent geologic history demonstrates that there are volcanic events many times more destructive than anything in recorded history.

5. The second most powerful destructive events have been tsunamis from earthquakes and volcanoes. Again, the danger of lost lives and property has been growing larger because more population has been crowding onto a more-vulnerable coast.

6. Cosmic impacts big enough to destroy civilization and even endanger humanity have been extremely rare—only happening at intervals of tens of millions of years. However, there is an unproved theory that smaller impacts could be associated with comet heating that caused conditions similar to volcanic winters.

7. Nuclear proliferation of more countries having enough nuclear bombs so that a nuclear exchange among two or more of them could cause significant fallout poisoning and a weaker nuclear winter of 1 or 2 years comparable to cold wet years after the Tambora eruption. Judging by the diabolic weapons of crossbows and gunpowder that were called too terrible to ever be used again, nuclear weapons will be used again.

As with concerns about food and energy, the dangers just described suggest their solutions. Next, things work best when there is profit available. Finally, some of those advances described earlier can be multiplied in use with the oceans.[702] The green wave in the ocean can, indeed stop global warming.[703,704]

The next two subchapters will deal with ways to avoid or at least mitigate these threats. They are roughly divided into the revolutions in shallow green water and deep blue water.

9.2 The Greenwater Revolution on the Continental Shelf

Aquaculture, not the Internet, represents the most promising investment opportunity of the 21st Century.

Peter F. Drucker (1909-2005)
Business writer, professor, and self-described social ecologist[705]

He was holding at bay the specter of famine which had confronted all earlier ages, but which would never threaten the world again while the great plankton farms harvested their millions of tons of protein, and the whale herds obeyed their new masters. Man had come back to the sea, his ancient home, after aeons of exile; until the oceans froze, he would never be hungry again ....

Arthur C. Clarke, 1957[706]

In in *Fishing: How the Sea Fed* Civilization, archeologist and historian Brian Fagan argues that fishing was an indispensable and often overlooked element in the growth of civilization.[707] Unfortunately, oceanic fishing hardly got beyond the hunter-gather stage, and that led to a stagnation of production as one fishing ground after another was fished out causing a population

---

[702] Brendan Smith, "The Coming Green Wave: Ocean Farming to Fight Climate Change," *The Atlantic*, Nov. 23, 2011. https://www.theatlantic.com/international/archive/2011/11/the-coming-green-wave-ocean-farming-to-fight-climate-change/248750/ (accessed Feb. 22, 2019)

[703] Brendan Smith, "The Coming Green Wave: Ocean Farming to Fight Climate Change," *The Atlantic*, Nov. 23, 2011. https://www.theatlantic.com/international/archive/2011/11/the-coming-green-wave-ocean-farming-to-fight-climate-change/248750/ (accessed Feb. 22, 2019)

[704] Nesar Ahmed, James D. Ward, Shirley Thompson, Christopher P. Saint & James S. Diana, "Blue-Green Water Nexus in Aquaculture for Resilience to Climate Change, *Reviews in Fisheries Science & Aquaculture*, vol. 26, no. 2, pp. 139–154, 2018. https://www.tandfonline.com/doi/abs/10.1080/23308249.2017.1373743?journalCode=brfs21 (accessed June 30, 2020)

[705] This Peter Drucker quote appears in many places, but I have never found a reference for the quote.

[706] Arthur C. Clarke, *The Deep Range* (novel), Harcourt, Brace & World, Inc., 1957.

[707] Brian Fagan, *Fishing: How the Sea Fed Civilization*, Yale University Press, Sept. 2017.

crash of the target species.[708] While wild catch has been stagnating since the mid-1990s, more people and more prosperity have been increasing the demand for seafood, and they have been getting it.[709]

How? More advanced aquatic farming (aquaculture) has been expanding more than 6% per year for two decades. Aquaculture has been expanding out from freshwater on the land into the oceans in a sustainable way.[710] Aquaculture has expanded from 4.7 million tons (4.2 metric tons) in 1980 to 80 million tons in 2016, which exceeded production from the wild catch.[711] Furthermore, this aquaculture revolution is expected to continue and even accelerate, first in the shallows of the continental slope and then on into deeper waters.

9.2.1 The Potential of the Continental Shelf and Barriers to Its Use

The continental shelf is the continuation of the continents that is under sea level. The shelf areas generally slope down at a shallow rate until they reach the continental shelf at roughly 100 meters (330 feet), and then the continental slope drops rapidly down into the deep ocean. The continental shelf is roughly 27 million square kilometers 10 million square miles) or three times the size of Australia.

The continental shelf zones comprise only 7.6% of the surface area of the world oceans, yet they provide 15–30% of the oceanic primary production. Even that is only a tiny fraction of what the shelf could produce. For food production and mitigation of global warming, those three Australias of shelf are handily right next to land, so the transportation cost to and from them is low. Also, being next to land and with land under the shallow water, the continental shelf tends to be more fertile because mineral nutrients are near the surface.

Yet with all that potential, the three Australias worth of sea area produce much less per unit area than hunter gatherers collected. On land, the transition from hunter–gatherer to farming made a hundredfold improvement. Then there was another hundredfold improvement as science and technology were applied to land farming over the centuries.

The same has been shown true in water farming of aqua + farming, hence aquaculture. Since antiquity, Chinese, Romans, Hawaiians, and many others gathered seaweed (sea vegetables

---

[708] Callum Roberts, *The Unnatural History of the Sea*, (originally Shearwater Books, 2007) Island Press, 2008.

[709] "Raising More Fish to Meet Rising Demand," World Bank, Feb. 6, 2014. https://www.worldbank.org/en/news/feature/2014/02/05/raising-more-fish-to-meet-rising-demand (accessed Nov. 10, 2020)

[710] John S. Lucas, Paul C. Southgate, Craig S. Tucker (editors), *Aquaculture: Farming Aquatic Animals and Plants*, 3rd Edition, Wiley-Blackwell, Jan. 22, 2019.

[711] Marta Correia, Isabel Costa Azevedo, Helena Peres, et al., "Integrated Multi-Trophic Aquaculture: A Laboratory and Hands-on Experimental Activity to Promote Environmental Sustainability Awareness and Value of Aquaculture Products," *Frontiers in Marine Science*, March 20, 2020. https://www.frontiersin.org/articles/10.3389/fmars.2020.00156/full (accessed Aug. 16, 2020)

as Judith Cooper Madlener insisted!) and later cultured them on mesh nets.[712] [713] They used cords hanging from rafts for shellfish and ponds or cages for fin fish. More advanced technology of the 2000s allows such production on an industrial scale. There are catfish farms, tilapia farms, and culturing of other species, often from facilities along the coast or even far inland.[714]

When the controlled water farming or aquaculture is taken out into the ocean, it becomes mariculture. However, there is a maricultural limitation in the United States and most other western countries. Most of this culturing is done in an artificial ocean-like environment or in pens or cages on or near the shore, not because technological limitations, but because of an institutional failing. As was mentioned in the seven deadly fins, waters off shore are often common property. They are owned by everybody; hence, they are guarded by nobody.

A common complaint for those harvesting wild plants, such as kelp, dulse, or nori, is the same as that of the fishers. If one hunter–gatherer does not take it, another one will. Just as fish are fished out, gathered sea plants are gathered out. It is another tragedy of the commons.

Government regulations often make things worse. A limited gathering period means that the hunter–gatherers risk their lives in high-speed work in that limited time. They may waste by-catch in their haste. They may gather sea plants too early. Even the actual culturing in shallows at the water's edge has problems in that pollution from fish culturing is more likely to be concentrated there, and pest predators (such as starfish, turtles, and sea urchins) are more likely to attack product fish and plants there.

For kelp production, California regulations have often been worse than useless. The regulators set limits on how much kelp could be trimmed in a given year, but they did not designate where. Thus, one harvester might cut half a kelp bed, which is no problem for the kelp. But then, another harvester might trim a half of a half of that same kelp bed the next week, which can seriously damage the bed. An even bigger danger to kelp is the spiny sea urchin; there should be

---

[712] Judith Cooper Madlener, *The* Seavegetable *Book*, Outlet, July 1, 1977.

[713] Thierry Chopin and M. Sawhney, "Seaweeds and Their Mariculture," *The Encyclopedia of Ocean Science*, Elsevier, Oxford, Editors: J. H. Steele, S. A. Thorpe, and K. K. Turekian, pp.4477-4487, Dec. 2009.
http://marineagronomy.org/sites/default/files/Chopin%20and%20Sawhney%202009%20Seaweeds%20and%20their%20mariculture%201.pdf (accessed July 3, 2020)

[714] Technically, culturing of fish and plants in ponds and tanks on land is aquaculture (water + culture), whereas culturing of such life forms in the ocean is mariculture (sea + culture). However, young plants and animals are now often cultured in controlled facilities and then transferred out to the ocean, so that blurs the distinction. Also, much of the literature is now using the one term of aquaculture. Similarly, more recent biology literature describes algae, from tiny one-celled pond scum up through giant bull kelp not as plants but as another order altogether, *protista*, because they have some non-plant features. My view is that if everything cultured in water is aquaculture, everything that looks like a plant ... is a plant. It's just too complicated otherwise.

a bounty on destroyed sea urchins. However, there is a specialty market for sea urchins, so the state of California has a bag limit on purple sea urchins as just another seafood.[715]

Many regulations are inimical to mariculture. Protected seals are multi-hundred-pound fish-eating machines that also attack fish nets and cages. Protected walruses dig up lobster pots, strip shellfish from culturing rafts, and occasionally heave themselves up onto small boats or shellfish rafts for picnics and napping causing serious damage to facilities. The regulators consider great white sharks to be "magnificent beasts" to be protected, whereas great whites consider fishermen and mariculturists in small boats as … lunch.

There are historical reasons why this institutional failing happens. Land farmers could mark and fence off their property exactly. They could even sleep next to their property lines and be there to defend their fields. In contrast, oceanic locations grew steadily less exact with distance from shore. Also, no one lived on a certain stretch of water; instead, they would periodically return to lobster pots, crab traps, set lines, or similar catching structures. Meanwhile, waters out of sight are out of security. While diving with sport divers, I was scandalized by some of my fellow divers who routinely stole from lobster traps.

The Industrial Revolution on land led much greater wealth and even food surpluses due to rationalized mass production. Meanwhile, oceanic production lagged and often declined. While great fleets went far away to fish exceptionally rich areas such as the Grand Banks to the east of Canada (until each was fished out),[716] coastal operations tended to be small operations.

When coastal areas were fished out, they reverted to a wilderness, largely just the playthings of the rich. The Rockefellers, Bushes, Kennedys, and now Obamas maintain their sea-view mansions. They pilot their sailboats and power boats out of their upscale marinas.

If someone proposes development within eye shot of their mansions, they pay for environmental protests. A classic case was a proposed wind farm off Cape Cod, Massachusetts. Wind farms are ordinarily dearly beloved by liberal environmentalists, but this wind farm would have been (distantly) visible from the Hyannis Port compound such as that of politically connected Kennedy dynasty that produced democrat representatives, senators, and even one president. With the help of neighbors such as the arch-conservative republican William Koch, Senator Edward Kennedy waged a 16-year bipartisan legal and political battle that ultimately prevented construction of the wind farm.

For all those reasons, the nearshore waters of the continental shelf are even less productive that of normal hunter–gatherer operations. That just makes it all the riper for a social and technological revolution.

---

[715] California Recreational Ocean Fishing Regulations: Invertebrate Fishing Regulations, California Department of Fish and Wildlife, updated on March 13, 2019. https://wildlife.ca.gov/Fishing/Ocean/Regulations/Sport-Fishing/Invertebrate-Fishing-Regs (accessed July 2, 2020)

[716] Kevin Cox, "Death of Newfoundland's Grand Banks," The Guardian, Dec. 17, 1999. https://www.theguardian.com/world/1999/dec/18/eu.politics (accessed Oct. 20, 2018)

9.2.2 Starting the Greenwater Revolution

The revolutionary advances in farming the continental shelf are starting most strongly with the countries of Asia. For centuries, people in these two areas have maintained extensive networks of sea vegetable culturing from the shoreline out from on the shelf floor to float depths several tens of meters from that floor. This culturing has advanced with the additions of

- Outboard motors to get from shore to the work areas and back in more practical times.
- Plastic nets and lines, for providing a place for seaweed and shellfish to grow and penning in fish.
- Plastic floats and stakes because the bamboo and other wood for these items is in increasingly short supply
- Scientifically designed nurturing facilities on shore to culture baby plants and animals.
- Entrepreneurial chemists and engineers finding markets for other uses of sea vegetables besides food, such as pharmaceuticals, plastics, cosmetics, and even leather substitutes.

These modest technologies have led to expansion of ocean aquaculture (mariculture) by 8.8% per year for the last three decades.[717] This growth is reflected in pictures of boats working rows of work areas stretching to the horizon. Because this revolution is largely happening in the tropics where workers are less subject to the chill danger of hypothermia, climate warming would just expand the area available to use these techniques.

But there is so much more. High-tech has arrived with fin-fish cage, shellfish racks, and sea plant holders.[718,719] Developments include

- Large floating net cages for fin fish
- Remote monitoring of the levels of temperature, oxygen, nitrates, pH, and carbon dioxide in the water to protect livestock health

---

[717] Pia Klee, editor in chief, *Rethinking Aquaculture to Boost resource and Production Efficiency: Sea and Land-Based Aquaculture Solutions for Farming High Quality Seafood*, Version 1.1, Rethink Water Network, Danish Water Forum and State of Green, 2017. https://stateofgreen.com/en/publications/aquaculture-rethinking-aquaculture-to-boost-resource-and-production-efficiency/ (accessed July 2, 2020)

[718] Alex Beyman, Mariculture: Farming the Ocean to Feed a Future Population of 10 Billion Humans," *Predict*, Nov. 12, 2018. https://medium.com/predict/mariculture-farming-the-ocean-to-feed-a-future-population-of-10-billion-humans-deb975abaa3d (accessed July 2, 2020)

[719] Suresh Kumar Mojjada, Imelda Joseph, P. S. Rao, et al., "Design, Development and Construction of Open Sea Floating Cage Device for Breeding and Farming Marine Fish in Indian Waters," *Indian Journal of Fisheries*, vol. 60, no. 1, pp. 61–65, 2013. http://eprints.cmfri.org.in/9383/1/164.pdf (accessed Aug. 14, 2020)

- Computer monitoring of operations to maintain supplies, schedule deliveries, and detect subtle changes (for instance, a slower rate of growth might indicate an infestation of sea lice)
- Integration of off-coast aquaculture with renewable energy systems
- Recycling techniques to minimize water use and/or pollution both on land tank culture and offshore.
- Robotic systems for identifying and controlling pests and dangerous predators

An ongoing project is the development of submersible cages, so that the cages can be submerged during storms. Furthermore, making all the infrastructure submersible would reduce potential storm damage. This is crucial for extending mariculture outside of protective bays and fiords. The protected waters are only a tiny part of the shelf. Thus, extending to the rougher (and usually deeper) waters multiplies the size of the potential commercial revolution not just in surface area, but in three dimensions. Various species thrive at different levels, so greater depth allows stacking of different maricultural operations at different depths.

Additional techniques can be adapted from other industries. Many of the revolutions in farming and energy can be applied directly to mariculture.

- Offshore wind farms may often be too expensive by themselves, but their power cables to shore can also include smaller side power lines to power the maricultural operations along the way. Moreover, the concrete and steel bases of the windmills can be safety shields against lackadaisical ship captains who sometimes drift out of their shipping lanes. Furthermore, more operations in a work area means more eyes and cameras to protect against theft, vandalism, and accidental ramming by ships.
- The minilivestock of Section 7.9—McBugs and McCrawlies—could provide food for the caged fish. Likewise, the cultured bacteria and yeasts of Section 7.8, Magic Shrooms—Bacteria and Fungi Alone Could Feed the World!—are already beginning to replace the increasingly scarce fish meal made from wild fish.
- Conversely, the unused plant and animal material from aquaculture could feed minilivestock or even conventional land livestock. (Several seaweeds introduced as supplements to cattle feed have been found to reduce methane emissions from the animals.)
- The renaissance in nuclear fission could provide cheaper (non-carbon-dioxide-producing process heat for cooking any organic material not used for food or fertilizer into chemicals or fuel.
- There have been factory ships to clean and process fishing catches for decades, but robotics could make them much smaller and more efficient to deliver products to market.

- On land, we have the rise of robo farm implements (Section 7.5). Robotic planters, weeders, and harvesters reduce human labor required. This would be even more important in ocean production in which cold water and limited visibility can reduce human productivity. Even better would be robo pest protection. Just imagine protecting your kelp fields from their bane, sea-urchins, using robo sea otters. How about robo cleaner fish to find and destroy sea lice, the bane of ocean-farmed salmon?

- Plant breeding can improve usefulness of sea plants. For instance, the warm-water sargassum sea plants can serve the same shallow water growing function as kelp does in colder waters … except that sargassum evolved in the open ocean of their namesake Sargasso Sea. Consequently, Sargasso do not have holdfasts (as kelp does) to keep it from drifting into shore. A little 21st Century plant breeding can fix that.

- The concept of using kelp or other sea plants for biofuel—and arresting global warming—is not new. Let us examine that concept in a few paragraphs.

Giant ocean energy and food farms are not a new idea. From 1929 until 2006 Kelco Corporation (later renamed ISP Alginates Inc.), operated out of San Diego California, using essentially giant floating lawn mowers to regularly harvest tons of giant bull kelp (*mycrocystis pyrifera*) for emulsifiers in food. (Regulatory difficulties and energy costs caused them to relocate in Ireland.)[720]

Howard Wilcox's 1975 *Hothouse Earth* described the U.S. Navy experimental effort to grow fields of kelp anchored with nets for giant kelp fields that would be harvested and bacterially decomposed into methane for fuel. This was to thwart the oil and gas exhaustion that was expected within a couple decades at that time and to avert the global warming that Wilcox urgently warned about in his book.[721] Unfortunately, an unknown ship drifted out of the ship channel and destroyed the demonstration field. (This being the 1970s there was no continuous remote sensing to identify the incompetent marauder and seek legal redress.)

Then, high oil prices caused production to increase enough to a glut, and energy prices collapsed along with the peak fossil fuels idea … twice, first in the 1980s from continued improvements in existing production fields and then in the 2010s from the Fracking Revolution.

Meanwhile, Wilcox's dire warning about possible catastrophic global warming became an accepted idea by the early 2000s. That led to Wilcox redux with a company that included Howard

---

[720] Michele Hackney, "Kelp Harvester Leaving," *San Diego Community Newspaper Group*, March 15, 2006. http://www.sdnews.com/view/full_story/298437/article-Kelp-harvester-leaving (accessed Nov. 4, 2019)

[721] Howard A. Wilcox, *Hothouse Earth*, Praeger Publishers, New York, 1975.

Wilcox's son proposing kelp farms tended largely by drone (robotic) craft.[722] In 2012, and a group of coauthors, proposed culturing much expanded beds of shallow-water large seaweeds such as kelp to draw down carbon dioxide from the atmosphere, provide seaweed (macroalgae) sufficient to be decomposed into carbon-neutral fuel (supplanting fossil fuel, and still have enough macroalgae to culture fish to feed large numbers of people.[723]

Environmentalist Tim Flannery popularized this concept in his 2017 *Sunlight and Seaweed: An Argument for How to Feed, Power and Clean up the World*, in which he proposed regional-sized areas of sea plant and animal production in 7% of the ocean surface, which sounds like the area of the continental shelves.[724]

Marine aquaculture/aquaculture is much more affordable on the continental shelf than in deeper waters, which have the following disadvantages:

1. The cables become much longer with greater expense, more chance of fouling on other cables, and greater unused areas to avoid cables from one facility entangling with others.

2. Power sources, maintenance support, and markets are farther away.

3. The deeper waters tend to have much lower levels of nutrients because the local sea floor is far from the sunlit zone where plants can grow.

4. Last and most important, deeper waters tend to be colder and to have greater ranges of pressure, both of which make it harder for human divers.

Still, as noted by N'Yeurt et al. and by Flannery, that 7% fraction of the ocean, just by itself, could provide nonfossil liquid fuels to run human society, food for ten billion people, and a drawdown of carbon dioxide to match present emissions from human technology. If successful, we could avoid the more heroic (or quixotic) aspects of the GND such as renovating all the houses on Earth, extermination of cattle, and grounding of most commercial aircraft.

However, the greatest caveat is that the time needed for that development. The three virtual Australias of the continental shelves are a new world much more so than Australia and the two American continents were new worlds to the first European settlers. Development of technologies and institutions for the immense potential of these virtual continents will require decades.

---

[722] Evan Ackerman, "Robotic Kelp Farms Promise an Ocean Full of Carbon-Neutral, Low-Cost Energy," *IEEE Spectrum*, March 15, 2017. https://spectrum.ieee.org/energywise/energy/renewables/robotic-kelp-farms-promise-an-ocean-full-of-carbon-neutral-low-cost-energy (accessed Dec. 9, 2019)

[723] Antoine de Ramon, N'Yeurt, David P. Chynoweth, Mark E. Capron, et al. "Negative Carbon via Ocean Afforestation," *Process Safety and Environmental Protection*, vol. 90, issue 6, pp. 467–474, Nov. 2012.

[724] Tim Flannery, *Sunlight and Seaweed: An Argument for How to Feed, Power and Clean up the World*, The Text Publishing Company, Melbourne, Victoria, Australia, 2017.

9.2.3 Organizing and Making the Greenwater Revolution

Dreams and potential are great, but the devil is in those clichéd details. As noted earlier, institutional organization is the key to developing the Greenwater Revolution. We have already seen how oceanic anarchy has led to the tragedy of the commons many times with no benefit for anybody who would try to husband stocks of fish or marine habitat. We have already seen how government regulation is often clumsy and often favors the rich.

Conversely, as noted earlier, herders and farmers owning their land make for tremendously more productivity. So too, it can be if continental shelf "land" can be owned and be for the use of the owners. The technology is now available to manage those boundaries. The Global Positioning System (GPS) that runs map applications for ships, cars, and airplanes can now delineate whose property one is on whether not just on land but now also at sea.

Technology is now emerging for comparable position finding under water and providing that capability with very small amounts of electricity.[725],[726] Such smaller and low-power instrumentation, combined with more affordable underwater drones becoming available, will allow detailed mapping and mapping of underwater land on the continental shelves.

Government regulations and balancing of conflicting interests will be even more important as ocean culturing becomes a major industry, but private property is the fuel for creating that industry to regulate. To understand how this can be, consider some of the events in a Greenwater Revolution.

- Major increases in sustainable fishing could be achieved simply by better management of fisheries (banning bottom trawling, catch limits, and limits on bycatch),[727] but that is only a start.

- People pay for property rights. This would be comparable to the enclosures and clearances of common lands in England and Scotland from the 1400s through the 1700s, which greatly increased productivity.[728] This would also be useful to the federal government and to coastal states. Since these governmental units have gone increasingly into debt, this is a powerful incentive to adopt such sales.

---

[725] Reza Ghaffarivardavagh, Sayed Saad Afzal, Osvy Rodriguez, and Fadel Adib, "Underwater Backscatter Localization: Toward a Battery-Free Underwater GPS," Massachusetts Institute of Technology, Cambridge, Massachusetts, 2020. http://www.mit.edu/~fadel/papers/UBL-paper.pdf (accessed Nov. 3, 2020)

[726] Daniel Ackerman, "An Underwater Navigation System Powered by Sound," *MIT News*, Nov. 2, 2020. https://news.mit.edu/2020/underwater-gps-navigation-1102 (accessed Nov. 3, 2020)

[727] Andy Sharpless, "How to Feed the World & Save the Oceans," *TEDxSF*, April 25, 2012. https://www.youtube.com/watch?v=tBolHk7_fxg

[728] Rögnvaldur Hannesson, *The Privatization of the Oceans*, The MIT Press, Cambridge, Massachusetts, 2006.

- Individuals (including corporations) with investments in continental-shelf land would have an incentive to argue against rich quasi-aristocracy onshore property owners for uses on shelf property.

- Production of food, fiber, and chemicals will usually be more profitable than synthesizing biofuel. As such, generating biofuel will always be the production of last resort. The fuels produced will probably be methane from either bacterial decomposition or synthesized from electrolyzed hydrogen combined with carbon, petroleum-like fluids from cultured algae, or ammonia again from electrolyzed hydrogen. Any of these routes to biofuel will be carbon neutral in that any carbon dioxide generated when they are burned as fuel will be from carbon dioxide that was drawn down from the atmosphere by the seaweeds and nuclear power or some form of oceanic power.

- For the deadly fin of overfishing and bad fishing, ownership of shelf land would have many benefits. It would immediately end bottom trawling in those areas because who would want every living thing on their sea bottom crushed. Meanwhile, the areas near plant and animal culturing would have protected areas for juvenile wild fish to hide from predators. Those culturing areas would have extra food from a relatively small percentage of plant material drifting away from the seaweed culturing areas. They would have spillage of food from fish cages. Production of farmed fish would be cheaper than fishing of wild species in distant waters, so the species ecology in unfarmed areas could return closer to the natural state.

- A major industry of sea-plant culturing would draw down carbon dioxide from atmosphere in the foods produced and in plant material settling on the shelf floor. Moreover, culturing of sea plants would be required to use the biological wastes/nutrients (especially nitrates and phosphates) generated by culturing of fin fish and shellfish. Plant and animal operations would of necessity be kept in balance for efficient operation. Thus, the deadly fins of global warming and increased stratification would be mitigated.

- For the deadly fin of increased carbon dioxide in the atmosphere contributing to global warming, waste shells from shellfish production would be another major calcium carbonate "sink" for some of that carbon dioxide in the air, drawing it down into those shells. Moreover, like an artificial tidal marsh or mangrove forest, an expanse of sea-plant production could take in much of the fertilizer runoff from land and protect sensitive ecosystems farther out to sea.

- For the deadly fin of trash and pollution, there is nothing like the power of aggrieved farmers, farm cooperatives, and their customers to enforce cleanup regulations.

- For the deadly fins of coastal development vulnerability to waves from hurricanes, tsunamis, and other causes, again, the collective assortment of fish cages, seaweed growing arrays, and windmill platforms would act as a coastal buffer providing some of the shoreline protection that was provided by the lost wetlands and mangrove swamps. Despite the history of floods and probable greater floods to come, people keep rebuilding.[729] Of course, the primary defense against waves is DO NOT BUILD YOUR HOUSES ON THE FLOOD PLAINS NEAR THE SHORE! Storm surges and tsunamis rapidly lose their power just a short distance from the water's edge.

Back at government operations and regulations, they just will not go away. Regulations will grow, but there will be revenues to generate smarter government regulation. Moreover, there will be a larger voter/political contributor base to assure government interest.

- Sea urchins are the most dangerous enemy of healthy kelp, and sea stars are the enemy of corals. Smarter regulation would put bounties on such creatures, not protections.

- Seawater nutrient levels should not be too little (weak sea-plant growth) or too much (possible poisonous algae blooms such as red tide). There could be major litigations asking for reduced emissions of nutrients from an operation or payment for releasing more nutrients.

- Taxpayers, state legislators, and congress people, stop subsidizing construction and periodic reconstruction in flood zones.

Perhaps the most important thing to remember about the Greenwater Revolution is that it is a fundamental change, just like the farming revolution. The farming revolution required centuries to evolve into a major part of human society,[730] and even at the faster pace of the Twenty-First Century, the Greenwater Revolution will require decades to happen. That is why we dare not expect it to stop global warming by itself. It must be one partial mitigation along with the other partial cures of revolutionary change in land agriculture and energy. Moreover, it will need more than 12 years to be felt significantly.

Major Things for Governments and Businesses to Do

- Government: Stop subsidizing construction in flood zones with flood insurance.

---

[729] Rona Kobell, "For Vulnerable Barrier Islands, A Rush to Rebuild on U.S. Coast," *Yale Environment 360*, Jan. 15, 2015. https://e360.yale.edu/features/for_vulnerable_barrier_islands_a_rush_to_rebuild_on_us_coast (accessed Jan. 4, 2020.

[730] This process is described in *An Edible History of Humanity* by Tom Standage, (originally published by Walker Publishing Co., New York) Bloomsbury USA, May 3, 2010.

- Government: Develop regulations for selling off, but then still monitoring, "land" on the continental shelves.

- Government: Require that significant amounts of the revenues derived from such sales be applied to aquacultural extension services analogous to those on land.

- Government: Develop regulations for balancing nutrients released (usually by animal aquaculture) with maintenance of areas containing plants and filter feeders.

- Government: Negotiate more stringent treaties with limitations on trash dumping at sea, bilge pumping at sea, and abandoning ghost nets at sea without paying a recovery fee.

- Busines: Develop geographic information systems and portable map applications for continental shelf locations analogous to the map applications in cell phones.

- Business: Assemble operations packages to support mariculture operations in an area by providing some combination of factory ship for processing of seafood produce, sanitary processing services (no human wastes in food producing areas), electrical support, maintenance support, and security.

What You Can Do

- Join the Seasteading Institute (https://www.seasteading.org/).

- Support ocean clean-up initiatives such as limited trash dumping or bilge pumping at sea. but not silliness such as forbidding plastic straws on land.

9.3 Marine Farming and Infrastructure for the New Atlantis

We must plant the sea and herd its animals using the sea as farmers instead of hunters. That is what civilization is all about – farming replacing hunting."

— Jacques Yves Cousteau

Aquaculture may be the future of seafood, but its past is ancient.

Kiona Smith[731]

Piles of ancient shells, fishbones, and fishing implements testify to fishing tens of thousands of years ago in inland fresh water and extended to the ocean. Fish farming or aquaculture is more recent, but even it started thousands of years ago. Culturing of carp in Chinese

---

[731] Kiona N. Smith, Aquaculture may be the future of seafood, but its past is ancient," *ARS Technica, Sept. 17, 2019.*
https://web.archive.org/web/20190917164315/https://arstechnica.com/science/2019/09/fish-farming-may-be-much-older-than-we-thought/ *(accessed Sept. 17, 2019)*

rice paddies started 8,000 years ago.[732]. Australian Aborigines dug fish ponds and channels 6,600 years ago,[733] Those practices preceded the pyramids of Egypt and the empire of Rome—and both of those cultures also practiced aquaculture.

However, land farming improved much faster than aquaculture. Worse, beyond the shallowest of waters, there was never anything but the hunter-gatherer technologies of hooks and nets—capture fishing. Worse, government controls were often nonexistent, ineffective, or a source of subsidies for overly large fishing fleets that crashed fishing stocks.[734] Only better fishing boats and progressively bigger engines of powered fishing fleets increased fishing production by going farther from home ports and operations farther down toward the bottom have kept wild fishery production increasing slightly since the 1990s. However, that has been at the cost of more fish populations being overfished and more population crashes such as the collapse of the cod stocks on the Grand Banks east of North America.

Meanwhile, aquaculture production has increased because aquaculture has returned as a major factor in food production. It has already surpassed wild-fish capture, and is expected to provide most of the expected increases in marine food production. (The distinction between freshwater, oceanwater, and artificial production environments of both in land operations has blurred distinctions among the three areas. For instance, hatcheries for both freshwater and seawater are usually on land.)

In the late 1900s, the traditional nets, fence operations spread into deeper waters with powered small boats and more rot-resistant materials such as plastic and synthetic fibers. Freshwater aquaculture increased, but the oceans are a much larger available production resource, still largely unused.

Agriculture on land usually only requires fences, farm roads, and sometimes irrigation for water. Ocean agriculture has unlimited water, but the infrastructure requirements are much more. There are three major materials costs.

- Keeping rafts, cages, culturing nets, and larger structures floating off the bottom to be get maximum light near the surface and avoid various predators and pests.
- Conversely, anchoring all this floating infrastructure so currents and waves do not drag the infrastructure away. The agro operators must either cable down to heavy anchors, or they must cable to attachments drilled into rocky sea bottom.

---

[732] T. Nakajima, M. J. Hudson, I. Uchiyama, et al., "Common Carp Aquaculture in Neolithic China Dates Back 8,000 years." *Nature Ecology & Evolution*, vol. 3, pp. 1415–1418, 2019. https://doi.org/10.1038/s41559-019-0974-3 (accessed Aug. 15,2020)

[733] Matt Neal, "Ancient Indigenous Aquaculture Site Budj Bim Added to UNESCO World Heritage List," ABC Net.AU, July 6, 2019. https://www.abc.net.au/news/2019-07-06/indigenous-site-joins-pyramids-stonehenge-world-heritage-list/11271804 (accessed Aug. 15, 2020)

[734] Richard Wellings, editor, *Sea Change: How Markets and Property Rights Could Transform the Fishing Industry*, The Institute of Economic Affairs, London, England, 2017.

- Connecting the floats and cables with ropes or cables.

Marine cages, with remote maintenance and operation, are one of the greatest sets of multiple inventions of the late 1900s and early 2000s.

- They prevent animals from swimming away and finding new homes far away from their intended destination in the super market.
- Food placed in cages is less likely to be taken by wild species that might (again) swim off to new locations far from the mariculturing operations.
- Cages protect the cultured species from becoming dinner for passing sharks and other marauders.
- The caged species can be watched for other threats such as diseases and parasites.
- Caged species can be harvested at optimal sizes and market timing rather than what showed up in the nets in a particular fishing run.

There have been several decades of growing size and complexity of the freshwater and oceanic cages.[735,736]

Another aquacultural concern—and a major advance being applied—is a build-up of excess nutrients or depletion of certain nutrients around a production area. Areas around fish cages accumulate organic material from food the animals miss and their excretions, which include organic material and compounds with phosphorus and nitrogen. This could encourage unhealthy conditions around the operation, such as the plankton species that cause the poisonous red tide or simply depleted oxygen as the organic materials decompose (which is also poisonous to the fish). These stressful conditions make fish prone to disease, and aquaculturists have often responded with heavy doses of antibiotics and pesticides—another environmental issue.[737] Conversely, plankton and seaweeds absorb and use nitrogen and phosphates from the water for their growth while depletion of these nutrient levels slows plant growth.

Traditional monoculture of one species tends to deal with these problems by spreading out in the ocean. They use "dilution to avoid pollution." However, this requires a great deal of space, and even the ocean has limits. Moreover, sprawling production thinly over large areas increases

---

[735] Ata Burak Cakaloz, "Fish Cage Construction," *FAO Regional Training on the Principles of Cage Culture in Reservoirs*, June 22–24, 2011, United Nations Food and Agriculture Administration, 2011. http://www.fao.org/fileadmin/templates/SEC/docs/Fishery/Fisheries_Events_2012/Principles_of_cage_culture_in_reservoirs/Fish_Cage_Construction.pdf (accessed Aug. 14, 2020)

[736] William See, Cage Culture - An Aquaculture Documentary" PT. Stargold Internusa Jaya, Jawa Barat, Indonesia, April 19, 2017. https://www.youtube.com/watch?v=YBQSFphHD34 (accessed Aug. 15, 2020)

[737] Joel K. Bourne, Jr., "How to Farm a Better Fish," *National Geographic Magazine*, May 29, 2014. https://www.nationalgeographic.com/foodfeatures/aquaculture/ (accessed Aug. 16, 2020)

the time and transportation costs of these operations that are already more expensive than agriculture on land.

Might these two contradictory processes be linked so that they work better together? Yes, oceanic coral reefs already combine plant and animal processes to multiply productivity in nutrient-scarce tropical waters. The corals are a number of filter-feeding animals that collect small bits of animal matter from the water. They also collect tiny algae (dinoflagellates) that remain inside the corals using coral waste products as fertilizer to photosynthesize extra food mass and free oxygen—both of which help feed the coral. Meanwhile, the algae get a bone-protected home. This mutually beneficial relationship is called symbiosis, and the giant clams of the coral reefs (*Tridacna gigas*) also host algae. [738]

The symbiotic model can be applied to aquacultural operations. The different levels of the food chain (trophic levels) into one area so each trophic level can support the other levels. A simple example is the oft-used polyculture in Chinese farming. The wastes from pigs fertilize algae in a nearby pond, the algae feed ducks, and the muck at the bottom of the pond is periodically dredged up and spread over the fields to fertilize the land crops.

A more impressive recent term is integrated multi-trophic aquaculture (IMTA). Different species at different levels in the food chain are raised together make a healthier and more productive environment. This creates a simpler version of the natural marine environment. in which: [739,740]

- Finfish and crustaceans get natural food and feed supplied. They are often called fed species. They, in turn, miss some of the food, and they excrete materials with phosphorus and nitrogen compounds.

- Filter feeders, such as clams, corals, mussels, and oysters, catch many of the tiny missed food particles to grow their own flesh and their shelves. They are often called extractive species. They too excrete compounds with nitrogen and phosphorus.

- Bacteria, worms, and sea cucumbers (an animal, not a vegetable) on the sea floor are another set of extractive species. They would probably not be kept in a controlled environment, but they build body mass and feed feeding other sea creatures, some of which can be taken as wild catch. Sea cucumbers (which look a bit like their namesake) are one of the most noticeable of such animals.

---

[738] Nicolette Perry, *Symbiosis: Close Encounters of the Natural Kind*, Chapter1, "The Coral Reef," Blandford Press, Dorset, United Kingdom, 1983.

[739] Thierry Chopin, Max Troell, Gregor K. Reid, et al., "Integrated Multi-Trophic Aquaculture," *Global Aquaculture Advocate*, pp. 38–39, Sept./Oct. 2010. https://www.researchgate.net/publication/284701399_Integrated_Multi-Trophic_Aquaculture (accessed Aug. 16, 2020)

[740] Amir Neori, "Can sustainable mariculture match agriculture's output? Global Aquaculture Alliance web page, Aug. 18, 2015. https://www.aquaculturealliance.org/advocate/can-sustainable-mariculture-match-agricultures-output/ (accessed Sept. 11, 2020)

- Ocean plants are the most important part of the food chain, the extractive species. They remove the phosphorus and nitrogen compounds and use them to grow more food. Reduction of those nutrients allows the maricultural operators to keep more fed and extractive species in a given area. They can also harvest seaweed directly for processing into food for the fed and extractive species, for human food, for chemicals, for and other uses. For the GND, ocean plants also provide a local drawdown of carbon dioxide from the atmosphere, thus contributing to the world total.

Multi-trophic Feed production from seaweeds deserves description in greater detail. Conventional fish farming operations of predator fish (such as salmon) use feed made from cheaper wild-catch fish, chiefly forage fish like sardines and anchovies. This is not sustainable or long-term affordable because the world is approaching peak wild catch with an ongoing series of fishery crashes, such as the collapse of the cod fishery, the collapse of the Monterey, California fishery, and several anchovy collapses of the Peruvian fishery. One proposed replacement is protein from land vegetable sources such as soy beans, but land protein sources lack the omega-3 fatty acid that is considered to be important for health and important to have a seafood taste.

Seaweed has omega-3 so it can be mixed with the land vegetable proteins to provide protein while still maintaining sufficient omega 3. Also, ocean plant materials can be fed to any number of worms (such as angle worms), insect larvae (such as for soldier ants), or insects (such as crickets), as was discussed in Subsection 7.9 Minilivestock: Chickens Down to Bugs, McBugs, McCrawleys, & McWigglies—Grazers of Small Prairies and Small Ponds. These minilivestock increase the protein to meat level (without the costs of buying and transporting land proteins), and, with one more trophic level up, the percentage of omega-3 is increased.

People can remove a prudent fraction of all these connected parts of the multi-trophic ecology for food and for other products. Meanwhile, the ecology draws down carbon from the atmosphere. Much of carbon is stored in the calcium carbonate ($CaCO_3$) of seashells.

9.3.1 It's Not Landfill Material, It's Floats for Ocean Farms

The deeper the water of agro–industrial process, the more material that is required. The materials for floats and anchors can become a major expense for oceanic aquaculture. However, they can also provide use for otherwise waste stream and (at the same time) a sink for materials that might otherwise be burned and contribute to emissions of global-warming carbon dioxide.

Remember the deadly fin of ocean pollution in 9.1.5?

The Brand-and-Benford proposal (mentioned earlier) to gather plant matter and drop it into the deep ocean for long-term carbon storage[741] had its problems—most importantly, it was cost for no return whatsoever. Conversely, American trash has more than 30 percent wood and paper

---

[741] Stuart E. Brand and Gregory Benford, "Ocean Sequestration of Crop Residue Carbon: Recycling Fossil Fuel Carbon Back to Deep Sediments," Environmental Science and Technology, vol. 43, no. 4, pp. 1000–1007, Jan. 12, 2009.

products, as well as 13% plastic.[742] Obtaining trash is not expensive. In fact, cities pay to have these materials disposed of.

The wood and paper float as long as they are sealed away from water. The plastic can provide sealing for the floats made from recycled wood and paper. Even better, the plastic itself floats. In fact, one environmental-activist group actually built the *Plastiki*, a 60-foot (18-meter) catamaran for which the buoyancy came using 12,500 recycled plastic bottles, ... and they sailed it from North America to Australia.[743]

We can apply the lesson from *Plastiki*, and go beyond *Plastiki*. The developed world has reached the level of prosperity in which trash is neatly packed into landfills—it's messy, but there is volume enough for thousands of years of trash. The poorer countries of the developing world have not all reached that level of prosperity to pack their trash to neat landfills, so their rivers often contain large amounts of floating trash ... just as the rivers of the now-developed world did decades ago. Worse, those nearly indestructible plastic items often float for years or even decades, often with damaging results to sea animals that eat these materials. However, the model of *Plastiki* can encourage these countries to gather wood, fiber, and other buoyant materials for floats and to gather plastic that can provide the sealing for those floats. The quest for floats might even pay for some of the cost for hunting floating trash in the open ocean. At the opposite end of the density spectrum, structural demolition waste could be part of the anchoring for ocean farm infrastructure.

Only in the worst areas of trash flow, might this be profitable. Still, gathering the trash in other areas would be LESS UN-profitable as those economies evolve to the point where they no longer emit massive flows of plastic wastes (with their unknown consequences) into the ocean.

Meanwhile, millions of tons of maricultural infrastructure will float up near the surface (in large objects so that the plastics will not be ingested by seal animals) where it can allow hundreds of millions of tons of carbon dioxide to NOT go up into the atmosphere.[744]

9.3.2 The First Stone of New Atlantis--Floating Concrete

Seaborne floating islands have been found in literature since Homer sang about the island of Aeolia in the *Odyssey* from the 700s B.C.[745] The evolution of reinforced concrete structures

---

[742] *National Overview: Facts and Figures on Materials, Wastes and Recycling*, web page, United States Environmental Protection Agency, undated. https://www.epa.gov/facts-and-figures-about-materials-waste-and-recycling/national-overview-facts-and-figures-materials (accessed Feb. 3, 2020)

[743] David de Rothschild, *Plastiki Across the Pacific on Plastic: An Adventure to Save Our Oceans*, Chronicle Books; March 16, 2011.

[744] Better methods of plastics recycling are being developed, particularly multiple-solvent systems for separating the multiple types of plastic in consumer products. Three of these methods being developed are discussed in: Theodore W. Walker, Nathan Frelka, Zhizhang Shen, et al., "Recycling of Multilayer Plastic Packaging Materials by Solvent-Targeted Recovery and Precipitation," *Science Advances*, vol. 6, number 47, Nov. 20, 2020. https://advances.sciencemag.org/content/6/47/eaba7599 (accessed Nov. 20, 2020)

[745] Homer (original author) and Robert Fagles (translator), *The Odyssey*, Penguin Classics; Reprint Edition, (originally released in the 700s B.C.) November 29, 1999.

has allowed such platforms to develop for various projects. One more generation of these structures should be sufficient to support major global-warming-mitigation projects that support themselves financially while helping mitigate global warming.

Yes, there is our old friend concrete again. While concrete production adds greenhouse-warming carbon dioxide to the atmosphere, concrete suitably modified for offshore structures can facilitate ventures to help cool the planet ... and make profits to pay for the venture along the way.

One concrete modification is borrowed from the natural rock type often thrown skyward from volcanoes. This rock is called pumice. It is filled with air pockets, so much so that it is lighter than water. After eruptions, pumice often floats in giant fields of floating rocks tens of miles across, and these rocks float for years until they eventually wash ashore when currents carry them to land.[746,747]

Pumice was the inspiration that led to floating concrete docks and ocean platforms with the following features that keep it buoyant.

- Light aggregate stones (such as pumice in particular) to replace ordinary stones or small pebbles in concrete because pumice is already lighter than water.

- Aluminum bronze strengthening rebar is much better than steel because of its high corrosion resistance. (At a slight additional weight penalty adding silicon carbide particles into the aluminum matrix makes a composite rebar that is much stronger and harder than aluminum bronze[748]—this is another example illustrating both revolutions of composites and nanotechnology.)

- Glass fiber-reinforced polymer rebar is totally immune to chloride attack, is lighter than aluminum bronze, and provides reinforcement strength roughly equivalent to steel rebar.[749]

- A structure of concrete chambers that can be filled with light-weight plastic foam can make a structure considerably more buoyant.

- For quick repairs, cement made with extra baked aluminum silicate clay can make a marine concrete (often called hydraulic concrete) that can set under water.

---

[746] Simon Refern, "Underwater Volcano Creates Huge Floating Islands of Rock, Disrupts Shipping," *Phys.Org.com*, April 25, 2014. https://phys.org/news/2014-04-underwater-volcano-huge-islands-disrupts.html (accessed Sept. 16, 2018)

[747] "Volcano's Legacy Still Washing up on Beaches," *PhysOrg.com* provided by the University of Queensland, Dec. 13, 2013. https://phys.org/news/2013-12-volcano-legacy-beaches.html (accessed July 1, 2020)

[748] Md. Habibur Rahman and H. M. Mamum Al Rashed, "Characterization of Silicon Carbide Reinforced Aluminum Matrix Composites," *Procedia Engineering*, vol. 90, pp. 103–109, 2014.

[749] Shahad Abdul Adheem Jabbar and Saad B. H. Farid, "Replacement of Steel Rebars by GFRP Rebars in the Concrete Structures," *Karbala International Journal of Modern Science*, vol. 4, pp. 216–227, 2018.

The advantages of such floating structures are that they can float indefinitely without constant monitoring and active bilge pumps to remove excess water leaking in. Besides the cost reduction of totally passive maintenance of buoyancy, this vastly increases the safety and durability of oceanic structures. The term "unsinkable" applied to ships like the Titanic was always a lie; these craft were only unsinkable as long as the bilge pumps and the various compartment bulkheads held. When they failed, the ship went down.

In contrast, floating-concrete ships and platforms may drift from their moorings and need to be retrieved, but they will never sink. They may pile up on the rocks if not rescued by tugs, but they will never sink. This provides a significant safety margin for maintaining people and infrastructure in deep water. Moreover, floating concrete structures can hold major loads. These features of high safety and high load capacity make possible a number of applications in deep water to be discussed later.

Floating concrete and steel rebar (also called ferroconcrete) structures have a long history. River barges and oceangoing vessels have sailed since the mid-1800s. They were most often built in wartimes when steel was in short supply. The United States built a number of concrete ships in both world wars of the 1900s.[750]

However, concrete ships required more labor for construction than steel. In operation, they were less profitable to run than steel ships for two reasons.

1. They were heavier than comparable steel ships, which increased fuel costs.
2. They had thicker hulls than steel ships. That translated into less profit-making cargo space or a wider ship that increased fuel costs, this time because of increased drag.

The high operating costs led most operators of concrete ships to sell them off to other owners who used them as floating docks and grounded or floating breakwaters. Some of them are still serving in those roles today.

The greatest use of floating, and sunken, concrete was in World War II. On June 4 of that year, Allied forces invaded German-occupied France. The Germans had their heaviest defensive forces around the deepwater port cities along the French coast because it would be insane to attempt an invasion without having a deepwater port.

The British had an insane response—they brought two prefabricated harbors to the invasion site. Allied forces landed on the Normandy coast 107 kilometers (66 miles) from the nearest major port of Cherbourg. While advance units attacked toward Cherbourg, the British built two long docks called mulberries with pontoons holding up a roadway to shore roughly 6 miles (10 km) of flexible steel roadways that floated on steel or concrete pontoons. The roadways were sheltered from the sea by lines of massive sunken caissons, scuttled ships, and floating breakwaters.

This great artificial harbor was assembled in 12 days. A storm destroyed one Mulberry, but the other held up for 10 months until enough conventional port capacity became available to

---

[750] A summary of the United States' concrete ships histories can be viewed at the Concrete Ships website (http://concreteships.org/ [accessed Nov. 3, 2020]).

cease its operation. This surviving Mulberry carried more than 2.5 million troops, 500,000 vehicles, and 4 million metric tons (4.4 million tons) of supplies.[751]

By the 1980s, very large concrete structures had already been used for permanently floating bridges and barges, and they were also used for temporarily floating structures that would be sunk and anchored in place such as drilling and production platforms, storage tanks, and lighthouses. There were floating cranes and floating breakwaters for marinas and design proposals for floating power plants (including ocean thermal energy conversion).[752]

Large floating structures have continued their evolution into the 2000s with well established guidelines for their use. As noted in 9.4.4, Dynamic Positioning of Floating Platforms, oil and gas platforms (with or without cabling to the sea floor) have grown into factory villages in the deep sea. There are floating offshore wind turbines,[753] large spar floating platforms,[754] proposed ocean thermal energy conversion,[755] and the wildly speculative but increasingly practical designs for ocean cities.[756] Wang Chien Ming, one of the co-editors of *Large Floating Structures: Technological Advances*, detailed a number of these concepts in a 2017 talk he gave for The Seasteading Institute (a group advocating ocean settlement).[757]

9.3.3 The Second Stone of New Atlantis—Regenerating Corals and Other Limestone-Accreting Species

Many ocean life forms use light from the sun and carbon dioxide from the air to make their food. Some, such as coral, also produce a byproduct of calcium carbonate ($CaCO_3$) stone, which they use to build a structure to protect themselves and build up toward the better sunlight near the

---

[751] Colin Flint, *Geopolitical Constructs: The Mulberry Harbours, World War Two, and the Making of a Militarized Transatlantic*, Rowman & Littlefield Publishers, Lanham, Maryland, Sept. 19, 2016.

[752] Floating Concrete Structures: Examples from Practice, Second Printing, VSL International Design Limited, Berne, Switzerland, July 1992. https://www.structuraltechnologies.com/wp-content/uploads/2018/02/PT_Floating_Concrete_Structures.pdf (accessed July 26, 2020)

[753] S. Useda, "Floating Offshore Wind Turbine, Nagasaki, Japan," *Large Floating Structures: Technological Advances*, C. M. Wang and B. T. Wang (editors), Springer, Singapore, pp. 129–155, 2015.

[754] J. Halkyard, "Large Spar Drilling and Production Platforms for Deep Water Oil and Gas," *Large Floating Structures: Technological Advances*, C. M. Wang and B. T. Wang (editors), Springer, Singapore, pp, 221–260, 2015.

[755] A. A. Yee, "OTEC Platform," *Large Floating Structures: Technological Advances*, C. M. Wang and B. T. Wang (editors), Springer, Singapore, pp. 261–280, 2015.

[756] Vincent Callebaut, "Lilypad: Floating Ecopolis for Climatical Refugees," C. M. Wang and B. T. Wang (editors), *Large Floating Structures: Technological Advances*, Springer, Singapore, pp, 261–280, 2015.

[757] Wang Chien Ming [identified as C. M. Wang in other documents], "Very Large Floating Structure Technology for Space Creation on the Sea," *International Conference on Floating Islands*, The Seasteading Institute, July 11, 2017. https://www.youtube.com/watch?v=x0WniV16evk (accessed July 26, 2020)

ocean surface. This is the age-old process that allows coral islands and reefs to hold their place just below the ocean surface even as the islands below them are slowly sinking.

Some of the most important carbon-fixing shallow-water communities are the coral reefs and sea grasses now endangered by pollution, coral and sand mining, overfishing of coral-protecting predators, global warming, and damage from boating. Government agencies and private groups have worked to protect the coral reefs by:

- Declaring large areas sanctuaries where there can be no development such as commercial fishing, dumping of dredging spoil, and mining of coral). These policies have helped. However, they do not protect these areas from nearby pollution from elsewhere, overfishing outside the sanctuaries, poaching inside the sanctuaries, and heat stress that might increase due to global warming.

- Culturing and replanting new corals to restock areas that have been killed.[758] However, the same conditions that killed corals in the dead zones will probably attack newly emplaced corals.

- Breeding corals and other carbon-fixing species that are more resistant to warmer waters and more polluted waters.[759] However, such breeding programs may require decades of time, and the carbon-fixing life forms may not have that much time before major extinctions occur.

- Emplacing artificial reefs of old cars, old tires, old ships, or (there it is again) concrete. The first three materials for artificial reefs have often caused environmental disasters when they collapsed and/or leaked poisonous materials into the ocean. Concrete has been more successful, especially with reef balls that are several feet (a meter) or more in diameter and have openings to a hollow interior where young fish can hide from larger predators. Also, concrete is somewhat inviting to colonizing sea life because it is essentially artificial limestone. However, the weight of concrete makes it expensive to transport and emplace at sea.

There is one more technique that has been used in small areas, electrical deposition (electrodeposition, also called electro-mineral accretion) of limestone. This process was developed from the cathodic protection of metals on ship hulls and oil platforms. A weak low-

---

[758] Sarah Emerson, "This Underwater Robot Will 'Squirt' Coral Larvae onto the Great Barrier Reef to Save It," *Motherboard*, Nov. 2, 2018. https://motherboard.vice.com/en_us/article/vba4nj/worm-brain-ai-park-car?utm_source=MIT+Technology+Review&utm_campaign=3a3dcbeceb-EMAIL_CAMPAIGN_2018_11_05_11_55&utm_medium=email&utm_term=0_997ed6f472-3a3dcbeceb-153929877 (accessed Nov. 5, 2018)

[759] National Academies of Sciences, Engineering, and Medicine, *A Research Review of Interventions to Increase the Persistence and Resilience of Coral Reefs*, The National Academies Press, Washington, D.C., 2019. https://www.nap.edu/login.php?record_id=25279 (accessed Nov. 26, 2020)

voltage electric current protects the negatively charged cathode from rust (hence, cathodic protection), and rusty areas are reduced to a non-rusty condition. Meanwhile, a corresponding piece of sacrificial metal (such as a block of iron) is the anode, and it corrodes much faster than it normally would.

In the 1970s, an architect named Wolf Hilbertz noted that such cathodic protection in the ocean also causes calcium carbonate ($CaCO_3$) to precipitate out of seawater and coat the negatively charged surfaces. Hilbertz and a number of other researchers developed a number of innovative restoration schemes built around the electrodeposition, and many of the researchers involved have participated in reef restoration projects as part of Biorock Technology Inc.[760]

- The current produces a hard limestone ($CaCO_3$) mineral variation called aragonite. The layer of aragonite increases the strength of the iron or steel new reef material.

- The current levels required to accrete aragonite are quite low—around a hundred watts, which is the amount of a single light bulb. Consequently, the needed electricity can be affordably supplied from solar panels or cables from shore, and the electrical levels are not dangerous to people or fish.

- There is no temptation to use stronger (more dangerous) levels of electrical current because such currents produce mineral form of $CaCO_3$ called brucite that is much softer than aragonite and easily flakes off, which makes it useless as a reef-building material.

- The steel frame can extend in much thinner structures per unit weight than concrete. This helps in three ways. First, the lighter materials can be more easily transported and assembled at the underwater site. Second, the thinner shapes can provide a larger surface area for culturing sea life. Third, the lighter thinner material can provide more complex interior spaces in which young fish can hide from predators, and provide more open areas through which water can flow. This last feature assures more nutrient flow through the reef and allows the reef to slow wave flow toward beaches and protect them from erosion with less risk to the reef than would come from trying to totally stop wave flow.

- Once the metal surfaces have an aragonite surface, this coral-like material supports growth of new corals and other sea life. Lack of such a sea life-welcoming surface has been a major failing of other artificial reefs made of sunken ships, sunken cars, sunken tires, and plastic.

- Biorock Technology claims that electrical environment at that surface and the more alkaline environment near it encourage significantly faster and more stress-resistant

---

[760] Thomas J. Goreau and Robert Kent Trench(editors), *Innovative Methods of Marine Ecosystem Restoration*, CRC Press, Taylor & Francis Group, Boco Raton, Florida, 2013. Thomas J. Goreau, *Biorock Benefits*, Biorock Technology Inc., Cambridge, Massachusetts, July 7, 2014. http://www.globalcoral.org/wp-content/uploads/2013/11/BIOROCK_TECHNOLOGY_july1114.pdf (accessed March 14, 2020)

sea life growth (particularly corals). The theorized reason is that it may allow such organisms to direct energy normally used in carbon fixing to faster growth and greater resistance to environmental stresses such as muddy waters, other kinds of pollution, and overly warm water temperatures.

- A final benefit is that an electrodeposition reef can simply re-accrete the aragonite coating on the steel if the structure is damaged, say by a boat crash or severe storm activity.

- The negative feature is that the steel will begin corroding if the cathodic protection of the electrical current ceases. (An aluminum structure would be much more corrosion-resistant than iron, but it is more expensive than iron might tempt recycling thieves.)

There are many shoreline people happy with their electrodeposition reefs. Unfortunately, most of the studies regarding such projects have been done by the electrodeposition researchers and entrepreneurs. There has been very little independent research comparing electrodeposition techniques with other reef-restoration techniques.[761]

9.3.4 The Seagoing Line Shack/Winnebago—We All Live in a Yellow Submarine

In 1957, a firm in Winnebago Cunty Iowa began manufacturing essentially trucks and trailers that were also motor homes used for camping without power, gas, or sewage lines (although they could attach to such lines).[762] The company took the iconic name of Winnebago. They and their competitors, developed mass production of these efficient-living-space recreational vehicles or RVs. Winnebago developed interior paneling that also served as insulation. All the manufacturers of large RVs incorporated standardized showers, foldout beds, sewage waste storage, and electrical power generators. Of course, these features became much more comfortable when the RVs parked in recreational camp grounds with electricity, potable water, and sewer lines.

The Winnebago model will be well used by mariculture systems at any serious distance from shore. Consider that small motor boats taking workers to mariculture sites generally move at speeds of 30 miles per hour (48 kilometers per hour) or considerably less for fuel economy. Hence, transporting workers one hour out and one hour back would probably be the cost limit for commuting workers, and this would be a major limitation on the extent of the possible maricultural zone.

Ranching, before the arrival gasoline-powered trucks, faced a similar situation. Cowhands often worked several days ride away from the ranch house, so nightly returns were impractical.

---

[761] Steve Baragona, "This Coral Restoration Technique Is 'Electrifying' a Balinese Village," *Smithsonian Magazine*, May 25, 2016. https://www.smithsonianmag.com/science-nature/coral-restoration-technique-electrifying-balinese-village-180959206/ (accessed March 13, 2020)

[762] Colin Ryan, "Legends - The History of Winnebago," *Motor Trend*, May 28, 2016. http://www.trucktrend.com/features/1605-legends-the-history-of-winnebago/ (accessed March 23, 2020)
Winnebago comes from a name for one of the first-nation tribes in the Midwest.

Instead, there would be a string of very simple shacks with rough bunks, a privy, a small fireplace, a covered open area on the side for horses to get out of the rain, and a lantern. These line shacks supplied places to sleep overnight within a couple hours ride from the work areas. The line shacks were especially useful during weather that was wet and/or cold.

The paradigm of the line shacks can be applied to maricultural operations too distant for daily commutes. However, the seagoing line shacks would need more than, bunks, a kitchen, and a privy for simple living during the work week. Some of the additional functions would be:

- Storage shed for storage of tools, as well as extra hoses, cables, floats, and netting to make minor repairs as necessary.

- Fence-line corners for connecting the cables for deepwater rows of mariculture floats, rafts, netting, and cages.

- Compressed-air source for adjusting levels of the floats so that the nearby mariculture infrastructure could be lowered below the surface area of severe wave action.

- Storage and wash-off area for scuba gear and cleaning of fish and vegetable matter for eating rather than transporting all food from shore.

- Storage of outboard motors and any instruments or other electronic gear used on the boats so that the boats could be stored under water.

- Wi-Fi connection via radio to nearby cell-phone towers to provide real-time operations data to the overall monitoring agency. (The towers would be another infrastructure expense but also another point for connecting cables.)

- Storage, maintenance of one or two aerial drones for overall monitoring and transportation of any extra needed materials to the work boats.

- Storage, maintenance of several underwater drones for checking and maintaining any deeper cages for animals and nets for culturing sea vegetables.

- Sewage treatment would probably be provided by a collection service visiting periodically. However, more distant and/or less used sites might have more exotic systems to cook or biologically process wastes to nontoxic materials that could be transported away. These two options are crucial for large areas of aquacultural operations producing safe food products on a large scale.

- Likewise, power could either be from a small generator on board or from external electrical cables connected to windmills or other types of power generation on the nearby continental shelf area.

- As noted in the next subsection, power generation for the seagoing line shacks and for the work boats would be provided burning compressed ammonia in an ammonia economy.

The seagoing line shacks would be much like small barges for maximum capacity anchored in place, and that would be their key difference from fishing boats and dive boats that motor into port during high winds and heavy seas. The seagoing line shacks would stay securely anchored like large bobbers. This can be a wonderful thing with stars reflecting off a calm sea, and foldout upper walk areas for extra space above and shade for the lower decks.

But the sea has storms with high winds and heavy seas. If you cannot run from a storm, what do you do? You follow another technology of the Great Plains, the storm cellar. When there is danger of a tornado, people retreat to the storm cellar below ground.

Similarly, the seagoing line shacks would submerge along with their associated maricultural lines, floats, rafts, cages, and nets to avoid bad weather and bad seas. Therefore, the seagoing line shacks would need to include a core 1-atmosphere submersible dry quarters with hull and hatches capable of surviving three additional atmospheres (3 bar), which would be a depth of roughly 100 feet or 30 meters. That would allow a much milder range of motion than at the surface. Such a submersible capability would entail

- A floatation/submergence tank directly below the main structure so that the seagoing line shack could submerge.

- Conversely, a considerable structural weight below the floatation/submergence tank acting as ballast so that the structure could never overturn.

- Hatches and structural body of the dry volume capable of handling three atmospheres of pressure (43 pounds per square inch, 3 kilograms per square centimeter).

- A truss structure with a buoyant foam-filled concrete column upward so that every foot lower into the water would meet more buoyancy. At the seagoing line-shack's design submergence depth. Buoyancy would match ballast so the seagoing line shack could sink no farther. The effect would be like a thin self-righting bobber used by sport fishermen (and fisherwomen).

- Snorkel (from German *Schnorchel* meaning nose or snout) air intake running up the column to provide air for running the electric generator and compressor, as well as for breathing, when the seagoing line shack needed to be submerged. (In World War II, German diesel-electric submarines used snorkels so that they could suck down air for their diesel engines and for breathing while running submerged.)

Major Things for Governments and Businesses to Do

- Government: Create a blue-ribbon panel to assess the issue of deterioration of concrete buildings due to corroding support rebar and suggest remedies. The GND proposal of renovating all buildings to increase energy efficiency would definitely have to wait until we can be sure that large structures with concrete will not require rebuilding within a few years.

- Government: Mandate that some of the revenue derived from underwater land on the continental shelves be provided to the United States Geological Survey for generation of maps detailing these areas.

- Government: Institute a federal waste disposal fee based on weight and increasing for toxicity. Thus, even plastic, wood, and paper (even though nontoxic) would have a charge. However, those using such materials from a waste stream would not face such a charge. This would encourage the use of paper and plastics for oceanic floats and demolition waste for oceanic anchoring.

- Busines: Build Winnebago-type anchorable work shelters (line shacks) for working in maricultural areas on the continental shelves.

- Business: Develop the technology and take waste plastic and wood or paper to make floats for aquaculture.

- Business: Develop maricultural-industrial parks that would be associated with offshore wind or other offshore power facilities. The maricultural-industrial parks would have seafood production of their own. They might also provide basic services and recruit entrepreneurs to run smaller operations.

9.3.5 The Third Stone of New Atlantis—Carbonated Igneous Rock for Power, Food, and Tsunami Defense

The edges of the continental shelf are another area where the aims of the GND can be met in an entirely different manner. Instead of replacing fossil fuel combustion with its resulting carbon dioxide emissions and other climate-warming factors, other processes can draw down carbon dioxide and store it to compensate for fossil-fuel combustion elsewhere.

The edges of the continental shelves are really the edges of the continents. At that edge of the continental shelf, the terrain of the continental slope drops off rapidly (on average) toward the abyssal depths of the sea floor. This means that energy and mariculture operations on those edges have the impracticalities of greater depth (about 100 meters or 330 feet) and greater distance from shore. This is particularly so for the Atlantic continental shelves of North and South America, some areas of which extend for many tens of miles (kilometers) from the land.

However, the diseconomies have some countervailing advantages for resource potential, environmental benefits, and governmental advantages. These countervailing advantages spring from energy concerns and climate mitigation

1. Thermoelectric (heat-driven) power plants at the edges of the continental shelves can access Immense amounts of presently scarce cooling water. Cooling in such sites can achieve greater power-plant efficiency.

2. Distance from land has legal and political advantages. Sites at some distance from land are less likely to be under the legal jurisdiction of local (city, county, and state)

governments. Sites at some distance from land are less likely to attract opposition from local land owners.

3. Building infrastructure on sites near the edge of the continental shelf has financial advantages.

4. Carbonation of rock on the continental shelf can provide permanent transformation and storage of carbon dioxide on a scale of billions of tons. Indeed, fossil fuel power plants at these sites can provide some of that carbon dioxide for rock storage.

5. Functioning power plants at these locations can supply electricity and logistical support for additional mariculture operations

6. Building infrastructure at such sites can reduce the dangers from possible mega-tsunamis by mining volcanic rock from potentially unstable volcanoes and using that rock to build islands

This major development scenario is described in greater detail in the following subsections.

9.3.5.1 Efficiency Advantage on the Edge

For accessing cooling, the thermal advantage of power plants near the steep drop of the continental slope starts with the present need; there is an increasingly severe shortage of cooling for land-based power plants. Roughly half of the fresh water used in the United States today is used for cooling power plants.[763] Moreover, the cooling potential of the land-based freshwater decreases tremendously in the summer. In the hottest months—often with the greatest air-conditioning loads— utilities must sometimes shut down thermoelectric power plants for lack of sufficient cooling water and/or water that is sufficiently cool.

In contrast, the deep ocean has a tremendous thermal inertia from eons of ice ages. Although the surface waters are generally warmed to some degree, the water temperatures become progressively colder with increasing depth and decline almost to freezing temperatures in the depths. There are well developed technologies for accessing pumping sewage and waste chemicals from the shore out to these cold deep waters (which was not necessarily a good thing).

- Starting at least since Santa Monica, California in 1910, cities all over the world have emplaced pipes to drain sewage and chemical wastes out to sea with the hope that these materials would become diluted and thus less of a danger. These pipes, called outfalls, vary in size from half a foot to 26 feet (15 to 800 centimeters). Outfall pipe materials have included polyethylene, stainless steel, carbon steel,

---

[763] "Power Plant Cooling and Associated Impacts: The Need to Modernize U.S. Power Plants and Protect Our Water Resources and Aquatic Ecosystems," NRDC Issue Brief IB:14-04-C, National Resources Defense Council, New York, April 2014. https://www.nrdc.org/sites/default/files/power-plant-cooling-IB.pdf (accessed April 12, 2020)

- glass-reinforced plastic, reinforced concrete, and cast iron. The longest of these are several tens of kilometers long.[764],[765]

- In 1930, Georges Claude built a pipeline off Matanzas, Cuba to bring cold deep water to the surface and generate power using the difference between that water and the warm tropical surface waters--what is now called ocean thermal energy conversion (OTEC).[766],[767]

- Offshore drilling for oil and gas became a major part of the industry in the last half of the 1900s, and with it, undersea pipeline technology has evolved to transport the production to shore. Two of the longest underwater pipelines are the 1,224-kilometer (760--mile) 48-inch (122-centimeter) Nord Stream gas pipeline across the Baltic Sea from Russia to Germany and the 1,166-kilometer (724-mile) 42-inch (107 centimeter) Langeled gas pipeline from waters off Norway to the United Kingdom.[768]

- Undersea pipelines also transport water. The 106-kilometer (66-mile) 1.5-meter (59-inch) water pipeline from Turkey to Cyprus is a suspended high-density polyethylene pipeline, moored to the ocean floor at a depth of 280 meters (919 feet) below the sea surface.[769] (Roman aqueduct builders from two -thousand years ago would have appreciated this work.)

For our immediate concern of mitigating potential global warming, these existing capabilities suggest that industrial concerns can move the needed materials around at affordable

---

[764] "Marine Outfall," *Wikipedia*, undated. https://en.wikipedia.org/wiki/Marine_outfall#cite_note-Database-2 (accessed April 18, 2020)

[765] International treaties since the 1990s have forbidden release of raw sewage and replaced it with some degree of sewage treatment, but it still includes large amounts of polluting organic matter (plus phosphates and nitrates that can become too much of a good thing). This is another issue for environmental reform to be dealt with elsewhere. Restaurants in Los Angeles, California often have comments to the effect that, "This establishment proudly serves locally caught seafood." I consider such words a warning.

[766] "History of OTEC," Offshore Infrastructure Associates, Inc. http://www.offinf.com/history.htm (accessed April 19, 2020)

[767] Willy Ley, *Engineers' Dreams: Great Projects That Could Come True*, Viking Press, 1954.

[768] "Underwater Arteries – the World's Longest Offshore Pipelines," *Offshore Technology*, Sept. 9, 2014. https://www.offshore-technology.com/features/featureunderwater-arteries-the-worlds-longest-offshore-pipelines-4365616/ (accessed April 19, 2014)

[769] N. Ağıralioğlu, A. Danandeh Mehr, Ö. Aakdeğirmen, and E. Taş, "Cyprus Water Supply Project: Features and Outcomes," *13th International Congress on Advances in Civil Engineering, 12-14 September 2018, Izmir/Turkey*, 2018. https://www.researchgate.net/publication/328042100_Cyprus_Water_Supply_Project_Features_and_Outcomes (accessed Nov. 3, 2020)

costs. Existing shoreline-based fossil-fuel power plants can ship carbon dioxide out to sites on the edge of the continental slope.

Similar pipes (with insulation) can pipe colder water up from the depths to power plants on the edge of the continental shelves. Furthermore, the additional cold water combined with the warm water flowing out of a power a bottoming cycle (similar to ocean thermal energy conversion power plant to get another 5% increase in electrical production) would yield cooler water at the outflow.[770] That, in turn, can provides more electrical power per unit of heat from the power plant. A slight increase in pumping costs allows the resulting exit water temperature to nearly match the existing water temperature so that the water will not harm the local sea life. Ideally, the operation can produce exit water just a little warmer than the existing surface water. Thus, the nutrients from the more fertile depths for increased plant life will stay near the surface to boost productivity in nearby waters. Furthermore, that slight temperature increases heat radiating to space and water evaporating from the surface and increasing clouds—slight cooling effects to subtract from global warming.

9.3.5.2 Legal and Political Advantages of the Edge

The legal and political advantages of infrastructure at the edge of the continental shelf start with the fact that individual states have legal control offshore out to 3 miles (almost 4 kilometers). Beyond that distance, the federal government must consult with state authorities and receive state reviews, but the federal government has ultimate control.[771] If there is a supportive federal government, it can proceed with fewer legal challenges from multiple government entities.

For public relations, the NIMBY, or not in my back-yard, principle, applies to major projects. Construction projects within those 3 miles (5 kilometers) are visible from shore and are more likely to draw protests from realtors and home owners, two of the more powerful political influence groups. Construction 10 or more miles (16 kilometers) is invisible to even most of the weekend boaters; hence, the political resistance is even less.

For international law, per the United Nations Law of the Sea, the exclusive economic zone of a coastal state extends out to 200 miles (320 kilometers) from the low-water line of the coast (although navigation can be closer.[772] Thus, particularly for the east coasts of the Americas, continental edge facilities might be built far from sight of land.

---

[770] Rodrigo Soto and Julio Vergara, "Thermal Power Plant Efficiency Enhancement with Ocean Thermal Energy Conversion," *Applied Thermal Engineering*, vol. vol. 62, pp. 105–112, 2014. (abstract and partial article available at https://pdfs.semanticscholar.org/38e7/9df70b2c5eada9fd39d6d63e4bb40ba69875.pdf (accessed June 30, 2020)

[771] Anna M. Phillips and Rosanna Xia, "Trump Might Limit States' Say in Offshore Drilling Plan. Here's How," *Los Angeles Times*, March 21, 2019. https://www.latimes.com/politics/la-na-pol-trump-offshore-drilling-states-coastal-act-20190321-story.html (accessed April 12, 2020)

[772] "Part V – Exclusive Economic Zone," Articles 56 and 57, *United Nations Law of the Sea*, United Nations. https://www.un.org/depts/los/convention_agreements/texts/unclos/part5.htm (accessed May 2, 2020)

Also, geographically, the 100-meter (325-foot) depth of continental shelf is just an average. A break at 200 meters (650 feet) would not be a problem as long as it were to happen within the exclusive economic zone.

9.3.5.3 Financial Advantages of Building on the Edge

We have already noted how thermoelectric power plants at the edge of the continental shelf can tap vast cooling and hence greater efficiency from access to cold deep-ocean waters. However, there is much more, and even more value streams are needed to compensate for the increased cost of construction and operation off shore. Fortunately, new islands and shoals can provide returns to help pay for their construction.

- Offshore power plants can free up high-value scenic properties of existing onshore power plants as they become worn out and obsolete.

- Considerably more new electrical generating capacity will be needed if the mainland economy is to transform into a more electrical and non-fossil economy with increased electrical demand of battery cars and less electricity production from coal-powered power plants.[773,774]

- Researchers have proposed bottoming cycles to use the extra low-grade heat to make power plants on the edge of the continental shelf more efficient.[775]

- Process plants that continue to produce fossil fuels (refineries) and plastics often also produce excess heat (that can run bottoming -cycle power plants).[776]

- Capturing carbon in rock (discussed in the next subsection) provides an easily monitored and verified storage method so that the stored carbon can be monitored for carbon offsets to draw carbon offset payments. This contrasts with various offset sales that are difficult to monitor and often fraudulent such as claimed

---

[773] S. Roussanalya, A. Aasen, R. Anantharaman, et al., "Offshore Power Generation with Carbon Capture and Storage to Decarbonise Mainland Electricity and Offshore Oil and Gas Installations: A Techno-economic Analysis," *Applied Energy*, vol. 233–234, pp. 478–494, 2019.

[774] Isao Roy Yumori, *Seaward Extension of Urban Systems: The Feasibility of Offshore Coal-Fired Electrical Power Generation*, PB-241346/6; Oceanic Foundation, Waimanalo, Hawaii; 1975.

[775] Lars O. Norda, Rahul Anantharaman, Actor Chikukwac, and Thor Mejdellc, "CCS on offshore oil and gas installation - Design of Post-Combustion Capture System and Steam Cycle," *13th International Conference on Greenhouse Gas Control Technologies, GHGT-13, 14-18 November 2016, Lausanne, Switzerland, Energy Procedia*, vol. 114, pp. 6650 – 6659, 2017.

[776] Yamid Alberto Carranza Sánchez and Silviode Oliveira, Jr., Exergy Analysis of Offshore Primary Petroleum Processing Plant with CO_2 Capture," *Energy*, vol. 88, pp. 46–56, Aug. 2015.

- Another approach in mitigating potential global warming is to replace fossil fuels with noncarbon synthetic fuels, such as hydrogen or ammonia (NH_3). Just recycling captured carbon dioxide into methane would drastically reduce net warming emissions into the atmosphere. Until recently, the energy to make such synthetic fuels often required even more heat for the processing, which requires even more energy from new solar, wind, or nuclear power using the century-old Haber-Bosch process that cooks nitrogen and hydrogen together at high pressure and temperatures above 350°C (660°F). However, ongoing advances in nano materials and computer calculations to analyze them as possible catalysts are producing more efficient processes. For instance, one approach is a catalyst that would allow the Haber-Bosch process to work at 50°C (120°F)[779] Another proposed process would use a photoelectrochemical process at ordinary room pressure to get ammonia from water and nitrogen (H_2O) and nitrogen (N_2).[780]

- Finally, building on presently unclaimed open continental shelf "land" has the financial advantages of no payments to existing owners (there is only the U.S. government), no demolition of existing structures (there are none), and no work-hour limits or other constraints due to surrounding property owners (there are ... well, you know the rest).

9.3.5.4 Carbonate Rock Synthesis for Carbon Capture and Storage

One serious proposal for reducing global-warming is to capture carbon dioxide emissions from burning fossil fuels before they reach the atmosphere. Less carbon dioxide entering the atmosphere should decrease possible greenhouse warming. The first versions of such proposals were to store the carbon dioxide in aquifers or geological traps underground. One problem with

---

[777] Nathaniel Gronwald, "Greenhouse Gas Emission Offsets May Be Fraudulent," *Scientific American*, June 14, 2010. https://www.scientificamerican.com/article/greenhouse-gas-emission-offsets-may-be-fraudulent/ (accessed April 26, 2020)

[778] Ryan Jacobs, "The Forest Mafia: How Scammers Steal Millions Through Carbon Markets," *The Atlantic*, Oct. 11, 2013. https://www.theatlantic.com/international/archive/2013/10/the-forest-mafia-how-scammers-steal-millions-through-carbon-markets/280419/ (accessed April 26, 2020)

[779] "Fueling the world sustainably: Synthesizing Ammonia Using Less Energy," *Phys.Org*, April 24, 2020. https://phys.org/news/2020-04-fueling-world-sustainably-ammonia-energy.html (accessed July 20, 2020)

[780] Li Yuan, "Scientists Propose Novel Electrode for Efficient Artificial Synthesis of Ammonia," *Phys.Org*, May 27, 2020. https://phys.org/news/2020-05-scientists-electrode-efficient-artificial-synthesis.html (accessed July 22, 2020)

that is that the carbon dioxide does not always stay captured.[781,782] Even minor faults (or forgotten unplugged wells!) could allow the expensively captured carbon dioxide to seep back into the atmosphere. Along the way, carbon dioxide under supercritical pressure and in the often-present underground water becomes a chemically active substance that can dissolve toxic substances and carry them up to the water table or even up to the air above the storage area if there were to be a leak.

Such leaks would usually be slow and expensive over time, but they might be quick over time. Such leaks have happened naturally. On the morning of August 22, 1986, people near Lake Nyos in Cameroon, Africa found that 1,700 of their neighbors and 3,000 domestic animals were dead. Lake Nyos is a small, but deep, volcanic lake (1.2 miles by 0.75 mile, 1.9 by 1.2 kilometers with volcanic seeps of carbon dioxide, methane, and/or sulfur dioxide in the deep waters while warm tropical waters prevent any convection to the surface.[783,784] Possibly, a landslide allowed a great overturning of the lake. Once released, the 80 million cubic meters (2.8 billion cubic feet) of compressed carbon dioxide erupted to the surface, but cooled as the gas decompressed just as gases cool in the decompression cycle of a refrigerator. The cold gases bubbling out of the water made a chilly poisonous fog that poisoned the people and livestock closest to the lake.

A similar escape of deliberately stored carbon dioxide might have a similar disastrous effect. Thus, the whole carbon-dioxide capture and storage effort might be less useful for storing carbon dioxide, more expensive, and more dangerous than originally hoped.

However, there are geologic processes that cause certain rocks to combine with carbon dioxide and form stable rocks containing the carbonate ion (CO_3) so there would be no chance of carbon escaping. Such drawdown of carbon dioxide into rock is already 30% of the natural flux of carbon dioxide out of the sky.[785] Many volcanic minerals contain oxides of the metals iron, magnesium, and calcium. When exposed to the elements, these metals naturally react with carbon

---

[781] Ariel Schwartz, "The Problem with Carbon Capture: CO2 Doesn't Always Stay Captured," *Fast Company*, Nov. 19, 2010. https://www.fastcompany.com/1704105/problem-carbon-capture-co2-doesnt-always-stay-captured (accessed July 3, 2020)

[782] Adriano Vinca, Johannes Emmerling, and Massimo Tavoni, "Bearing the Cost of Stored Carbon Leakage," *Frontiers in Energy Research*, May 15, 2018. https://www.frontiersin.org/articles/10.3389/fenrg.2018.00040/full (accessed July 3, 2020)

[783] David Bressan, "The Deadly Cloud at Lake Nyos," *Forbes*, Aug. 21, 2019. https://www.forbes.com/sites/davidbressan/2019/08/21/the-deadly-cloud-at-lake-nyos/#2e248c6f5dbf (accessed Aug. 18, 2020)

[784] John Misachi, "What Was the Lake Nyos Disaster?" *World Atlas.com*, Aug. 22, 2019. https://www.worldatlas.com/articles/what-was-the-lake-nyos-disaster.html (accessed Aug. 22, 2020)

[785] Klaus S. Lackner, Christopher H. Wendt, Darryl P. Butt, et al., "Carbon dioxide disposal in carbonate minerals," *Energy*, vol. 20, issue 11, pp. 1153–1170, 1995.
Sandra Ó. Snæbjörnsdóttir, Bergur Sigfússon, Chiara Marieni, et al., "Carbon Dioxide Storage Through Mineral Carbonation," *Nature Reviews Earth* & Environment, vol. 1, pp. 90–102, Jan. 20, 2020.

dioxide in the air to make carbonates such as limestone, dolomite, and magnesite. However, this natural process may take years, decades, or even millennia.

The proposed quicker climate-warming mitigation is to fracture deposits of such rock in place underground, pump carbon dioxide into them, which can cause the carbonation reaction to happen within months. Test applications of this rock carbonation has been done in Iceland and Washington state in the U.S.,[786] and it has been proposed in a number of other geologic provinces.[787]

Unfortunately, this process as used is expensive ($50 to $100 per ton of carbon stored), and it provides no return for the money spent other than possible carbon offset payments, which would probably not be enough. The example from Iceland has a power-plant source of carbon dioxide near the volcanic rock. The convenient juxtaposition of carbon dioxide source and favorable volcanic minerals is rare, so most applications would be more expensive. Moreover, there is one additional problem/cost when capturing carbon dioxide by transforming oxide minerals into carbonates; the land above the carbon capture swells up, leading to fracturing, minor earthquakes, and possible damage to roads and other infrastructure. Considering all that, such rock carbonation would probably not be extensively used.

However, there have been proposals to use volcanic rock from nearby deposits on land or subsea and carbonate these rocks away from existing infrastructure.[788] Such volcanic rocks are common in many areas on the continental shelves and nearby land areas. As noted above, pipeline technology is well established for moving fluids from existing onshore power plants, refineries, and chemical plants. Carbon dioxide has often been pipelined from production sources to oil fields where the carbon dioxide is injected to enhance oil recovery for decades.[789] Furthermore, carbon dioxide compresses to a liquid at roughly room temperature, so liquid CO_2 pipeline or tanker transportation should be much cheaper than transporting gaseous natural gas. Thus, it should be

---

[786] B. Peter McGrail, Herbert T Schaef, Frank A Spane, et al., "Field Validation of Supercritical CO2 Reactivity with Basalts," *Environmental Science and Technology Letters*, vol. 4, no. 1, pp. 6–10, Nov. 18, 2016. https://pubs.acs.org/doi/pdf/10.1021/acs.estlett.6b00387 (accessed June 1, 2020)

[787] By Eli Kintisch, "Underground Injections Turn Carbon Dioxide to Stone," *Science*, June 10, 2016. https://www.sciencemag.org/news/2016/06/underground-injections-turn-carbon-dioxide-stone (accessed April 12, 2020)

Kimberly M. S. Cartier, "Basalts Turn Carbon into Stone for Permanent Storage," *EOS*, vol.101, March 20, 2020. https://eos.org/articles/basalts-turn-carbon-into-stone-for-permanent-storage (accessed April 12, 2020)

[788] Sandra Ó. Snæbjörnsdóttir, Bergur Sigfússon, Chiara Marieni, et al., "Carbon Dioxide Storage through Mineral Carbonation," *Nature Review*, vol. 1, Feb. 2020. https://www.nature.com/articles/s43017-019-0011-8.pdf (accessed April 13, 2020)

[789] Michael E. Parker, James P. Meyer, and Stephanie R. Meadows, "Carbon Dioxide Enhanced Oil Recovery Injection Operations Technologies," *Energy Procedia*, vol. 1, pp. 3141–3148, 2009. https://core.ac.uk/download/pdf/82017119.pdf (accessed Nov. 3, 2020)

possible to bring carbon dioxide and volcanic oxides together to produce artificial carbonate rocks much like the industrial processing of concrete.

Indeed, there is a concrete analog from ancient history. More than two-thousand years ago, the Romans invented hydraulic concrete that could set under water. One of their most impressive constructions was the harbor of Caesarea Maritima. The location was the east end of the Mediterranean Sea in the present area of northern Israel. Herod the Great, king of the vassal state of Judea, partnered with the Romans in building a major artificial harbor in an area that had no good natural harbors. This benefitted both Rome and Judea for two reasons. First, the area was the western end of several camel caravan routes bringing exotic goods from the Orient; these routes had more rainfall and resulting forage than routes going to Egypt. Second, pesky Parthian armies sometimes also invaded from the east, so the northern harbor provided a major staging point for Roman reinforcements to unload heavy military gear and drive the Parthians back.

The Judeo-Roman construction effort entailed massive limestone mining, tree felling to bake limestone into lime for the cement, baking the limestone into lime, digging up volcanic ash to mix with the lime, transporting those materials to the site, quarrying, breaking the local rock for aggregate to use in the concrete construction, and finally the construction itself.

The construction included a Roman city, an aqueduct to bring needed water from the nearby highlands, and (last but not least) the city's harbor of Sebastos (Greek for Augustus, the name taken by Octavian Caesar, the first Roman emperor). The Judeo-Roman contractors built two jetties—breakwater walls—from the sea floor to high enough above the water to protect against storm waves. On the top of each was a row of warehouses and a road for transporting materials to and from the docks. The larger southern jetty ran west and then north for nearly a half kilometer (a quarter mile). According to the one surviving historical account from Josephus, the southern jetty extended down about 20 fathoms (120 feet or 35 meters) at its deepest point.[790] The shorter northern jetty went west and almost closed a square, leaving a passage into the totally storm-sheltered harbor.

The contractors built all this in 7 years with only hand-rowed boats, gang rope-pulled cranes, and free divers to put in box-like concrete forms, fill them with hydraulic concrete, let each block set, and repeat … until the greatest harbor construction project of antiquity was completed.[791]

Despite earthquakes, tsunamis, and subsidence associated with a nearby earthquake fault, the harbor of Sebastos functioned well for several hundred years. It allowed the Romans to stave off those pesky Parthians, it allowed the Romans to bring in heavy siege machinery and crush the several Judean rebellions against Rome, and it provided a logistical point for several crusader

---

[790] "Ancient Caesarea Harbor: Did King Herod build the most magnificent harbor in the ancient world?" *Bible History*, undated. https://www.bible-history.com/archaeology/israel/3-caesarea-ancient-bb.html (accessed May 3, 2020)

[791] Robert Courland, Chapter 3: "The Gold Standard," *Concrete Planet: The Strange and Fascinating Story of the World's Most Common Man-Made Material*, Prometheus Books, 2011.

invasions seeking to reconquer and hold the birthplace of Christianity (until a Moslem ruler eventually ordered the jetties destroyed to prevent future sea invasions).

This harbor construction was more than two thousand years ago. Since then, the British built their more ambitious World-War-II Mulberry harbors in less than two weeks, and the advances in marine construction from World War II until now are probably as great as that the Judeo-Roman construction techniques to that of the British in 1944. there has been evolution of container ships for moving materials around the world, powered work boats, powered cranes, powered dredges, explosive quarrying, powered grinding of stone into pebble-sized stone for aggregate, pumped concrete, hardhat diving, scuba diving, robotic submarine equipment, and offshore drilling (which can supply steel anchors on the sea floor). Construction technologies are thousands of times more capable than those available during the time of Herod the Great and Augustus. With those thoughts in mind, a development scenario for continental-shelf carbonate rock storage on that average depth of 100 meters (325 feet) is as follows.

- Drilling platforms provide anchoring pivot points for corners and intermediate support points for concrete walls from the sea floor to higher than wave-breaking height above sea level, which would be 100 meters (330 feet) plus another 10 meters (31 feet) for enclosing a desired area. (Of course, the rebar must be bronze–aluminum, basalt fiber, or some other chloride-resistant material). The walls might enclose a square of 2×2 kilometers (1.3 miles square) for later containment of volcanic oxide rocks.

- Vertical stand pipes to the sea floor and a grid of pipes covering the enclosed sea floor are emplaced for later dispersing of carbon dioxide into the carbon dioxide carbonation site.

- Volcanic rock gravel is dumped into the enclosed area on a grid of pipes rather than drilling multiple bore holes for the earlier technique of underground injection. The site can also take concrete, stone, and brick demolition waste that might also garner payments for avoided landfill dumping fees. It might also receive ash from coal combustion or steel-mill slag, which (besides having the value of avoided landfill fees) can also help form hydraulic concrete.

- At the top of the enclosed material, a collection of gas gathering pipes is emplaced, so that carbon dioxide reaching that point can be refluxed back to the input pipes at the bottom.

- Next, workers add a cap rock of concrete at the top to contain the pressure of the heated gases—that comes next.

- The operators pump water from the top of the heap. Then, they pump carbon dioxide, a little oxygen, and a limited amount of ground-up organic material (such as agricultural waste or yard waste) into the bottom array of dispersing pipes. The organic matter feeds composting bacteria that attack the rock oxides directly, generate heat to speed the rock carbonation process, and produce more carbon dioxide. In addition, this material is carbon-containing, so its injection into the containment area is also carbon storage. Theoretically, there might be enough low-grade heat to run a bottoming-cycle power plant.

- A hot damp, but still gaseous, area near the top attains sufficient heat for the carbonation process to become exothermic; that is, the process produces more heat than it consumes.

- With sufficient heat and pressure, the volcanic metal oxides can combine with carbon dioxide orders of magnitude faster than through fissures in rock. Only in this instance, the flow of carbon dioxide will accrete new carbonate minerals filling in gaps between rock down from the active zone building a giant monolithic block.

- If there is excessive carbon-dioxide pressure buildup at the top of the heap, the operators can pump it out of array of gathering pipes at the top and return it through the lower array.

- Some billions of tons of carbon combining with rock later, the concrete walls will allow accretion of a stone mesa fully capable of maintaining itself even without the concrete walls. This is island building on a geological scale.

- Such islands can be constructed with appropriate patterns of gaps on the top to provide protected anchorages for power ships energized by natural gas, nuclear, or even coal. These mobile power plants can be anchored and generate power. When they become obsolete or require major repairs, they can simply be towed away and replaced by newer better units.

- Any coal-fired or gas-fired plants can be additional sources of carbon dioxide for insertion in the island-building subsea landfills.

- An additional way to reduce the cost of carbon capture is the Allam cycle now working in a 50-megawatt demonstration power plant. The Allam cycle has two innovations. First, it burns natural gas with only oxygen instead of the usual use of ordinary (ambient) air, which is 80% nitrogen. The combustion of $2CH_4 + 3O_2 \rightarrow 2CO_2 + 2H_2O$ means there are no extra steps of separating out oxides of nitrogen and purifying the carbon dioxide, both of which have been major parasitic power drains until now. Second, the exhaust is high-pressure (supercritical) carbon dioxide, which has higher efficiency and requires much less volume than a conventional steam turbine (and there still might be enough heat for a secondary steam turbine).[792] Of course, with the cold deep waters available on the edge of the

---

[792] David Roberts, "The Carbon Capture Game Is about to Change," *Vox*, June 1, 2018. https://www.vox.com/energy-and-environment/2018/6/1/17416444/net-power-natural-gas-carbon-air-pollution-allam-cycle (accessed April 26, 2020)
Rodney J. Allam, "NET Power's CO2 cycle: the breakthrough that CCS needs," *Modern Power Systems*, July 10, 2013. https://www.modernpowersystems.com/features/featurenet-powers-co2-cycle-the-breakthrough-that-ccs-needs (accessed April 26, 2020)

continental slope, there could be an additional bottoming-cycle power plant to harness another 5 or 10%.

9.3.5.5 Using Power Plants on the Edge Can Be Carbon Neutral or Carbon Negative

As noted above, fossil-fuel power plants near the rock-forming island-building landfills would not emit any greenhouse-warming carbon dioxide—it would all be captured by rock oxides transforming into carbonates. Thus, they would be a climate neutral.

Even better, these facilities could import carbon dioxide from elsewhere for carbon capture, thus rendering themselves carbon sinks to reduce the net global emissions of carbon dioxide into the atmosphere and building those artificial islands. As such, they would be carbon negative and conceivably able to claim any carbon credits that that individuals, countries, or international agencies might pay.

Furthermore, such islands would provide infrastructure for continued extension of mariculture into deeper water. Again, the ecology of production would be best served by a balanced mix of plants, filter feeders, and animal species—wastes from animal species providing material with phosphorus, nitrogen, and organic material with plants absorbing and using those materials. As a side benefit, the plants would draw down carbon dioxide from the atmosphere and further increase the carbon-negative effect.

9.3.5.6 Defense Against Tsunamis

Last, but not least, rare but real giant tsunami waves do happen in the ocean (and several were mentioned in Subsection 9.1.7.). The largest waves come from island volcanic lava flows overbuilding steep slopes as they cool at the water line. Those overbuilt slopes eventually collapse, sometimes producing tsunami waves higher than anything in our recorded history.[793]

Such collapses are highest close to the collapse. For instance, the active volcanoes in the Hawaiian chain have dropped sizeable collapses at intervals of thousands of years. Some of these collapses have shoved thousand-foot (thee-hundred-meter) waves up nearby islands.[794]

Fortunately, the wave energies are massively diluted as they spread out with long distances over the ocean. For instance, a wave coming from the Cumbre Vieja volcano on Los Palmas Island in Azores could be mountainous as it left the island, but it might only have 10–25 feet (3–8 meters) height when hitting the beaches to the west in North America, the Caribbean, and South America.[795] This could still do massive damage, but it would be much less of a danger. (Granted,

---

[793] Adrienne LaFrance, "The Most Destructive Wave in Earth's (Known) History," *The Atlantic*, October 23, 2015. https://www.theatlantic.com/science/archive/2015/10/traces-of-an-ancient-mega-tsunami/411970/ (accessed April 14, 2020)

[794] Becky Oskin, "Landslide-Driven Megatsunamis Threaten Hawaii," *LiveScience*, Dec. 6, 2012. https://www.livescience.com/25293-hawaii-giant-tsunami-landslides.html (accessed June 1, 2020)

[795] Steven N. Ward, "Cumbre Vieja Volcano -- Potential Collapse and Tsunami at La Palma, Canary Islands," Paper number 2001GL000000, American Geophysical Union, 2001. https://websites.pmc.ucsc.edu/~ward/papers/La_Palma_grl.pdf (accessed June 1, 2020)

there have been disputing estimates on the chances of one large collapse versus a number of smaller collapses, which would produce multiple correspondingly smaller tsunami waves.)

Such waves (and they come in a train of several waves) would generally only penetrate a mile or two (2 to 3 kilometers) inland on low-relief terrain (including river valleys), but this could be a major issue for cities built directly on a low shore facing the ocean such as Miami Beach. The initial high breaking waves expend energy in that breaking and in becoming a temporary flood flowing inland and then back out to sea. The damage would be beach communities, ports, and shoreside power plants and fuel gathering points.

Moreover, there are some coastline features that might amplify an incoming chain of tsunami waves. One such feature is the New York Bight. The new Jersey shore (facing east) runs roughly south to north. Then, Long Island runs west to east, forming a triangle with the New Jersey shore. The point of this triangle has a gap going toward Manhattan Island and Staten Island. This triangle has demonstrated a giant funnel effect for storm surge from occasional hurricanes making landfall in that area, such as the 1893 hurricane, also known as the Midnight Storm, and "Super Storm Sandy" in 2012.

Likewise, a tsunami generated by a volcanic collapse in the Canary Islands might well be amplified in that funnel. Depending on the tide heights at impact, the damage on the New Jersey shore, New York City, Long Island and on up the Hudson River could be immense.[796]

Defense against such rare but immense waves could be done in any of three methods:

1. Building carbon-capture blocks on the flanks of dangerous volcanoes such as Cumbre Vieja in the Canary Islands or Mauna Loa in the Hawaiian Islands. Such carbonate blocks or massifs would be built by removing some of the unstable volcanic rock on the slopes of the volcanic island. That would decrease the mass of expected mass flow into the ocean. Moreover, the massifs would be expected to block or at least separate the flow of a collapse flow into separate directions that would be less and would interfere with each other causing the net tsunami wave train generated to be much less energetic. The limitation would be the cost of excavating the volcanic rock, grinding it, transporting it to the offshore site, transporting waste carbon dioxide from distant industrial sites, and cooking the volcanic rock with carbon dioxide in a confined space. The positive cash flows would be carbon credits, energy, or chemical production using the excess heat from rock carbonation, and byproduct mariculture based around those operations. The additional limitation would be that there are thousands of volcanoes that have been quiet for thousands of years that can unexpectedly roar to life.

2. Building carbon-capture blocks (massifs) at the edge of the continental shelves offshore from vulnerable areas. Such blocks would not totally stop a chain of tsunami waves, but it would drain much of the wave energy and scatter wave energy

---

[796] Autumn Heisler, "A Mega-Tsunami Is Coming; Can the East Coast Even Prepare? *Risk & Insurance*, July 30, 2018. https://riskandinsurance.com/mega-tsunami-wipes-out-east-coast/ (June 1, 2020)

in multiple (often interfering) directions (often with interfering wave patterns that would reduce wave energy further). For sites near developed cities in North America and South America, the cost of carbon dioxide and other materials to build the blocks would be much less. Likewise, any byproduct energy, chemical, and/or food production would be close to potential markets.

3. (An entirely different approach) Trimming ultramafic rock from the sides of such volcanoes, grinding the rock as in the other two, and then spreading it around in chemically and biologically active areas for the oxides to convert into carbonates, thus drawing down carbon dioxide from the atmosphere. The areas to spread to would be beaches (most active), nearby shallow oceans (not as active but larger area), and agricultural fields (nearly as active as a beach, useful as a fertilizer, but more expensive because of transportation costs).[797] The Vesta Project (https://projectvesta.org/) spun off from the original research report referenced above. They claim to be within the cost of carbon offsets. It would be much better if they could find a way to be profitable.

9.3.6 What to Do

Conclusion: This has been a partial summary of ocean risks and remedy proposals associated with the three virtual Australias of the world's continental slope. The risks noted in this presentation were overfishing, global warming, and more damage from the other two. They really are a threat to our lives and our civilization. The proposed greenwater revolution is as big a change as the switch from the hunter-gatherer existence to farming.

Major Things for Governments and Businesses to Do

- Government: Legislate carbon capture and storage compensation for capturing carbon dioxide by transforming silicates to carbonates by using volcanic silicates in carbonation for island building, for soil fertilization, or simply as green sand to take in carbon dioxide with extra payment for removing such rock that maybe unstable near island volcanoes.

- Business: Develop business proposals for island building using carbonation of volcanic silicate rocks.

---

[797] R. D. Schuiling and P. L. de Boer, "Rolling stones; Fast weathering of olivine in shallow seas for cost-effective CO2 capture and mitigation of global warming and ocean acidification," *Earth Systems Dynamic Discussions*, vol. 2, pp. 551–568, 2011.

What You Can Do

- Support proposals for silicate carbonation for island or shoal building using carbonation of volcanic silicate rocks.

9.4 A Step Even Further: The Blue-to-Green Revolution

A thousand years after Plato wrote about Atlantis sinking beneath the waves, straggling bands of people began entering what is now central Mexico. They called themselves Azteca (Aztecs). They said they were poor because their land, *Aztlan*, had sunk in a flood. The locals allowed them to be vassals in a poor area around a lake with an island in the middle.

Since they needed more farmland and there was not enough land on the island, the Aztecs made more farmland with a little-known agricultural technique that was called *chinampas* or floating gardens. These humanmade islands could be either of two things:

1. Lake-bottom mud piled up in beds above the water surface in shallow areas.

2. When the shallow areas were filled, the Aztecs floated large woven reed mats piled with more mud from the lake bottom. These little islands could be anywhere in the lake. Fast growing willow trees were planted so the root systems would grow to the bottom and anchor the floating islands and brace the raised beds.[798,799]

The food production from the *chinampas* made the Aztec economy a powerhouse that fed great armies to dominate most of what is now Mexico until the Spanish *conquistadores* arrived.[800]

The *chinampa* farming revolution of the Aztecs is an example of what people can do. Now, after another five-hundred years since the Spanish conquest of Mexico, people are just starting another agricultural revolution the Greenwater Revolution on the continental shelves described in the previous subsection. That will be a stretch with the social and technological advances described earlier in this book. However, even the greenwater continental shelves will not be enough for an expanding population.

Fortunately, the ocean can support much more.[801] Further into the future and more technologically ambitious is the concept of the Blue-to-Green Revolution for entering the deeper waters that are less fertile, so they have less green. Consequently, those areas have blue water.

---

[798] "Farming," *Aztec Empire* (website), date continually updated. http://aztec.com/page.php?page=home2 (accessed June 29, 2020)

[799] Louis Werner, "Cultivating the Secret of Aztec Gardens," *Americas*, vol. 44, issue 6, pp. 6–16, Nov.–Dec. 1992.

[800] Alfred Aghajanian, *Chinampas: Their Role in Aztec Empire -- Building & Expansion*, IndoEuropeanPublishing.com, July 2018 (first published 2006).

[801] Carlos M. Duarte, Marianne Holmer, Yngvar Olsen, et al., "Will the Oceans Help Feed Humanity?" *Bioscience*, vol. 59, pp. 967–976, Dec. 2009. https://www.researchgate.net/publication/222092668_Will_the_Oceans_Help_Feed_Humanity/link/09e4150bf72de6e39d000000/download (accessed Sept. 10, 2020)

Turning blue waters green would multiply productivity for two reasons. First, the blue waters are essentially deserts with very little life per unit volume of water, so they have more to gain. Second, the deep waters have that third dimension of slowly moving currents that can fill in with more layers of more complex life processes than in shallower waters.[802]

Converting patches of bluewater nutrient-starved desert into greenwater oases requires upwelling of nutrients to create green zones. Just as importantly, it requires maintaining control and maintenance of those newly enriched areas.

In contrast with the 7.6% of the ocean that is continental shelf, the other 92.4% (about 11 times more) of the ocean is much deeper, averaging a depth of approximately 14,000 feet (4,300 meters or 2.6 miles). These deeper waters are generally bluewater desert. These are areas far from land nutrients and far above nutrient-rich cooler—and hence denser—waters below. Consequently, these nutrient-starved waters have less green plant matter (plankton) causing them to be clear and blue. There is much less life in these waters and very little drawdown of carbon dioxide from the atmosphere. A comparable low fertility of surface waters also happens in many areas of Arctic Ocean where the nutrient-rich waters are actually slightly warmer Gulf Stream waters from the Atlantic water that are overlain by freshwater from the many large rivers flowing north to the Arctic Ocean. Even though colder, the fresh waters are still less dense than the salty warm waters below. Again, the result is diminished green algae, diminished productivity, and diminished drawdown of carbon dioxide from the atmosphere.

There are nutrient exceptions. In certain areas, combinations of winds cause nutrient-rich waters to upwell from the depths. These areas are only about 1% of the ocean surface, but they supply nearly 50% of the world's wild fishing catch. (Except for around Antarctica and around the Arctic Sea, most of the continental shelf areas have been mostly fished out—otherwise, they would be much more productive.) Waters west of Peru, around the Canary Islands, and west of California are some of these rich areas.

There are several largely developed or nearly developed technologies that can open the blue waters to becoming green.

1. Fertilizing the bluewater deserts and the ammonia economy at sea
2. Motor ships
3. Totally Buoyant construction
4. Dynamic positioning
5. Syntactic foams
6. Robo farmers (again, but colder and under pressure!)
7. Transportation advances

---

[802] Marc Gunther, "Can Deepwater Aquaculture Avoid the Pitfalls of Coastal Fish Farms?" *YaleEnvironment360*. Jan. 25, 2018. https://e360.yale.edu/features/can-deepwater-aquaculture-avoid-the-pitfalls-of-coastal-fish-farms (accessed Aug. 10, 2020)

8. Virtual presence, telebusiness,

9.4.1 Fertilizing the Bluewater Deserts and the Ammonia Economy at Sea

How would one bring the fertility of an upwelling to the bluewater deserts? The easiest method is placing nutrients in those surface waters, much like a farmer putting fertilizer on a field. We shall examine that first, and creating an artificial upwelling will be described a little later.

Artificial fertilizer increased the productivity of land agriculture to feed many times more people. However, scientists discovered that these bluewater deserts in deep water already had sufficient nutrients to sustain much more plant matter than what existed. They wondered why these areas were so bluewater barren.

Then, oceanographer John Martin theorized that bluewater areas lacked trace amounts of iron. That was important to him because he wanted an explanation for the ice ages. He thought that dry periods with lots of desert dust would supply that iron, make large fertile areas, draw down carbon dioxide out of the atmosphere … and chill the planet even more. (Ironically, the glacial advances of severe ice-age cold locks up water in ice and causes many areas to become desert and yield more airborne dust, thus making the ice age worse.) At one scientific meeting, Martin famously proclaimed:

"Give me half a tanker of iron, and I'll give you the next ice age."[803]

…

Fortunately, he was exaggerating … we think. In fact, we'd be happy with just LESS WARMING, LOTS MORE FOOD, AND EVEN SOME CARBON-NEUTRAL FUEL.

People have proposed fertilizing ocean areas and harvesting some of the resulting increased marine production for fuel production that would be carbon neutral or even carbon negative

---

[803] "Other Lines of Evidence Inform the Debate on Ocean Iron Fertilization" *Oceanus Magazine*, January 9, 2008. https://www.whoi.edu/oceanus/feature/lessons-from-nature-models-and-the-past/ (accessed March 22, 2020)

because it would be produced by ocean plants drawing carbon dioxide down from the sky.[804,805,806,807,808]

Along the way, such fertilizing operations could pay for themselves by generating marketable products. In 2012, the Native American Haida tribe of British Columbia, Canada did just that. They spread 120 tons (109 metric tons) of iron sulfate into nearby waters in the northeast Pacific Ocean to fertilize a plankton bloom and ultimately a better salmon catch. (Iron sulfate [$Fe_2(SO_4)_3$] is water soluble and stays in the surface waters, whereas solid-iron particles sink to the bottom before they can cause much fertilization compared to the initial nutrient input.)

The 2012 salmon catch in the northeast Pacific more than quadrupled from 50 million the previous year to 226 million.

Many environmental activists denounced the Haidas' ocean fertilization success as a dangerous gamble. However, many GND boosters have said that humanity would be doomed unless global warming were controlled soon. Following that rationale, the GND resolution calls for replacing cattle ranching with vegetarian foods, replacing most airline travel with a new system of electrified railroads, rebuilding all structures for high efficiency, and many other fabulously expensive ventures. Even with all those changes, the climate effects still might be enough.

It seems that the initially proposed GND is already a gamble: Betting the farm while drawing for an inside straight. Compared to that wild bet, a little ocean fertilization is minor risk, especially because natural fertilization is a known occurrence. The Haida experiment is similar to an exceptional harvest associated with a 1958 volcanic eruption in the Gulf of Alaska that supplied a load of volcanic dust onto the nearby waters—natural fertilization.[809] (The volcanic dust was largely in the form of metal and silicon oxides. As with pure iron, noted above, they tended to sink quickly. However, the massive tonnage of an eruption meant that enough iron would have dissolved to cause a plankton bloom.)

---

[804] C. M. Moore, M. M. Mills, et al., "Processes and Patterns of Oceanic Nutrient Limitation," *Nature Geoscience*, vol. 6, pp. 701–710, Sept. 2013.

[805] John H. Martin and Steve E. Fitzwater, "Iron Deficiency Limits Phytoplankton Growth in the North-East Pacific Subarctic, *Nature*, vol. 331, pp. 341–343, Jan. 28, 1988. doi:10.1038/331341a0

[806] Guillem Chust, J. Icarus Allen, et al., "Biomass Changes and Trophic Amplification of Plankton in a Warmer Ocean, *Global Change Biology*, DOI: 10.1111/gcb.12562, May 7, 2014.

[807] Hugh Powell, "Fertilizing the Ocean with Iron: Should we Add Iron to the Sea to Help Reduce Greenhouse Gases in the Air?" *Oceanus*, vol. 46, no. 1, Jan. 2008. http://www.whoi.edu/oceanus/viewArticle.do?id=34167&sectionid=1000 (accessed Nov. 11, 2013)

[808] Eli Kiintisch, Hack the Planet: Science's Best Hope—or Worst Nightmare—for Averting Climate Catastrophe, John Wiley & Sons, Inc., Hoboken, New Jersey, 2010.

[809] Robert Zubrin, "The Pacific's Salmon Are Back — Thank Human Ingenuity," *National Review*, April 22, 2014. https://www.nationalreview.com/2014/04/pacifics-salmon-are-back-thank-human-ingenuity-robert-zubrin/ (accessed July 20, 2020)

9 Fears and Fixes from Sea and Sky

Back at use of presently existing waters, Howard Wilcox's kelp farms in the 1974 Hothouse Earth[810] were just a proposed expansion of the existing bull kelp (*macrocystis pyrifera*) beds off the North American West Coast by emplacing nets for attaching kelp where the water was too deep for the kelp holdfasts or where the bottom was too smooth or sandy for the help holdfasts to grip. These waters are already highly fertile because the south-flowing California current causes upwelling all along that coast, and Wilcox proposed kelp farming on the continental shelf areas because it would be easiest in those areas.

Likewise, Tim Flannery's more ambitious 2017 *Sunlight and Seaweed*[811] describes only use of continental-shelf areas because operations are much easier there than in the ocean deeps. He refers to a mere 7% (and a fraction) of the ocean surface.

Then, there is the other 92+% of the ocean, more than 12 times the potential of the continental shelf for increased production and drawdown of carbon dioxide from the atmosphere. Actually, the potential is even more because deep waters are usually bluewater desert while continental-shelf waters are at least slightly green because of sea-floor nutrients only a few meters below. Despite most of the ocean being bluewater desert, plankton in the ocean still produce half of the primary food production (and carbon dioxide drawdown) for the planet.[812]

Furthermore, the humble phytoplankton (tiny floating algae much smaller than macroalgae kelp) have more anti-greenhouse powers available to them than just fixing carbon out of the sky.

Phytoplankton also outgas a number of other byproducts besides oxygen. One of those outgassed chemicals fits with the Gaia Hypothesis of James Lovelock and Anne Margulis, by which the biosphere might act almost like a powerful godlike being (Gaia) capable of compensating for changes that might damage the system.[813] Building from that principal, several scientists proposed that under heat or high irradiance, phytoplankton might produce chemicals that would provide more shade and cooling.[814] The scientists had the surnames of Charlson, Lovelock, Andrea, and Warren—hence, from the first letters of those surnames, they hypothesis has often been called the CLAW hypothesis.

The CLAW hypothesis notes that phytoplankton produce dimethyl sulfide (DMS) as a byproduct of their growth, and they produce more DMS when water temperatures and/or received sunlight are elevated. Some of the DMS reaches the atmosphere where it eventually oxidizes and produces sulfur dioxide (SO_2) and various sulfates. These chemical species are anti-greenhouse

---

[810] Howard A. Wilcox, *Hothouse Earth*, Praeger Publishers, New York, 1975.

[811] Tim Flannery, *Sunlight and Seaweed: An Argument for How to Feed, Power and Clean up the World*, The Text Publishing Company, Melbourne, Victoria, Australia, 2017.

[812] Paul G. Falkowski and John A. Raven, *Aquatic Photosynthesis*, Second Edition, University Press, Princeton, New Jersey, 2007.

[813] James E. Lovelock and Lynn Margulis, "Atmospheric Homeostasis by and for the Biosphere: the Gaia Hypothesis," *Tellus. Series A*, vol. 26 (1–2): pp. 2–10, Stockholm: International Meteorological Institute, doi:10.1111/j.2153-3490, February 1, 1974.

[814] R. J. Charlson, J. E. Lovelock, M. O. Andreae, and S. G. Warren, "Oceanic Phytoplankton, Atmospheric Sulphur, Cloud Albedo and Climate," *Nature*, vol. 326 (6114), pp. 655–661, 1987.

chemicals in that they reflect visible sunlight back into space but allow infrared light to pass through, leading to a cooling effect. Because greater heat increases this cooling effect, the phytoplankton produce a negative climate feedback.

For future reference, increased productivity of sea plants could be another way to increase the CLAW effect and draw down more carbon dioxide from the sky, both of which would decrease global warming.

As noted earlier, one can avoid putting greenhouse-warming carbon dioxide into the atmosphere by burning hydrogen instead of the carbon or hydrocarbons of fossil fuels. Unfortunately for climate concerns, the present production of hydrogen by splitting it from coal or natural gas just emits the carbon dioxide somewhere else, in the hydrogen-generation process.

C (coal) + steam $2H_2O$ →
yields CO_2 (carbon dioxide) + $2H_2$ (hydrogen gas))

or

CH_4 (methane, natural gas) + steam $2H_2O$ →
yields CO_2 (carbon dioxide) + $2H_2$ (hydrogen gas))

The limitation is that the "clean" hydrogen is produced at the cost of producing an even larger amount of climate-changing "dirty" carbon dioxide elsewhere. Even the electrolyzing from wind or solar costs the energy from coal and other sources to mine and refine steel, mine and bake limestone to make concrete, rare earths, and other exotic materials used in solar and wind generators. Furthermore, the climatically clean hydrogen is hard to store. It is the least dense fuel there is, so stored hydrogen is not useful unless tremendously compressed (expensive and dangerous) or compressed and highly chilled into liquid hydrogen (even more expensive and dangerous).

The more practical GND alternatives are synthetic hydrocarbons made by combining hydrogen with waste carbon dioxide and making methane … or even some heavier hydrocarbon, such as gasoline that packs tremendously more energy per unit volume than hydrogen.

The best alternative fuel from a GND standpoint is ammonia (NH_3). It has the advantage of zero carbon. As noted earlier, ammonia can be produced by either heat or electricity from nuclear fission, solar, and wind; but where might it be used?

Ammonia can be liquid at room temperature and at much lower pressure than hydrogen requires. It has the disadvantage of only being two thirds the density of gasoline, so there would be some loss of land-vehicle trunk space to accommodate larger fuel tanks. Also, the emissions from ammonia combustion--oxides of nitrogen and unburned ammonia—need expensive emissions control systems, which are expensive for relatively small personal vehicles.[815]

---

[815] Pavlos Dimitriou and Rahat Javaid, "A Review of Ammonia as a Compression Ignition Engine Fuel," *International Journal of Hydrogen Energy*, vol. 45, issue 11, pp. 7098–7118, Feb. 28, 2020.

These issues are less or nonexistent at sea where commercial freighters and tankers emit roughly 3% of the world's carbon-dioxide emissions (roughly a billion tons (metric tons).[816] Shipbuilders are already designing ammonia-powered ships.[817,818] Ammonia fueling for ships has three crucial additional advantages for humanity and for a better GND.

1. Ships can easily carry additional space for the less-energy-dense ammonia fuel.

2. Ships can use stronger tanks and refrigeration to maintain ammonia in the more compact liquid form.

3. Ships can afford complex injection systems in the large mostly diesel engines (also called compression ignition engines) that are the most common power plants for commercial ships.[819]

4. The emissions from combusting ammonia fuel can serve as plant-fertilizer at sea rather than pollutants. Ammonia is a common fertilizer on land, and it is highly soluble in water, so it can increase productivity at sea. Wet scrubbers on land often use calcium hydroxide ($Ca(OH)_2$, also known as slaked lime or limewater) or powdered limestone to remove oxides of nitrogen from exhaust gas flows.[820] Furthermore, adding limestone to water bodies to reduce acidification is already a well-established practice.[821]

But the ammonia would not be ejected from the smoke stack. It would be mixed with input seawater and pumped out the aft generating a slight increase in thrust ... and a thoroughly mixed flow of water.

Of course, the as mentioned earlier trace amounts of usable iron are the most significant limiting factor for fertility of the bluewater deserts. Thus, the combustion propwash exhaust would

---

[816] Henrik Selin and Rebecca Cowing, "Cargo Ships Are Emitting Boatloads of Carbon, and Nobody Wants to Take the Blame," *Phys.org*, Dec. 18, 2018. https://phys.org/news/2018-12-cargo-ships-emitting-boatloads-carbon.html (accessed July 28, 2020.

[817] Jasmina Ovcina, "Consortium Sets Sights on Commercializing Ammonia-Fuelled Ships" *Offshore Energy*, Oct. 21, 2020. https://www.offshore-energy.biz/consortium-sets-sights-on-commercializing-ammonia-fuelled-ships/ (accessed Oct. 27, 2020)

[818] Naida Hakirevic, "NYK, JMU and ClassNK Ink R&D Deal to Commercialise Ammonia-Fuelled Ships, Aug. 12, 2020. https://www.offshore-energy.biz/nyk-jmu-and-classnk-ink-rd-deal-to-accelerate-use-of-ammonia-as-marine-fuel/ (accessed Oct. 27, 2020)

[819] Stephen H. Crolius, "Literature Review: Ammonia as a Fuel for Compression Ignition Engines," web page, Ammonia Energy Association, March 23, 2020. https://www.ammoniaenergy.org/articles/review-of-ammonia-as-a-ci-fuel-published/ (accessed July 28, 2020)

[820] A. Saleem, "Flue Gas Scrubbing with Limestone Slurry," *Journal of the Air Pollution Control Association*, vol. 22, issue 3, pp. 172–176, March 15, 2012.

[821] Louis A. Helfrich, Richard J. Neves, and James Parkhurst, *Liming Acidified Lakes and Ponds*, VCE Publication 420-254, Virginia Cooperative Extension and Virginia Polytechnic and State University, Blacksburg, Virginia, May 1, 2009. https://www.pubs.ext.vt.edu/420/420-254/420-254.html (accessed July 28, 2020)

also receive some iron sulfate to multiply the fertilization. The sum of these nutrients would be trails of green plankton in the bluewater desert. They would slow global warming in the following ways.

1. There would be roughly a billion tons (metric tons) per year less global-warming carbon dioxide going into the atmosphere. This would decrease the net carbon dioxide emissions, which would still be immense.

2. The plant plankton (phytoplankton) would draw down carbon dioxide from the air as part of their growth process. Various predators, starting with slightly larger zooplankton on up to baleen whales, would eat the phytoplankton (or eat smaller predators). All these sea actors would drop excretions and dead individuals toward the sea floor, thus further reducing the net carbon dioxide flow to the atmosphere. (Critics of ocean fertilization have claimed that predator populations would eventually increase in number, eat all the phytoplankton, and exhale (gill out?) most of the carbon drawn down as carbon dioxide that would bubble up to the surface and return to the atmosphere. However, passing ships are an occasional thing, so phytoplankton blooms would return to bluewater desert before predator populations could peak—leaving excess planktonic material to settle downward.

3. The phytoplankton would decrease ocean acidity, helping protect corals and other calcium-carbonate-fixing species such as corals.

4. As noted earlier as the CLAW Hypothesis, phytoplankton blooms release chemicals with a shading anti-greenhouse effect, thus slowing global warming.

5. All these first four items are benefits that might be worth paying for as carbon credits.

A similar, but more modest benefit could be attained in those already somewhat fertilized continental-shelf waters. These waters already have most of their trace iron, but there are other benefits related to the ammonia economy.

The outboard motors of small boats already expel their exhausts into the water to suppress noise, and the same has also been done compressors, and generators at sea. When such equipment exhausts into the water, any inefficiencies and initial running while the engine is still cold releases unburned gasoline or diesel fuel, which is harmful for marine life. These engines burning ammonia would release oxides of nitrogen and even some ammonia itself. These compounds would be pollutants in the air and land world. However, in the watery world, all these exhaust products are fertilizer for mariculture. The managers of the various operations need only balance the increased ammonia and oxides of nitrogen with more sea plant growth to use, and hence consume, those nutrients.

The ammonia economy can improve performance of mariculture on the continental shelf where the waters are already at least mildly fertilized with nutrients. It will be much more useful for mariculture in the deep waters that are now bluewater deserts, almost entirely devoid of nutrients.

When mariculture advances to work in the deep ocean, there is another massively larger way to fertilize those waters. As noted in 9.1.3 Reduced Ocean Productivity Due to Increased Stratification, upwelling is the natural way that oceans bring nutrients to the surface. Thus, it is not an unnatural fertilizer input subject to international legal controls. There have already been experiments and many more proposals for generating areas of upwelling, most often with wave-powered platforms pumping water up from about 300 meters (1000 feet).[822]

Another proposed approach that has been proposed for decades would be to use ocean thermal energy conversion (OTEC) for causing upwelling of nutrients.[823] The water flowing up in an OTEC riser pipe would be substantial because there is only a small temperature difference between the cold deep waters and the warm surface waters. Consequently, there must be a great flow to compensate. A 100-megawatt OTEC power plant might have a flow comparable to the Nile River. That would carry a great amount of nutrients.

Unfortunately, for GND climate concerns, water pumped by OTEC, wind, or any other source have a problem with the physics of heat and density. Cold water pumped from the depths is colder than the surface waters; therefore, when released at the surface, it immediately plumes back down. That applies totally for wind power. The effect would be less for an OTEC power plant that mixes warm surface water with the cold as part of the power cycle, but by the nature of OTEC, the released water must be colder than the surface water. Thus, the nutrients would plummet away from the sunlit zone needed by plants.

More heat must be supplied to warm the nutrient-rich water enough so that it is warm enough to stay at the surface. Also, higher temperatures would increase the efficiency, and thus profitability, of OTEC power plants. The waste heat from any other type of thermal power plant could provide such heat. Handily, there are reasons to propose doing just that, and there have been several proposals.

Researchers from Korea Electric Power corporation (KEPCO) did preliminary designing for combined ocean thermal energy conversion (C-OTEC). The C-OTEC would use the latent heat of the steam exhausted into the condenser of a power plant as a heat source. The condenser steam can always be maintained at around 32 °Celsius (90°Fahrenheit), which is the temperature of saturated steam when it is condensed. In contrast, most tropical surface waters generally do not

---

[822] "Artificial Upwelling: Current Efforts and Anticipated Impacts of Intermingling the Ocean," *Geoengineering Monitor*, Oct. 24, 2019. http://www.geoengineeringmonitor.org/2019/10/artificial-upwelling-current-efforts-and-anticipated-impacts-of-intermingling-the-ocean/ (accessed Oct. 1, 2020)

[823] C. K. Liu, "Ocean Thermal Energy Conversion and Open Ocean Mariculture: The Prospect of Mainland-Taiwan Collaborative Research and Development," Sustainable Environment Research, vol. 28, pp. 267–273 2018. https://reader.elsevier.com/reader/sd/pii/S2468203918300645?token=D6147A7A32F856D2E6F8129CF20B46A5C76FD05F5143DDFB9BB25C27354E0A88242BE259CE482BB1CDA89C90A7F33F19 (accessed Oct. 1, 2020)

exceed 27°C (81°Fahrenheit) yearlong. The benefits from the KEPCO proposal would be that the powerplants would have increased electricity production and that their water exhaust not damage local sea life because of excessive heat.[824] The benefit for ocean fertilization would be that the nutrient-rich water released would be just warm enough to remain at the surface.

Engineers proposed a similar system to use excess heat from a Malaysia-Thailand Joint Authority (MTJA) gas production platform. Again, the OTEC production cycle of cold deep water to chill low-grade heat would be used. Only in the Malaysia-Thailand case, the proposal described it as geo-thermal ocean thermal energy conversion (GeOTEC).[825]

Researchers have also proposed OTEC as a bottoming cycle for nuclear power plants for the same reasons of increased electricity production and avoidance of thermal problems from cooling-water emissions.[826,827]

Finally, the benefits of oceanic fertilization can be tremendously greater when such combined (hybrid) systems of OTEC topped with an additional system go to sea. That is our next subsection.

9.4.2 Motor Ships and Factory Ships for Power and Infrastructure at Sea

Now comes the hard part, paying for fertilizing areas of bluewater by making a profit so that you can keep doing it.

That is where motor ships (also referred to as plantships) come in. Motor ships have been around for a thousand years. Medieval motor ships in rivers harvested the power of flowing rivers with paddle wheels most commonly to grind grain. There have been seagoing processing plants

---

[824] Hoon Jung and Jungho Hwang, "Feasibility Study of a Combined Ocean Thermal Energy Conversion Method in South Korea, *Energy*, vol. 75, pp. 443–452, Oct. 2014. https://www.sciencedirect.com/science/article/abs/pii/S0360544214009360 (accessed Nov. 4, 2020)

[825] N. H. Mohd Idrus, M. N. Musa, W. J. Yahya, and A. M. Ithnin, "Geo-Ocean Thermal Energy Conversion (GeOTEC) Power Cycle/Plant," *Renewable Energy*, vol. 111, pp. 372–380, Oct. 2017.

[826] Nam Jin Kim, Kim Choon Ng, and Wongee Chun, "Using the Condenser Effluent from a Nuclear Power Plant for Ocean Thermal Energy Conversion (OTEC)," *International Communications in Heat and Mass Transfer*, vol. 36, issue 10, pp. 1008–1013, Dec. 2009. https://www.sciencedirect.com/science/article/abs/pii/S0735193309001900 (accessed Nov. 5, 2020)

[827] Bin-Juine Huang and Ho-Tsen Lee, "Feasibility Analysis of an OTEC Plant as the Bottom Cycle of the Third Nuclear Power Plant," *Journal of the Chinese Institute of Engineers*, vol. 16, 1993. https://www.tandfonline.com/doi/abs/10.1080/02533839.1993.9677555 (accessed Nov. 5, 2020)

(or factory ships) to process whale and fish catches since the 1800s and seagoing power plants to provide electric power out to remote islands and coastal areas since the mid-1900s.[828,829,830]

The venerable electric-machinery company Siemens has designs for a combined cycle (gas turbine and then steam) natural-gas-fired power plant that could deliver more than 1300 megawatts of electricity from a power plant anchored within a breakwater.[831] Mitsubishi[832] and Kawasaki[833] have also proposed barge-mounted generating plants burning liquified natural gas (LNG).[834] Such power plants can be as much as a third more efficient than diesel generators.

For nuclear energy, the *Akademik Lomonosov* began delivering 70 megawatts of electrical of electrical power from a breakwater protected nuclear-fission power ship to a remote Siberian city beside the Arctic Ocean in December 2019.[835] There are even proposed designs for coal-fired barge power plants.[836]

The advantage of power ships and factory ships is that they can be built in a shipyard and sailed or towed to the point of use; this can be much cheaper than building on site. Likewise, factory ships as canneries, or chemical processing centers, can move to or moved to where the resource is for the most efficient processing. The limitation for these near-shore floating power

---

[828] Dag Pike, "The Future of Factory Ships," *The Motorship*, March 28, 2011. http://www.motorship.com/news101/ships-and-shipyards/the-future-of-factory-ships

[829] Patrick Jannsens, "Floating Power Plants: A New Solution to an Old Problem," *Gastech Insights*, Aug. 1, 2018. https://gastechinsights.com/article/floating-power-plants-a-new-solution-to-an-old-problem (accessed Jan. 19, 2019)

[830] "Floating Power Plant Market Research Report- Forecast to 2023" (summary of full report), Market Research Future, Oct. 2017. https://globenewswire.com/news-release/2018/09/26/1576340/0/en/Floating-Power-Plant-Market-Valuation-to-Touch-USD-1-769-7-Mn-by-2023-Registering-10-35-CAGR-Says-Market-Research-Future.html (accessed Jan. 19, 2019)

[831] *Floating Power Plants*, Siemens Power and Gas Solutions, Sept. 2018. https://new.siemens.com/global/en/company/stories/research-technologies/energytransition/the-future-of-energy-seafloat.html (accessed July 3, 2020)

[832] Sonal Patel, "Novel Floating Power Plants on the Horizon," *Power*, Dec. 2, 2018. https://www.powermag.com/novel-floating-power-plants-on-the-horizon/ (accessed April 4, 2020)

[833] Kenji Asada, "Kawasaki Heavy develops floating gas-fired power plant," *Nikkei Asian Review*, Dec. 3, 2018. https://asia.nikkei.com/Business/Companies/Kawasaki-Heavy-develops-floating-gas-fired-power-plant (accessed April 4, 2020)

[834] "Floating Power Plants: Mobile Power Generation for Island Living," *Forbes*, Aug. 18, 2017. https://www.forbes.com/sites/mitsubishiheavyindustries/2017/08/18/floating-power-plants-mobile-power-generation-for-island-living/#4232248d7e37 (accessed April 4, 2020)

[835] Kenneth Rapoza, "Russia's First Floating Nuclear Power Plant Turns On, Set to Replace Coal," *Forbes*, Dec. 19, 2019. https://www.forbes.com/sites/kenrapoza/2019/12/19/russia-first-floating-nuclear-power-plant-turns-on-set-to-replace-coal/#25f514da1e3d (accessed April 4, 2020)

[836] "Floating Coal Power Plants," web page, SeaPower Inc. USA, 2019. https://www.seapower-inc.com/ (accessed April 4, 2020)

plants and floating factories is that, for developed areas, onshore construction is fairly cheap with trucked-in supplies and nearby motels for various construction contractors. Consequently, on-site construction is the norm in developed areas.

However, as noted earlier, there are significant advantages in accessing cold water available at the edge of the continental shelf. These advantages are even greater in the bluewater deeps, and they have been proposed, and even attempted, before.

In the 1930s, Georges Claude, built a motor ship using a concept that could be applied nicely to fertilizing bluewater deserts. Claude (who invented neon lights and was called the French Edison) knew that deep water in the ocean is always nearly freezing cold, while tropical surface waters are tens of degrees warmer. [837] This temperature difference is not as much as in steam engines, which have temperature differences in hundreds of degrees (so efficiency is low), but the total potential energy of that ocean temperature differences is huge. Claude hoped to produce major amounts of electrical power in the South Atlantic and use it to make ice to sell in Brazil. A surprise hurricane (due to the rudimentary forecasting of the time) and exhausted capital during the Great Depression of the 1930s doomed Claude's venture, but the concept was plausible even with the technology available at that time.

During the two Energy Crises of the 1970s, Claude's concept got significant funding from several governments under the name of ocean thermal energy conversion or OTEC.[838] Unfortunately for those development efforts, the limitation of small temperature differences between the hot and cold temperatures were noted again. The potential energy for tapping was massive, but the slim efficiency of 5% or less might easily become some minus percentage of loss by just a thin layer of marine slime on the heat exchangers or by accretion of material inside the pipes.[839] Last and worst, that vast source of potential energy was generally far, far away from cities and industrial sites of developed-world countries in temperate climates whose governments were funding the research. Thus, OTEC-research funding dwindled down, and then collapsed entirely when the price of oil collapsed … again (!) in the 1980s.

Successfully using the potential energy of cold deep ocean water, requires additional revenue streams and/or additional uses. For our global-warming climate concern, those deep waters also have all the minerals to make green waters greener and turn bluewater desert green without need of shipping in fertilizers. The plants of such waters going greener could draw more

---

[837] Walter E. Pittman, Jr. "Energy from the Oceans: George Claude's Magnificent Failure." *Environmental Review: ER*, vol. 6, no. 1, pp. 2–13, spring1982. www.jstor.org/stable/3984046 (accessed Nov. 5, 2020)

[838] Clarence Zener, "Solar Sea Power," *Physics Today*; pp. 48–53, Jan. 1973; (This is a classic paper that helped restart OTEC development efforts).

[839] R. Paul Aftring and Barrie F. Taylor, "Assessment of Microbial Fouling in an Ocean Thermal Energy Conversion Experiment," *Applied and Environmental Microbiology*, vol. 38, no. 4, pp. 734–739, Oct. 1979.

carbon dioxide drawn from the atmosphere so there would be less global warming.[840] Furthermore, as noted in the last subsection, the green phytoplankton in such areas would also release sun-shading sulfates for additional cooling.

Those cooling processes are both great, but there are still two barriers. First, the mixed cold-and-hot water expelled from an OTEC is still cooler (and hence heavier) than the hot surface water, so that cooler water still tends to plummet back down below the sunlit (euphotic) upper waters where green plants can grow. The remedy from the previous subsection is a hybrid of OTEC and some other energy source—preferably noncarbon—to provide the energy to pump nutrient-rich water up from the depths, warm those nutrient-rich waters sufficiently to remain in the sunlit zone, and provide some additional money to pay for that large powership/plantship.

Second, if one creates a new upwelling with its riches of seafood, who benefits? Remember the tragedy of the commons? One entity building an OTEC facility (costing tens or hundreds of millions of dollars) would increase the fish catch for fishing boats from everywhere.

A motor ship can transcend all those problems and do much more. Let's start with the power plant. With a pipe to colder water, the temperature difference for the engine will be greater so the motor ship's engine will be even more powerful. After cooling the motorship's engine, the formerly cold bottom water can be released at a temperature slightly warmer than the surface water. Slightly warmer water is slightly lighter water; consequently, that mineral-rich water stays in that sunlit zone where plants can grow instead of sinking back into the dark waters below.

But there's more. With a heat source and a power plant, you can run a cannery (factory ship) and provide a base for fishing boats just as mobile canneries or factory ships work in the world's oceans today. Better yet, you can run power lines and compressed air lines to satellite stations maintaining facilities as far as several tens of miles (kilometers) away to string fish cages, shellfish rafts, and nets for culturing sea plants. Better yet again, the motor ship can provide the maintenance yard for work boats, the storage areas for produce and extra equipment, the entertainment centers for worker relaxation, the loading dock for importing and exporting materials, and a landing deck for short-takeoff or vertical-takeoff aircraft.

As noted earlier regarding power plants on the continental shelf, such plants could also synthesize ammonia fuel for work boats and for remote power systems such as on seagoing line shacks. There have already been such design proposals decades ago.[841]

Most importantly, the motor ship and its cabled-in satellite islands could essentially fence their perimeter to control access by potential seafood poachers and bottom-trawling vandals. It would be analogous to the barbed wire fencing of the Great Plains in North America during the 1800s.

---

[840] John S. Corbin, William A. Brewer, and Gerald S. Clay, *Mariculture and Ocean Thermal Energy Conversion, State-of-the Art Assessments: A Technical Supplement to Ocean Leasing for Hawaii, Hawaii Aquaculture Development Program*, 1981.

[841] W. Avery and D. Richards, "Design of a 160 MW OTEC Plantship for Production of Methanol," *Oceans*, vol. 15, pp. 746–750, Aug. 1983.

Jules Verne's 1895 fictional *Propeller Island* was a gigantic motor ship that carried its rich dilettante owners around the Pacific.[842] The fictional concept was plausible at the time but not financially practical.

Now, the technology has evolved to make such large craft financially practical, but not for sailing around the seas. These energy-and-mariculture islands would serve the world by holding station in one place and cabling power to a large number of smaller surrounding islands (an artificial archipelago) to produce energy, produce food, and hold control of the wealth produced. (See 9.4.4, Dynamic Positioning of Floating Platforms.) Various floating oil and gas platforms provide the prototypes for much larger central-island (platform) for maricultural archipelago to be described later.

The motor ship / central island of the archipelago would need to be sufficiently large and sturdy enough to hold position and to remain functional despite dangers such as hurricanes and rogue waves. (Tsunami waves would not be a threat because they only build to great height when they come to shallow water.) Such a craft would be too large and too far from shore for running to port. Moreover, abandoning the maricultural lines, nets, and stations would be a costly sacrifice.

9.4.3 Additional Technologies Needed for Archipelagoes

A number of existing and developing technologies are needed to make the islands and their surrounding archipelagoes practical, and the most important of these technologies are described in the following additional subsections. These technologies are

1. Totally buoyant construction
2. Dynamic positioning of floating platforms
3. Syntactic foams
4. Transportation advances
5. Virtual Presence, telebusiness, and robo workers in distant seas

9.4.3.1 *Totally Buoyant Construction*

As noted in 9.3.2 Floating Concrete, buoyant structures are a recent update to the hydraulic concrete of Roman days. Floating concrete structures can never spring a leak and sink. This is important for building floating islands and for maintaining use of them for decades. Such floating island structures would have a number of complexities.

- They also need structural support that would be maintained by steel, other metal, or composite skeleton materials.

---

[842] Jules Verne, [editions of the novel have been variously titled as] *Propeller Island*, *The Floating Island*, *The Pearl of the Pacific*, or *The Self-Propelled Island*, 1895.

- Metal skeleton supports would need active galvanic protection from corrosion.

- Shipyard construction might not be large enough. In many cases, shipyard-constructed modules would be barged in and attached to one or more structural members.

- Fail-safe propulsion would be needed in four or more directions for maintaining position (see next subsection). Hence, the island would need a backup power/thrusting system in addition to the main power plant.

9.4.3.2 Dynamic Positioning of Floating Platforms

As noted, the main platform (and any other platforms holding position) would need to hold position (station keep) to connect with maricultural facilities and all additional infrastructure. In shallower and more protected waters, existing large offshore structures, such as drilling platforms, can be held up by jack-up legs or dynamic tensioning with anchor chains in slightly deeper waters. However, increasing depths eventually make jack-up legs too expensive and too prone to failure. Likewise, anchoring cables in progressively deeper waters may become too massive to work.

Fortunately, geologists and oil drillers both wanted development of that different method for station keeping, dynamic positioning.

Science sometimes moves technology development. In 1909, Croatian seismologist Andrija Mohorovičić noticed two different types of earthquake waves that seemed to indicate the depths of the upper rock crust and the lower mantle rocks. The boundary between the two layers was later named after him as the Mohorovičić discontinuity or the Moho.

By the 1950s, geologists proposed a program to drill to the Moho, on through it, and down to the mantle to find out what made up the Moho and the mantle. Because the Moho is only 5 to 10 kilometers (3–6 miles) below the ocean floor, versus 20 to 90 kilometers (10–60 miles) beneath typical continental crusts, they proposed a drill ship that could hold position long enough to reach the mantle.

Also, in the 1950s, offshore oil and gas drillers had gotten steadily more ambitious setting drilling towers and then anchored drill rigs in progressively deeper water in the Gulf of Mexico. Eventually, they reached the practical limits mentioned earlier for jack-up rigs and anchor chains or cables. The drillers also wanted a drill ship that could hold position.

Joint funding efforts from the scientists and the drillers led rig builders to develop drill ships with powerful directable propellers at each of four corners. Position sensing was first done with radar on the surface and sonar pointing to the ocean floor. Later, position-sensing data from the Global Positioning System (GPS) of satellites was added. The rig developers also added power, crew quarters, and production facilities on these independent facilities.[843],[844] By 2016, the deepest well was being drilled more than two miles (three kilometers) below the surface of the ocean, and in a water depth of 3,400 meters (11,156 feet).[845] Today, there are hundreds of such rigs operating in the world, so the technology is well established. Furthermore, the rigs are still evolving to facilities that are bigger and even more capable

As for the geologist and their Moho Project, it never completed that deep hole to the mantle, but the technology provided shorter samples around the world that revolutionized geology in many ways including proof of drifting-plate tectonics, climate changes, and details about the asteroid or comet impact that finished off the dinosaurs. Also, geologists did discover a few places on Earth where shifting plates brought portions of the Moho upward enough to outcrop on the surface.)[846]

In those deeper waters (more than 1,500 meters (4,900 feet), the semisubmersibles or drillships are usually maintained at the required drilling location using dynamic positioning. Dynamically positioned drill rigs combined with power-plant ships have all the technologies that George Claude wanted for his ocean thermal energy. Most importantly, the drillers have well developed procedures for attaching and maintaining long strings of pipe and hanging them far into the depths. Sucking cooling water from the depths would require much greater pipe diameters, but the pipes would not need to hold drill tools or recycle driller's mud.

Another industry that may be using floating rigs is the space launchers. Thanks to the revolution of reusable boosters, entrepreneurs could actually realize the dream of industrializing Earth's near orbital space. However, one environmental downside of the budding industry is that daily launches or even multiple daily launches—and booster returns—can become a major noise irritant in the local area around the launch site. For that reason, Elon Musk's SpaceX is designing an offshore platform to launch and recover launchers from their new launch site on the Texas Gulf Coast. The floating launch facilities will likely be about 300 meters (1000 feet) long and 100 meters (330 feet) wide with a displacement of several tens of thousands of tons. Such a platform would be ten times larger than SpaceX's football-field-sized drone ships, which are already larger than a city block. Furthermore, SpaceX has suggested that the platform design would be repeated

---

[843] "Special Anniversary - The History of Offshore: Developing the E&P Infrastructure," *Oil & Gas Journal*, Jan. 1, 2004. https://www.offshore-mag.com/home/article/16756956/special-anniversary-the-history-of-offshore-developing-the-ep-infrastructure (accessed June 2, 2020)

[844] Sean, "A Brief History of Dynamic Positioning," *gCaptain*, Aug. 25, 2009. https://gcaptain.com/history/ (accessed June 2, 2020)

[845] Mike Schuler, "Maersk Drillship Spuds World's Deepest Well," *gCaptain*, April 1, 2016. https://gcaptain.com/maersk-venturer-begins-drilling-worlds-deepest-well/ (accessed June 2, 2020)

[846] "What We've Learned from 50 years of Ocean Drilling," *EarthSky*, Oct. 2, 2018. https://earthsky.org/earth/50-years-of-ocean-drilling-findings (accessed June 5, 2020)

for other launch sites.[847] The result would be more experience in building and operating very large high-tech ocean platforms.

Continued construction of floating platforms by drillers and space launchers should help assure that these structures will remain a vibrant and evolving technology capable of providing a dynamically positioned (or tension moored) central island. As noted earlier, it would provide everything needed for maintaining a network of mariculture operations: a stable attachment comparable to a real island: at least a semi-protected harbor for receiving and sending out ship cargoes, a stable harbor for support craft larger than outboard boats, a local repair yard, a fuel station, and electricity. It would essentially be a small floating industrial city.

9.4.3.3 Syntactic Foams

Extending energy and mariculture operations into the deeper waters has several difficulties.

- Steel cables become impractically heavy and could fail under their own weight while plastic lines can be severed by sharks grinding their teeth on the lines, which they tend to do.

- Bringing the exceptionally cold (but nutrient rich) water from the depths can be defeated by warmth conducting through the walls of the pipes bringing it up.

- Likewise, any mariculture or other work in those colder deep waters must deal with cold conducting in and reducing the effectiveness of batteries and people.

Syntactic foams can help with these difficulties. Syntactic foams are another composite. They are tiny bubbles (hollow spheres) composed of material such as glass, titanium, steel, or silicon carbide inside a matrix of some other material such as plastic or another metal—for example, titanium bubbles inside the lower temperature melting aluminum and likewise, glass beads inside resin. Syntactic foams increase the strength of their matrix material around them while providing a lower density, sometimes even lighter than water. Most importantly, the bubbles make the syntactic foam a good insulator.[848]

Such properties make syntactic foams ideal for easing the difficulties with structures of diving equipment such as submersibles, buoys, pipelines, and cables (for protection, buoyancy,

---

[847] Eric Ralph, "SpaceX Starship Will Need Fleet of Ocean Spaceports, Says Elon Musk," *Teslarati*, June 16, 2020. https://www.teslarati.com/spacex-starship-floating-spaceports/ (accessed July 17, 2020)

[848] "Metal Foams May Be Effective for Fire and Heat Protection," *Engineering 360*, April 7, 2016. https://insights.globalspec.com/article/2452/metal-foams-may-be-effective-for-fire-and-heat-protection (accessed Jan. 5, 2020)

and insulation).[849,850] The benefits of adding syntactic foams provides a mirror image to the problems of the earlier list.

- Steel cables can have a buoyant coating to eliminate the mass issue while polymer cables can have a protective metallic coating.

- Metal cold-water pipes can have a temperature-insulating coating that increases the buoyancy of the pipe (usually to neutral buoyancy). Conversely, a polymer pipe might have a protective metallic syntactic foam covering that would provide insulation, provide greater strength, and reduce the pipe's buoyancy to neutral or slightly negative.

- Insulation is important to slow loss of heat from batteries and personnel to the outside near-freezing temperatures of the depths. With that insulation, it will be much easier to use of the greatly increased third dimension of mariculture in the blue waters, fish cages hundreds of feet below the surface with thousands of feet more below to disperse organic waste products.

Syntactic foams are now exotic materials only used in fancy scientific research craft, but that must change for the Bluewater-to-Greenwater Revolution. As with many other technologies, 3D printing is decreasing costs and increasing capabilities. Conventional syntactic-foam part fabrication uses injection molding to produce individual parts that are then secured together with adhesives or fasteners to create complex-shaped parts in submersibles or other infrastructure. However, 3D printing allows fabrication of complex syntactic-foam parts of various shapes in a single form without the failure points of the adhesives and/or fasteners.[851]

---

[849] Gary Gladyz and Krishan K. Chawla, "The Space Between: To the Bottom of the Sea Using Buoyancy and Voids," *Scitechconnect*, Sept. 4, 2014.
http://scitechconnect.elsevier.com/bottom-sea/ (accessed March 14, 2020)

[850] Kimiaki Kudo (January 2008). "Overseas Trends in the Development of Human Occupied Deep Submersibles and a Proposal for Japan's Way to Take," *Science and Technology Trends Quarterly Review*, no. 26, pp. 104–123, Jan. 2008.

[851] Daphne MacDonald, "3D-Printed Syntactic Foams Might Bring Submarines to New Depths," *Engineering.com*, Feb. 22, 2018.
https://www.engineering.com/DesignerEdge/DesignerEdgeArticles/ArticleID/16463/3D-Printed-Syntactic-Foams-Might-Bring-Submarines-to-New-Depths.aspx (accessed March 14, 2020)

Ashish Kumar Singh, Balu Patil, Niklas Hoffman et al., "Additive Manufacturing of Syntactic Foams: Part 1: Development, Properties, and Recycling Potential of Filaments," *Additive Manufacturing of Composites and Complex Materials*, pp. 303–309, Jan. 24, 2018.

Ashish Kumar Singh, Brooks Saltonstall, Balu Patil, et al., "Additive Manufacturing of Syntactic Foams: Part 2: Specimen Printing and Mechanical Property Characterization," *Additive Manufacturing of Composites and Complex Materials*, vol. 70, pp. 310–314, Jan. 16, 2018.

9.4.3.4 Transportation Advances

The blue waters tend to be both far from land and more prone to high waves. These more severe conditions can best be served by transportation advances that have not been generally used before or are just being developed.

First, each central island would need at least industrial-sized cranes. One would be as used an offshore drill rig for sending and maintaining a big insulated pipe deep into cold waters (although no drilling bits or drillers mud recycling required). The second crane would pluck large items such as generators, construction supplies, and industrial modules from supply ships and return product containers to those ships.

Each central island would need a small breakwater to protect a couple tugboats to help keep freighters and tankers steadier when offloading and onloading. The breakwater area would also shelter several work boats for distributing materials via work boats to smaller facilities throughout the archipelago and could dock a couple of collection boats at a time bringing in mariculture production.

Even better, a bigger central island would have a breakwater-protected area big enough to accommodate at least one oceangoing ship for the most efficient delivery of supplies and equipment with pickup of produce.

Depending on the business plan, the breakwater-protected area might have tankers to bring in compressed natural gas or for pickup of synthetic fuel if the main power plant is nuclear fission. Production of synthetic fuel is low return on investment, but is steady money on a long-term contract.

Moving workers and visitors in and out within a single day would require air transportation. The best such transport would be the something along the lines of the original vertical takeoff and landing (VTOL) craft, the helicopter. Unfortunately, helicopters are expensive, relatively slow, and limited in range. Meanwhile, conventional aircraft require long runways that would be very expensive to build for a deepwater facility. Think of an aircraft carrier large enough to accommodate Boeing 737s.

Short takeoff and landing (STOL) aircraft provide greater speed and shorter runways.[852] As a practical matter, the airport for a large central island would probably be best sited on its own dynamically positioned platform to avoid large aircraft at STOL speeds flying near the central island.

Fortunately, the much more affordable alternative is a turboprop that can launch vertically as a helicopter and then tilt its wings into a conventionally winged craft to travel faster and farther than a helicopter, and then tilt its wings back again to land vertically like a helicopter. The Bell Boeing V-22 Osprey is an American multi-mission, tiltrotor military aircraft with both vertical

---

[852] Graham Warwick, "Is Super-STOL A Viable Alternative to Electric VTOL?" *Aviation Week*, May 13, 2020. https://aviationweek.com/aerospace/aircraft-propulsion/super-stol-viable-alternative-electric-vtol (accessed June 4, 2020)

takeoff and landing (VTOL), and short takeoff and landing (STOL) capabilities. It is designed to combine the functionality of a conventional helicopter with the long-range, high-speed cruise performance of a turboprop aircraft.

The Osprey has been pioneered for military use, but a civilian version would be ideal for transporting people and high-priority mail and parts to bluewater central facilities. There is already a smaller commercial version, the Leonardo AW609, that can carry 6,000 pounds (2,721 kilograms) or nine passengers at a speed of about 300 miles per hour (480 kilometers per hour) versus 140 miles per hour (225 kilometers per hour) for a helicopter. More importantly, it has a range of 700 nautical miles versus 360 for a helicopter (1130 versus 480 kilometers).[853,854]

However, the AW609 is too small and expensive except for passengers and small high priority cargo. There must be an Osprey-sized transport for both cargo and passengers at more competitive prices. Cargo deliveries would have standardized containers smaller but patterned after the standardized shipping containers that transformed world transportation by ship, train, and truck.[855]

Still, another possibility is that marine transportation might go forward to the past. In July 2020. the Aviation Industry Corporation of China (AVIC) demonstrated the first public flight of the AG600 Kunlong, a four-engine turbo-propeller amphibious aircraft that can carry 50 passengers as far as 2700 miles (4300 kilometers).[856] Its cruising speed is 500 kilometers per hour (310 miles per hour). It can take off and land in adverse weather conditions with a wave height of 2 meters (6.5 feet).

The AG600 is half the size of Howard Hughes' "Spruce Goose" (which only flew once), and the Kunlong is now the world's largest type of operational seaplane. As such, it will have much greater potential than helicopters for moving personnel and high-priority parts to distant ocean platforms.

---

[853] Jeremy Bogaisky, "After 24 Years, the Civilian Version of the Marines' V-22 Osprey Tiltrotor Is Finally Nearing Takeoff," *Forbes*, May 9, 2020. https://www.forbes.com/sites/jeremybogaisky/2020/03/08/aw609-leonardo-marines-v-22-osprey-tiltrotor/#7d118e8d5354 (accessed June 6, 2020)

[854] Thomas Pellini, "The First Civilian Version of the Half-Plane, Half-Helicopter V-22 Osprey Will Soon Be Available to Buy. Here's How It Could Revolutionize Air Travel," *Business Insider*, May 29, 2020. https://www.businessinsider.com/leonardo-aw609-civilian-v-22-osprey-details-photos-2020-3 (accessed June 6, 2020)

[855] Marc Levinson, *The Box: How the Shipping Container Made the World Smaller and the World Economy Bigger*, Princeton University Press, Jan. 27, 2008.

[856] Bryan Hood, *Robb Report*, "The World's Largest and Newest Seaplane Completed Its Maiden Voyage over Water Last Weekend," *Robb Report*, July 28, 2020. https://robbreport.com/motors/aviation/china-giant-seaplane-kunlong-flies-2938880/ (accessed July 31, 2020

9 Fears and Fixes from Sea and Sky

9.4.3.5 Virtual Presence, Telebusiness, and Robo Workers in Distant Seas

Living space on ocean habitats will always be more expensive than housing on land. Thus, development of ocean habitats will drive greater automation just as the limited number of workers in the early United States led to a more automated, and thus more productive, economy that led the world by the early 1900s.

Robotic maritime systems will be a key enabling technology in growing the Bluewater-to-Greenwater Revolution of offshore energy, fishery production, and mariculture industries. In Subsections 7.5. Rise of the Robo Farm Machines and 9.2.2 Starting the Greenwater Revolution, I noted the benefits of robo farming equipment for both land agriculture and the sea agriculture of mariculture.

These advantages are multiplied in the blue waters where people will be more expensive to house and where cold high-pressure waters will be harder on them for actual work in the water. Everything learned in the earlier robo land and near-shore operations will apply in the new realm, and even better technologies will be needed to succeed in an environment expensively far from shore, with more extreme wave action and with operations extending down to areas of near-freezing temperatures and high pressure. Some of those robotic technologies include the following:

- More distant open-water mariculture needs to reduce costs by using robotic vehicles to move fish cages at a slow pace, monitor for pests such as sea lice, and clean fish cages.[857]

- Small robo vehicles provide continuous monitoring inside fish cages without the cost or disturbance to the fish of human divers.[858],[859]

- Small uncrewed helicopters (such as the Camcopter S-1) can make high-priority deliveries much more cheaply than a piloted and fully instrumented conventional helicopter as well as the drones more common aerial survey work.[860]

---

[857] Dee Ann Divis, "The Tide Rises for Open Ocean Farming," *Inside Unmanned Systems*, Feb. 26, 2016. https://insideunmannedsystems.com/the-tide-rises-for-open-ocean-farming/ (accessed June 5, 2020)

[858] M. Kruusmaa, R. Gkliva, J. A. Tuhtan, A. Tuvikene, and J. A. Alfredsen "Salmon Behavioural Response to Robots in an Aquaculture Sea Cage," *Royal Society Open Science,* vol. 7, issue 3, March 1, 2020. https://royalsocietypublishing.org/doi/pdf/10.1098/rsos.191220 (accessed June 5, 2020)

[859] Lisa Jackson, "Rise of the Machines: Aquaculture's Robotic Revolution," Global Aquaculture Alliance web page, Feb. 13, 2017. https://www.aquaculturealliance.org/advocate/rise-of-the-machines-aquacultures-robotic-revolution/ (accessed June 5, 2020)

[860] Fabienne Lang, "Camcopter S-100 Helicopter Drone Revolutionizes Oil Rig Deliveries," *Interesting Engineering*, Sept. 3, 2020. https://interestingengineering.com/camcopter-s-100-helicopter-drone-revolutionizes-oil-rig-deliveries?_source=newsletter&_campaign=vMY2oEBM3Wj0D&_uid=LDdwpX6Re1&_h=9616d98f82a7a631519bae2822a86b243f923bbc&_nt=l76bw0&utm_source=newsletter&utm_medium=mailing&utm_campaign=Newsletter-03-09-2020 (accessed Sept. 3, 2020)

- Robo inspectors for oil and gas platforms are faster than humans and can operate in harsh unsafe conditions including drones for flying overhead and crawling robots for inside pipes and chambers.[861]
- A South Korean firm developed a type of remote operating vehicle to destroy jelly fish, which have increasingly become dangerous pests (probably due to loss of controlling predators).[862]
- Robo divers have been proposed for collecting and processing the sea urchins that attack the holdfasts of kelp.[863]

The point of the above list is that many of the needed robo technologies exist or are being developed.

Likewise, many ancillary services will be automated or performed remotely from onshore facilities (onshored). Handily, this fits with a long-term trend in many industries

Even before the 2020 Corona Virus Pandemic there was an existing trend toward internet shopping, business management, and entertainment. Many people already worked remotely from home, held virtual business meetings remotely from home or office, and had remote social gatherings.[864] The widespread pandemic lockdowns only accelerated the trend.[865]

Long-term bluewater operations will minimize on-site personnel while maximizing production not just because of health concerns but more importantly because of distance and limited space for housing personnel. Hence, administrative functions such as personnel, payroll, ordering, accounting, and customer service can all be "onshored" with remote electronic

---

[861] Tsvetana Paraskova, "Robots and Drones Are Changing the Offshore Oil Industry," OilPrice.com, May 6, 2018. https://oilprice.com/Energy/Energy-General/Robots-And-Drones-Are-Changing-The-Offshore-Oil-Industry.html (accessed June 5, 2020)

Liz Hampton, "In U.S. Gulf, robots, drones take on dangerous offshore oil work," *Reuters*, May 3, 2018. https://www.reuters.com/article/us-world-work-oildrones/in-u-s-gulf-robots-drones-take-on-dangerous-offshore-oil-work-idUSKBN1I4100 (accessed June 5, 2020)

[862] Drew Prindle, "Meet South Korea's Autonomous Jellyfish-Murdering Robots," *Digital Trends*, Oct. 8, 2013. https://www.digitaltrends.com/cool-tech/jellyfish-murdering-robots/ (accessed June 6, 2020)

[863] Jeff Kart, "Startup Says Robot Divers Can Help Remove Urchins, Restore Oceans, *Forbes*, Feb. 19, 2019. https://www.forbes.com/sites/jeffkart/2019/02/19/startup-says-robot-divers-can-help-remove-urchins-restore-oceans/#3d2c30c95cb4 (accessed June 5, 2020)

[864] Sean Peek, "Communication Technology and Inclusion Will Shape the Future of Remote Work," *Business News Daily*, March 18, 2020. https://www.businessnewsdaily.com/8156-future-of-remote-work.html (accessed June 6, 2020)

[865] Jared Spataro, "Remote Work Trend Report: Meetings," Microsoft 365 web page, April 2, 2020. https://www.microsoft.com/en-us/microsoft-365/blog/2020/04/09/remote-work-trend-report-meetings/ (accessed June 6, 2020)

connection to the oceanic facilities. Mechanical checks and repairs can be trained and/or assisted remotely.[866] Even secretarial functions can be moved off site with virtual assistants.[867]

The key technologies for extending this internet revolution to oceanic facilities are fiber-optic cables, cell-phone repeaters, drone repeaters, and fleets of low-altitude satellites that will make the entire planet—and especially the archipelagoes—wireless high-fidelity (Wi-Fi) accessible. The future possibilities include high-altitude drone virtual repeaters for Wi-Fi throughout the archipelago. Flying drone hot spots were proposed in 2014.[868] By 2018, a Drone America website was advertising an autonomous airborne repeater.[869] Within a decade or two, there will be ubiquitous 5G and perhaps even 6G for bandwidth sufficient to remotely connect in many more ways.

9.4.4 Putting It All Together–A Complete Archipelago

Beginning in the early 1400s, the Europeans began the Age of Discovery during which they explored, traded, and often colonized all over the world. They could have started their voyages much sooner. For centuries, they had possessed compasses for direction, jib sails for tacking into the wind, and improved astrolabes for knowing one's latitude; yet they waited.

Why? Because sailing off across a vast ocean was an entirely new concept with many risks, real and imaginary. Hence, they waited and waited.

It is much the same for the worldwide land-based society hardly imagining the impending extension of human agriculture and industry onto the ocean even as the technologies to do so become steadily more practical. Undersea pipelines, offshore oil and gas production platforms, robotics, next-generation small modular reactors, and many other technologies are coming to the fore.

A tropical oceanic archipelago would combine all those technologies and more for ocean agriculture (mariculture) and industry while also reducing global warming. Let's examine a scenario about what an oceanic archipelago might be like and its issues.

---

[866] Courtney Linder, "This Robotic Arm Can Lend a Helping Hand with Repairs," *Popular Mechanics*, Nov. 17, 2019.
https://www.popularmechanics.com/technology/robots/a29712497/robot-repair-arm/ (accessed June 6, 2020)

[867] Chau Lim, "Trend of the 2020s: Fewer Full-time Workers, More Virtual Assistants," Virtualdoewell.com, Oct. 4, 2019. (accessed June 4, 2020)

Fiona Swaffield, "Virtual Assistants: The Growing Trend of Outsourced Assistance for Entrepreneurs," *Entrepreneur*, March 24, 2016.
https://www.entrepreneur.com/article/272879 (accessed June 4, 2020)

[868] Michael Peck, "That Drone Is a Wi-Fi Hotspot," *War is Boring*, April 8, 2014.
https://warisboring.com/that-drone-is-a-wi-fi-hotspot/ (accessed July 3, 2020)

[869] "Solving Your Line of Sight Radio Connection Issues," *Drone America*, 2018.
http://www.droneamerica.com/daconnectxtr (accessed July 3, 2020)

9.4.4.1 Location

The first key feature of a tropical oceanic archipelago is that it be in tropical deep bluewater desert, far from land. This provides multiple benefits:

- Great depth allows for greater maricultural production using multiple species at multiple levels.

- Great depth allows culturing the full range of animal species from tropical in the surface waters down to cold-water species in the depths. (Of course, plants would be limited to tropical species because the sunlit surface waters would only be tropical.)

- Great depth allows sinking organic material to spread widely before sinking into the depths. This would be used to reduce the net flow of global-warming carbon dioxide into the atmosphere while avoiding an excess flow of organic matter in any one area, which could cause a dead zone from too little oxygen.

- Changing an area of bluewater desert to greenwater causes the maximum change in global-warming factors. This is a major plus from a GND standpoint.

- An area of bluewater desert has no unique environment compared to other millions of square miles of bluewater desert. Thus, it will be more difficult for environmental activists to challenge such a facility as destroying an irreplaceable habitat.

- Operators can more easily defend a remote location against demonstrations.by nongovernmental organizations.

- There will be no developed land up-current within a set distance from the archipelago. An archipelago will be first of all an agricultural area, so several days of sunlight and marine bacteria-attacking organisms between nearby settlements and the archipelago will increase food safety and decrease the arrival of marine pests such as star fish, crown-of-thorns seastars, and sea lice.

- A favorable location for a tropical archipelago will be along one of the five great subtropical ocean gyres that move in stately circles around the Indian, North Pacific, South Pacific, North Atlantic, and South Atlantic Oceans. This will allow low-nutrient bluewater desert at the up-current side (such as coral reefs) and high-nutrient fertilized greenwater flowing in a steady plume down current for continued maricultural applications. Also, it will help prevent over fertilization that could happen with slow or no currents closer to a gyre center, in which a fertilized plume might circle back to its source.

- There would be no coral islands or coral shoals within a set distance down-current from the archipelago. Corals evolved to thrive in very low nutrient levels, and much of the recent damage to coral areas is because of excess nutrients from flows of sewage,[870] fertilizer runoff,[871] livestock runoff, and dumped dredging spoil[872] (which was still being dumped in the Great Barrier Reef Marine Park in Australia as of 2019).[873] An archipelago will be a super nutrient source, so it should not be located up-current and near existing coral reefs. Also, proximity to an existing island ecology would increase the danger of escaped farm animals polluting the gene pool of an existing island ecology.

One additional feature that would help an archipelago would be if at least, the portions of an archipelago complex with riser pipes bringing up deep water were located above a seamount, which is a comparatively isolated elevation of 1,000 meters (3,200 feet) above the ocean floor.[874] (A related term of guyot [or tablemount] is a seamount that has a level top; however, the three terms are often intermixed.) These features are generally foot notes in oceanic texts. However, they are an important feature for siting archipelago complexes. their relative shallowness makes mooring platforms to them much easier and hence more affordable than the average 3,000 to 6,000 meters [10,000 to 20,000 feet) in the abyssal plains, which are the most common feature of the ocean. This is particularly so for guyots that might afford the opportunity for a number of mooring locations because guyots often spread over a wider area than a seamount peak. There are at least 10,000 seamounts in the world's oceans, and many more are yet to be discovered.

Siting for an archipelago complex would need a certain minimum water depth for three reasons.

1. Greater distance above bottom-dwelling pests (such as coral-attacking crown-of-thorns seastars, coral-choking seaweed, and kelp-attacking sea urchins) can make it easier to avoid or at least control them.

2. Greater distance from the surface goes from the tropical shallow waters to polar cold and dark at depth. Consequently, Fish cages at greater depths can provide

---

[870] Stephanie L. Wear and Rebecca Vega Thurber, "Sewage Pollution: Mitigation Is Key for Coral Reef Stewardship," *Annals of the New York Academy of Sciences*, May 8, 2015. https://nyaspubs.onlinelibrary.wiley.com/doi/full/10.1111/nyas.12785 (accessed July 30, 2020)

[871] Jennie Mallela, Stephen E. Lewis, and Barry Croke, "Coral Skeletons Provide Historical Evidence of Phosphorus Runoff on the Great Barrier Reef," *PLOS One*, Sept. 27, 2013. https://journals.plos.org/plosone/article?id=10.1371/journal.pone.0075663 (accessed July 30, 2020)

[872] Martin LaMonica, "Dredge Spoil Linked to Coral Disease, WA Study Shows," *The Conversation*, July 16, 2014. https://theconversation.com/dredge-spoil-linked-to-coral-disease-wa-study-shows-29265 (accessed July 30, 2020)

[873] "Great Barrier Reef Authority Gives Green Light to Dump Dredging Sludge," *The Guardian*, Feb. 19, 2019. https://www.theguardian.com/environment/2019/feb/20/great-barrier-reef-authority-gives-green-light-to-dump-dredging-sludge (accessed July 30, 2020)

[874] Harold D. Palmer, "Seamounts (Including Guyots)," *The Encyclopedia of Oceanography: Encyclopedia of Earth Sciences Series*, Vol. I, Rhodes W. Fairbridge, editor, *Reinhold Publishing Corporation, New York,* pp. 782–786, *1966.*

habitat for cold-water species, so fish culturing will not be limited to tropical species.

3. The ultimate purpose of each archipelago is to produce not just a profitable set of sea animal and plant maricultural products, but an overabundance of organic matter such that a significant amount can sink into deep water and reduce the net flow of carbon dioxide into the atmosphere. If these organic materials landed in warm shallow waters they would eaten or decompose, and much of the organic matter would be "recycled" back to the atmosphere as respired carbon dioxide. Worse, too much organic material sinking into a small (shallower and warmer) area would generate a low-oxygen dead zone such as the dead zone at the mouth of the Mississippi River. A deeper basement under the archipelago would allow the plume of excess organic matter to spread wider and decrease in concentration.

9.4.4.2 Power and Chemical-Production Platform Technology Developments from the Oil and Gas Industry

As noted earlier, offshore oil and gas platforms have a history approaching three-quarters of a century. During that time, they have grown steadily larger, more capable, and able to float rather than be grounded on the seabed.

Operators of these platforms have developed many of the technologies and procedures that will be needed for an archipelago agricultural complex and are developing still more. Some of the existing larger and more sophisticated oil and gas platforms and their key features include:

- Hibernia (1997) is a 37,000-metric-ton (41,000-ton) integrated topsides facility mounted on a 600,000-metric ton (660,000 ton) gravity base structure, meaning that it sits on the seabed east of Newfoundland, Canada. Hibernia does well drilling, intake of fluids, separating out of water and natural gas, reinjection of the natural gas underground, and pumping of the oil on to shore. Hibernia has been producing since November 1997.[875]

- Petronius A (2000) is a giant Chevron–Marathon free-standing tower 130 miles (208 kilometers) southeast of New Orleans. It is located in water depths of 1754 feet (535 meters). The tower is 2,010 feet (610 meters) in total height (taller than the Eiffel tower) with 256 feet (78 meters) above the water.[876]

- Perdido (2010) is a Shell, Chevron, British Petroleum, and other partners single-column floating oil-and-gas drilling and production platform in the Gulf Coast between the United States and Mexico. "The Perdido hull and spar weigh in at 55,000 metric tons (60,000 tons). The main column, or spar, was built in a Finnish shipyard, and it now bobs 2,450 meters (8,040 feet) above the seabed. It gathers

---

[875] *Hibernia*, Hibernia website, undated. https://hibernia.ca/ (accessed Aug. 1, 2020)

[876] "Into the Abyss – The World's Deepest Offshore Oil Rigs," *Offshore Technology*, Oct. 11, 2011. https://www.offshore-technology.com/features/featureinto-the-abyss-the-worlds-deepest-offshore-oil-rigs/ (accessed Aug. 1, 2020)

oil from many wells (eventually 35) and forwards the oil to via pipeline to shore on the seabed. It has about 270 staff living on the platform and on the nearby floating hotel, referred to as a flotel.[877] The Perdido began operations in March 2010.[878] It is the first major platform held in place by polyester tensioning lines instead of steel.[879]

- Mars B/Olympus (2014) is a Shell 140,00-ton (127,000-metric-ton) tension-legged oil gathering facility located approximately 210 kilometers (130 miles) south of New Orleans in 3,100 feet (945 meters) of water. It started pumping in January 2014.

- Berkut (2015), on the Russian Pacific Coast near Sakhalin Island, is 200,000 tons (180,000 metric tons) grounded on the seabed in 48 meters (156 feet) of water. In January 1915, it began processing oil from a number of wells and forwarding the oil to via a pipeline to shore. It is designed to withstand ice 3 meters (10 feet) thick. The Berkut is owned by a consortium of the U.S. major ExxonMobil (30%), Japan's Sodeco (30%), Russia's Rosneft (20%) and India's ONGC Videsh (20%).[880]

- Stones (2016) is a Shell floating production, storage, and offloading (FPSO) rig. 320 kilometers (200 miles) southwest of New Orleans, Louisiana. It is the deepest operating project, reaching a depth of 2,900 meters (9,500 ft) to the seabed below. Stones started production in September 2016. The Stones is actually a ship that can detach from the oil penstock and sail to calmer waters during storms or to a shipyard for refitting.[881]

- Prelude (2018) is a 488-meter (1,605-foot) long floating liquid natural gas (FLNG) platform anchored 475 kilometers (295 miles) from Broome, Australia. It gathers, liquifies, and stores natural gas for pickup by tankers. It is two-thirds owned by

---

[877] Rob Almeida, "A Quick Tour of Perdido, the World's Deepest Offshore Production Platform," *gCaptain*, Sept. 12, 2012. https://gcaptain.com/quick-tour-perdido-worlds-deepest/ (accessed Aug. 1, 2020)

[878] Brett Clanton, "Shell's Perdido Blazes a Deep-Water Trail," *Houston Chronical*, April 18, 2010. https://www.chron.com/business/energy/article/Shell-s-Perdido-blazes-a-deep-water-trail-1619664.php (accessed Aug. 1, 2020)

[879] Daniel Canty, "Perdido's Journey," *Oil & Gas Journal*, April 14, 2009. https://www.oilandgasmiddleeast.com/article-5284-perdidos-journey (accessed Aug. 3, 2020)

[880] "Biggest Oil Rig Ever: 200k-ton Sakhalin Giant Begins Production," *RT.News*, Jan. 20, 2015. https://www.rt.com/news/224371-oil-rig-berkut-extraction/ (accessed July 31, 2020)

[881] Stones Deep-Water Project, website, Shell, undated. https://www.shell.us/energy-and-innovation/energy-from-deepwater/shell-deep-water-portfolio-in-the-gulf-of-mexico/stones.html (accessed Aug. 1, 2020)

Shell with additional partners, and it was built in a South Korean shipyard.[882] Prelude's gas production started in 2018. Unfortunately, the 600,000-ton (540,000-metric-ton) barge has had production problems, and it may be difficult for Shell and its partners to recover the (probably $14 billion) invested in the project.[883]

- Appomattox (2019) is a 125,000-metric-ton (138,000-ton) Royal Dutch Shell platform in water 7,400 feet (2,200 meters) off the Louisiana coast that gathers oil from a number of wells and forwards the oil via a pipeline on the seabed going to shore. Because it was built during a time of low oil prices, the designers worked especially hard to lower costs. The Appomattox commenced operations early and under budget in 2019[884] For instance, the Appomattox was the first platform in the Gulf of Mexico to use the especially-efficient technology called combined cycle power generation with not only a gas turbine but with the turbine hot exhaust supplying heat for a steam-turbine-driven electric generator.[885,886]

These structures have the size, the complexity, and the experienced organizations to expand enough to support major oceanic complexes, such as archipelagoes, which would include the following functions:

- Serve as the cannery ship (also known as mother ship) that cleans, processes, freezes, cans, or even prepares for live shipment all plant and animal produce for export to shore. The cannery ship was a key innovation for more profitable whaling and fishing fleets. It will be even more important for the archipelagoes, which will need greater efficiency to compensate for the high costs for operating in deep waters distant from port.

- Electrolyze or thermally crack water to generate hydrogen.

- Use hydrogen and more energy to synthesize ammonia for fuel in the archipelago with the advantage that exhausts from ammonia power would generally fertilize the

---

[882] Peter Williams and Peter Milne, "Shell's Prelude, the World's Biggest Vessel, Sets Sail for WA's North West," *The West Australian*, June 28, 2017. https://thewest.com.au/business/oil-gas/shells-prelude-lng-vessel-sets-sail-ng-b88521899z (accessed July 31, 2020)

[883] Tim Treadgold, "Shell's $12 Billion LNG Experiment Becomes a Big Headache," *Forbes*, June 23, 2020. https://www.forbes.com/sites/timtreadgold/2020/06/23/shells-12-billion-lng-experiment-becomes-a-big-headache/#7d9aebce1107 (accessed July 31, 2020)

[884] David Blackman, "Shell Appomattox Project: Ahead of Time and under Budget," *Forbes*, May 24, 2019. https://www.forbes.com/sites/davidblackmon/2019/05/24/shell-appomattox-project-ahead-of-time-and-under-budget/#514882e3a576 (accessed July 31, 2020)

[885] Thomas Francis, "Appomattox: the Energy Project That Defied the Odds," Shell web page, May 19, 2019. https://www.shell.com/inside-energy/appomattox-the-energy-project-that-defied-the-odds.html (accessed July 30, 2020)

[886] *Appomattox*, Shell web site, undated. https://www.shell.com/about-us/major-projects/appomattox.html (accessed July 31, 2020)

water rather being toxic as unburned hydrocarbons are. It might also export ammonia to shore, depending on the business plan.

- Cook and purify chitin from shrimp and other crustaceans for export to shore or further processing with seaweed to synthesize plastics for lines, floats, and structural materials.

- Grind and use seaweed for the culture of various worms, insect larvae, or insects to generate high-protein material to use as livestock feed—all protein for carnivorous species, mixtures of protein and plant for omnivores, and entirely plant matter for herbivores.

- Collect, concentrate, and use carbon dioxide that will outgas from deep chilling water brought up from the riser pipe.

- Last, but not least, process some of the seaweed and livestock produce for serving to the archipelago crew.

9.4.4.3 Upwelling Riser/Penstock

The key function of the archipelago is to transform an area of tropical nutrient-poor bluewater desert into a highly productive greenwater agricultural area that draws carbon dioxide down from the sky and pays for the process by yielding food. This is a farming operation comparable to irrigation except that nutrient levels in water, rather than water supply, are the limiting factor.

The tropical bluewater areas have limited living matter because they are far from rivers carrying nutrients from the ocean and because hot tropical surface waters inhibit upwelling of nutrient-rich waters in the depths, which are colder and thus denser than the surface waters. The goal is to bring nutrient-rich water up to the sunlit zone (also called the euphotic or epipelagic zone).

The sunlit zone is the layer from the surface down to depths of roughly 200 meters (660 feet). Going down from the surface, nutrient levels increase steadily to a maximum concentration at depths of roughly a thousand meters (3300 feet), and this concentration stays roughly the same all the way to the sea floor. That 1,000-meter (3,300-foot) depth is also the point where there is absolutely no surface light (the midnight zone where the only light is bioluminescence from life forms that glow). From the tropical ocean surface to the top of the midnight zone averages 80°Fahrenheit down to 39°Fahrenheit (27°Celsius down to 4°Celsius). Thus, the top of the midnight zone should be adequate to supply both cold water and nutrients for surface operations.

The only limitation on that depth estimate also comes from that water-irrigation analog. When irrigators suck a large amount water up from a well, the water level around each pumping well intake sinks in a cone-shaped depression in the aquifer. Although water flow in the deep ocean is much faster than in a land aquifer, there would be some lowering of the high-nutrient layer to farther down from the ocean surface. Consequently, an additional riser pipe length of a couple hundred meters (660 feet) would be prudent. This is easily within the capabilities of the present technologies of oil and gas platforms.

These advanced platforms would amaze and gratify George Claude if he were here to see them. They could easily raise the cold deep waters that he sought to bring up for ocean thermal energy conversion power on the Cuban coast and in the South Atlantic.

The riser or penstock for bringing up cold water to the platform will be very much like the flow of petroleum, water, and natural gas coming up from the wells to a platform's processing plant. The difference will be that the pipes will have a larger diameter and they will need insulation to keep the deep waters cold for more power from energy conversion, more chilling for food processing, and cooling the overall exhaust water so that it is not so hot as to be bad for sea life.

The larger diameter pipe would have to the same potential problems that have confronted designers of ocean thermal energy conversion power plants. They would tend to be heavier, and even a riser of 4 meters (13 feet) in diameter would lose significant amounts of efficiency-making cold from heat conducting into the pipe as water flowed upwards.

Fortunately, there are two solutions to both of these concerns. First, the pipe would not need to serve the drilling tasks of carrying drilling tools and pumping up drilling mud along with cuttings from the drill.

Second, the riser pipe could have the syntactic foams discussed in subsection 9.4.2.3 wrapped around them for significant insulation … and also additional strength. The term foam is a somewhat misleading because it implies something like a pile of soap bubbles. Actually, the foam may be fabricated as a soft gooey mixture, but it hardens into solid material that also has air-containing beads that make it an insulating material. Also, the proper bead material can both make the riser pipe stronger and (most important for a pipe of 1,200 meters) a lighter material so that the pipe might present minimal weight to the platform or even have a net neutral buoyancy. Furthermore, the material is a composite so adding ceramic or metal fibers into the matrix material could add even more strength.

Syntactic foams have already been used for coating deep-sea pipelines. In those cases, the syntactic foam was used to keep cold out of the oil being transported so that the oil would flow better, but the principle is just reversed to maintain cold within the pipe. A 2019 patent on syntactic foam and pipelines with such foams refers back to a number of related patents,[887]

9.4.4.4 The Power Plant(s)

The original researchers and questing entrepreneurs who worked for ocean thermal energy conversion always had the problem of the parasitic loss from pumping water from the depths to the power plant. That issue never changes for a large enterprise that needs to make a profit even

---

[887] Thomas C. Mirossay and William J. Benton, *Syntactic Foam Compositions, Pipelines Insulated with Same, and Method*, United States Patent Application US 2011/001734.0 A1, Jan. 27, 2011.
https://patentimages.storage.googleapis.com/bf/2f/7f/1d412cca14e2e1/US20110017340A1.pdf (accessed Aug. 2, 2020)

if the power plant has a heat source much greater than the temperature of the surface water. Whatever heat source is used, it will have its own expenses that must be covered.

The immediately available power source is natural gas, which has again gone to very low prices—this time because of the Fracking Revolution. With such low prices, liquified natural gas (LNG) tankers have begun to ply the world's waters. As noted in 9.4.2, Motor Ships and Factory Ships for Power and Infrastructure, there have already been several proposals from major power-system builders for floating natural-gas power plants.

The biggest advantage would be that a floating gas-fired power plant would not need any nuclear design certification—no nuclear. As such, a gas-fired power plant would be an ideal demonstration facility for the full-up nuclear archipelago.

It would still need to transcend the major challenges of operating a major floating power and agricultural enterprise in the deep ocean. Therefore, there are four ways to wring the maximum possible amount of electricity from the heat source that apply first with the fossil-fuel demonstration and later with nuclear energy.

1. A gas turbine engine: Like a jet engine in the sky, a gas turbine can run at least at temperatures of 1,600°Celsius (2,900°Fahrenheit).[888] That is the hottest heat engine available, and that makes the gas turbine the highest efficiency topping cycle of about 55% efficiency.

2. A steam (lower-temperature) cycle: The gas turbine exhaust can heat a water boiler to raise steam for a steam turbine, and that can boost the combined heat-to-electricity efficiency up to 65%. This combined cycle power generation with gas and steam turbines is used in many power plants. The Appomattox floating oil gathering platform is the first combined cycle power plant on a floating platform.

3. A supercritical carbon dioxide cycle (sCO$_2$): As described in subsection 8.7.5, supercritical carbon dioxide might be used as another step down in temperature from the gas and steam turbines or as a more compact replacement for the steam turbine. In either case, it can get that 10% of the steam turbine plus maybe another 5% increase in efficiency.

4. A low-temperature (bottoming) cycle: Finally, the absolute bottom cycle would be the one of the low-temperature cycles from the ocean thermal energy conversion (OTEC) proposals. These would be either open cycle or closed cycle. Closed cycle is just liquid to gas and through a turbine, only using a fluid such as ammonia that boils at a lower temperature. Closed cycle decreases air pressure to a partial vacuum so the warmer water goes to steam and then cool water in pipes condenses

---

[888] "MHI Achieves 1,600°C Turbine Inlet Temperature in Test Operation of World's Highest Thermal Efficiency "J-Series" Gas Turbine," No. 1435, press release, Mitsubishi Heavy Industries, May 26, 2011. https://www.mhi.com/news/story/1105261435.html (accessed Aug. 2, 2020)

it again. The side benefit is that the open cycle produces fresh water. If used, an OTEC-style bottoming cycle might yield another several percent of efficiency.

Even that would not be the end of the heat use because some of the heat would provide process heat for chemical processing, hot-water cleaning, canning or freezing of harvested produce, and HVAC (heating, ventilation, and air conditioning. At the same time, more of the cold deep water would boost freezing of maricultural and chilling for air conditioning on the platform. Some heat would also be used for chemical processing, but that is another subsection.

In addition, one more use for the parasitic pumping of water is that water flow from the platform can be directed in several directions to provide thrust for some part of the station-keeping thrust for the platform if the platform is using dynamic positioning.

Even if the platform were to be moored to the sea bottom, backup propulsion for dynamic positioning is needed ... just in case there are any problems with the mooring. Likewise, the power must have backup so that the platform can continue operations if the main power system is down for any kind of maintenance.

Finally, as noted earlier, nutrient-rich water released from the power plant can be exhausted at a temperature slightly warmer than the local surface water. The whole complex of the archipelago finally leads to the GND goal of a massive area of fertile upwelling that help slow global warming.

Back at the heat source, nuclear fission is the ultimate goal as the heat source for two reasons. First, the purpose of the GND is mitigating global warming, and less burning of hydrocarbon fuels means less global-warming carbon dioxide entering the atmosphere. Second, as described in 8.3, The Return to Atom Splitting Power with Better Next-Generation Nuclear Fission Reactors, fission could become cheaper than fossil fuels.

Once again regarding development time, the widespread use of fission is not a technology that will be immediately implemented. Then, extending it to the deep sea is another delay of some years or decades. Many tools must be used until better tools become available—many of them years after the initial 12-year GND proposal.

9.4.4.5 Rationale and Details for Mariculture as the Key Activity Rather Than Fuel Production
Ocean thermal energy conversion proposals always had the primary goal of generating energy. Then, the various researchers and entrepreneurs sought ways to use the energy. The reality is just the opposite.

Mariculture is the key to successful (that is, profitable) archipelago operation. It is key because foods sell at more per unit weight than fuels and chemicals. Consider that that a gallon of gasoline (3.8 liters) might rise back to $4 a gallon ($1.06 a liter) and perhaps twice that price in Europe. With gasoline weighing about 0.71 kilograms per liter, which is 2.7 kg per gallon or

9 Fears and Fixes from Sea and Sky

6 pounds per gallon.[889] That is price of about $0.67 per pound ($1.47 per kilogram. In contrast, consider the typical prices for a number of frozen maricultural food products tabulated in 2020.[890]

- Abalone: $55 per pound ($121 per kilogram) with abalone steaks higher
- Clams: $3 per pound ($6.60 per kilogram)
- Crab legs: $10 per pound ($22 per kilogram)
- King crab: $30 per pound ($66 per kilogram)
- Lobster: $8 per pound ($18 per kilogram)
- Mussels: $2.50 per pound ($5.50 per kilogram)
- Oysters: $8 per pound ($19 per kilogram)
- Salmon: $4 per pound ($9 per kilogram)
- Scallops: $6 per pound ($13 per kilogram)
- Shrimp (frozen jumbo, with the shelled peeled): $11 per pound ($24 per kilogram)

Food items yield returns several times that of fuels. Thus, food production is the first item to be emphasized—the "low-hanging fruit."— rather than energy production.

That said, a deep-ocean maricultural complex must be highly productive to compensate for the expenses associated with structures able to withstand the ocean environment, housing a significant number of crewmembers, and shipping everything needed and everything produced to and from land.

A different concern is that expensive commodities are expensive because they are rare. The prices of abalone and king crab will drop when a major maricultural complex, such as an archipelago, goes into high production. The price would (and probably will) plummet when several such complexes were to put these items on the market. However, the expenses of these complexes would still remain the same—an age-old problems for farmers if the harvest is … too good.

Enterprise survival dictates that a complex must generate as wide an assortment of products as possible with its standard tool set. Production might be salmon one day and lobster the next, but it would all be processed in a single cannery. Live abalone, live king crab, and live coral-reef exotics would go out on the same aircraft.

---

[889] Maureen Shisia, "How Much Does Gasoline Weigh? World Atlas website, Oct. 25, 2017. https://www.worldatlas.com/articles/how-much-does-gasoline-weigh.html (accessed Aug. 4, 2020)

[890] "How Much Is It," web page, undated. https://www.howmuchisit.org/ (accessed Aug. 4, 2020)

The major potential production areas would be an artificial coral reef area using existing low-nutrient bluewater, upwelling fertilized tropical greenwater, and deeply submerged temperate-water shellfish and finfish production using

9.4.4.6 The Archipelago Coral Reef as a Production Area

The first production area of the archipelago will be a cultured coral reef on the up-current side of the complex. Water there will still be the clear bluewater low-nutrient condition favored by corals.

An artificial coral reef does not need a stone sea floor to hold up many tons of coral-made stone, so the support does not need to be massively heavy. It only needs to be a thin layer of growing corals on a metal screen parallel to the ocean surface, floats and adjustable air tanks underneath, and a frame down to ballasting weights below (to prevent the reef from flipping). It can ride at 10 meters (33 feet) below the sea surface so scuba divers could work for several hours without decompression time. Such an artificial structure is much more expensive than a natural coral reef, but it has key advantages:

1. The corals and associated other reef life forms are protected from bottom-living pests (such as sea urchins and crown-of-thorns seastars) by vertical distance from the sea bottom and by robotic hunters if they do rise to the reef.

2. an archipelago reef can mitigate heat stress of its corals at dangerous temperature of 31°Celsius (88°Fahrenheit) or greater in several ways. First, the reef pieces are buoyant structures that can be sunk a short distance down into cooler waters. As with submersible nets and lines for culturing the fish and plants mentioned elsewhere, partial flooding of the buoyancy tanks can reduce buoyancy for corals to submerge farther to essentially a cool basement. A depth of 20 meters (65 feet) is still in the sunlit zone, but the light is slightly dimmed (in the clear blue waters) and the water is slightly cooler. That is all the corals need. Furthermore, corals only need short pulses (2 hours per day) of cooler water to prevent damage.[891] Second, an archipelago can run power lines from the generating station on the central platform with power sufficient to spray ocean water high into the air for evaporative cooling in the air above the corals. Furthermore, the cooled mist generated would also form reflective clouds to make shade for more coral cooling and a slight cooling of the world.[892]

3. Control of nearby waters will prevent overfishing of coral-protecting fish that eat coral-choking algae and coral-attacking pests.

---

[891] Yvonne Sawall, Moronke Harris, Mario Lebrato, et al., "Discrete Pulses of Cooler Deep Water Can Decelerate Coral Bleaching During Thermal Stress: Implications for Artificial Upwelling During Heat Stress Events," *Frontiers in Marine Science*, Aug. 28, 2020. https://www.frontiersin.org/articles/10.3389/fmars.2020.00720/full (accessed Oct. 3, 2020)

[892] Tim Smedley, "How Artificially Brightened Clouds Could Stop Climate Change," *BBC Future*, Feb. 25, 2019. https://www.bbc.com/future/article/20190220-how-artificially-brightened-clouds-could-stop-climate-change (accessed Oct. 21, 2020)

4. As noted in 9.3.3. The Second Stone of New Atlantis—Regenerating Corals and Accreting Limestone, a low electrical current on a metal framework can boost coral growth by cathodic accretion (and protect the coral-culturing structure).[893]

5. The archipelago will be far enough away from other reefs that they will not give any diseases to the archipelago corals or associated fish. Likewise, the archipelago reefs will be distant from the pests and diseases of existing reefs.

6. The archipelago can acquire and/or breed corals resistant to heat. There are already known heat-resistant corals in the eastern Pacific,[894,895] the Persian Gulf,[896] the Red Sea,[897] and Samoa.[898] There are already major programs to breed corals more resistant to higher levels of both heat and carbon dioxide.[899,900]

7. Last, but not least, as entirely new artificial coral reefs, these areas would not have any new animals or plants to protect. All species on them would be artificially produced. Hence, any produce from such reefs should not be subject to any rules about endangered.

---

[893] Thomas J. Goreau and Wolf H. Hilbertz, *Method of Enhancing the Growth of Aquatic Organisms, and Structures Created Thereby*, US Patent 5543034A, Aug. 6, 1996. https://worldwide.espacenet.com/patent/search/family/023479064/publication/US5543034A?q=pn%3DUS5543034A (accessed Aug. 8, 2020)

[894] Mauricio Romero-Torres, Alberto Acosta, Ana M. Palacio-Castro, et al., "Coral Reef Resilience to Thermal Stress in the Eastern Tropical Pacific," *Global Change Biology*, DOI: 10.1111/gcb.15126, April, 21, 2020.

[895] "Some Tropical Coral Reefs Resilient to Rising Temperatures – May Show '"The Key to Survival for Future Reefs'," *SciTechDaily*, Aug. 3, 2020. https://scitechdaily.com/some-tropical-coral-reefs-resilient-to-rising-temperatures-may-show-the-key-to-survival-for-future-reefs/ (accessed Aug. 6, 2020)

[896] Michael Casey, "Hot Water Corals in the Persian Gulf Could Help Save the World's Reefs," *Scientific American*, March 3, 2015. https://www.scientificamerican.com/article/hot-water-corals-in-the-persian-gulf-could-help-save-the-world-s-reefs/ (accessed Aug. 7, 2020)

[897] Thomas Krueger, Noa Horwitz, Julia Bodin, et al., "Common Reef-Building Coral in the Northern Red Sea Resistant to Elevated Temperature and Acidification," *Royal Society Open Science*, vol. 4, issue 5, May 17, 2017. https://royalsocietypublishing.org/doi/full/10.1098/rsos.170038 (accessed Aug. 7, 2020)

[898] Megan K. Morikawa and Stephen R. Palumbi, "Using Naturally Occurring Climate Resilient Corals to Construct Bleaching-Resistant Nurseries," *Proceedings of the National Academies of Sciences of the United States of America*, vol. 116, no. 21, pp. 10586–10591, May 21, 2019. https://www.pnas.org/content/pnas/116/21/10586.full.pdf (accessed Aug. 7, 2020)

[899] Warren Cornwall, "Researchers Embrace a Radical Idea: Engineering Coral to Cope with Climate Change, *Science*, vol. 363, issue 6433, pp. 1264–1269, March 21, 2019. https://www.sciencemag.org/news/2019/03/researchers-embrace-radical-idea-engineering-coral-cope-climate-change (accessed Aug. 7, 2020)

[900] Nerissa Hannink, "Breeding Baby Corals for Warmer Seas," *Phys.org*, March 26, 2019. https://phys.org/news/2019-03-baby-corals-warmer-seas.html (accessed Aug. 6, 2020)

Coral reefs can contribute to archipelago revenues in three ways: 1) sale of live coral exotics (corals and fish) for aquaria; 2) sale of coral-reef species; and 3) sale of wild catch fish harvested throughout the archipelago that that proliferated because their young had a protective reef habitat.

Sale of sustainable live coral is already an existing industry as evidenced by advertising websites such as that from Bali Coral Farm[901] and Oceans, Reefs & Aquariums.[902] Colorful coral-reef fish and their living coral surroundings have increasingly graced high-tech saltwater aquaria of high-end customers. These items yield hundreds of dollars per pound. When operated sustainably, they help protect wild reefs that are still being grievously damaged when people gather fish by cyanide poisoning or dynamiting the reefs and gathering the stunned but still-living fish.[903]

Culturing of corals is also used for regenerating reefs that have been damaged by storms and bleaching events associated with hot spells. Coral Maker, a collaboration between the California Academy of Sciences and Autodesk is developing a 3D technique for building the coral skeleton bases and a robotic arm to insert a number of coral fragments (also called seed plugs) into the bases. The robot arm allows planting thousands of seed plugs in a day, and the pre-made skeleton base allows the coral to reach adulthood, not in the present 3 to 10 years, but in 6 to 18 months.[904] This is just one of many research and development efforts.

An archipelago reef will not be able to compete with the large areas available around real islands, but those islands cannot protect the much more aquarium-valuable small fish from local predators and fishers. Furthermore, many of the governments of these areas allow culturing of corals but forbid taking of the colorful coral fishes. The artificial reef with total control of an isolated strip reef with predator control, should be able to get a large number of such fish and at larger sizes than seen is often scene around today's wild coral reefs.

Chinese diners have shown a taste for reef fishes, particularly on special occasions. In 2014, the napoleon wrasse was the most expensive at 800 to 2000 yuan for 500 grams (a little more than a pound for 130–326 U.S. dollars). Next in popularity were mouse groupers at 700–1500 yuan

---

[901] "Coral Farming – Land or Water?" Bali Coral web page, July 31, 2020. https://balicoralfarm.com/coral-farming-land-or-water/ (accessed Aug. 8, 2020)

[902] "ORA," Oceans, Reefs, & Aquariums home page, undated. https://www.orafarm.com/ (accessed Aug. 8, 2020)

[903] Jennifer C. Selgrath, Sarah E. Gergel, and Amanda C. J. Vincent "The Effects of Dynamite Fishing on Coral Reefs," Shifting gears: Diversification, Intensification, and Effort Increases in Small-Scale Fisheries (1950-2010)," *PLOS One*, March 14, 2018. https://journals.plos.org/plosone/article?id=10.1371/journal.pone.0190232 (accessed Aug. 5, 2020)

[904] Erica Argueta, "Dying Coral Reefs Are Being Saved by Automation," *C/Net*, July 30, 2020. https://www.cnet.com/news/dying-coral-reefs-are-being-saved-by-automation/ (accessed Aug. 8, 2020)

(114–244 U.S. dollars) for that pound, and scarlet leopard coral groupers at 500–1000 yuan (81-162 U.S. dollars) for that pound.[905]

Environmentalists have proposed strict worldwide enforcement to prevent such dining. However, forbidden fruit is often worth even more to buyers. Production from a place that did not even exist before should be legal. Moreover, its sale should be able to protect wild stocks of reef fish by being cheaper and being certified as checked for health issues known to be in wild reef stocks.

Another species ripe for mariculture is the giant clam (*Tridacna spp.*). These magnificent creatures are the largest bivalves in existence. Some have been documented at weighing 500 pounds (230 kilograms). Although 90% of the weight is in the shell, the 10% that is flesh is largely protein and is an expensive delicacy. The giant clam is near to being endangered because of overfishing, illegal trade, poaching, habitat degradation, sedimentation, and pollution. The giant shell is also an exotic sales item.[906]

There are still other product species. Conches supply food, but their most valuable product is the exotic shells that grace mantles and other displays. These shells may be worth tens of dollars, or if an unusual configuration, they may be worth hundreds or even thousands of dollars.

9.4.4.7 Microalgae (Phytoplankton), Zooplankton, Macroalgae (Seaweeds), and animals That Feed upon Them

Down-current from the coral reef will be upwelling water from the power plant. This fertile water will provide the nutrients for a steady bloom of phytoplankton (that is, tiny drifting plant plankton). Along with the phytoplankton will be zooplankton (floating animals bigger than the phytoplankton that feed on the phytoplankton). More conventional-looking plants (macroalgae or seaweeds) will also flourish while attached to nets and cables. Some of the researchers who added iron-salt fertilizer to blue waters commented about a resulting smell hours later like that of new-mown hay and a water surface that had become green.

Both the tiny drifting phytoplankton microalgae and the larger macroalgae draw down carbon dioxide from the atmosphere and produce plant matter for the tiny zooplankton and larger marine animals. Both the tiny plants and the tiny animals provide food for bivalve filter feeders such as clams, abalone, mussels, oysters, and scallops.

The name plankton is derived from the Greek *planktos*, meaning *errant*, and by extension, *wanderer* or *drifter*. The plankton largely live up to their name, drifting horizontally with the

---

[905] Michael Fabinyi, "Seafood Banquets Put Tropical Reef Fish at Risk," *China Dialogue*, Oct. 30, 2014. https://chinadialogue.net/en/food/7442-seafood-banquets-put-tropical-reef-fish-at-risk/ (accessed Aug. 8, 2020)

[906] Ping Sun Leung, Yung C. Shang, K. Wanitprapha, and Xijun Tian, *Production Economics of Giant Clam (Tridacna species) Culture Systems in the U.S.-Affiliated Pacific Islands*, Publication No. 114, Center for Tropical and Subtropical Aquaculture, April 1994. https://marine-aquaculture.extension.org/wp-content/uploads/2019/05/Poduction-Economics-of-Giant-Clam-Culture-Systems-in-the-US-Affiliated-Pacific-Islands.pdf (accessed Aug. 8, 2020)

current (although zooplankton often migrate vertically to hide in darker waters during the day). This horizontal drift means that the plankton produced in the richest upwelling carries its load of fertility for days down-current and some that reach the bottom will be spread out enough to prevent dead zones from too much organic matter.

Phytoplankton blooms in high-nutrient areas are often limited by one of the most ferocious predators on the planet, copepods. Fortunately for humanity, copepods are a centimeter (less than half an inch) in length. This is still much larger than most of the phytoplankton so fine mesh netting can catch them while letting the smaller algae pass through. Besides maintaining high phytoplankton productivity by culling copepods and other zooplankton, the netted zooplankton can also be processed into fish pellets to feed cultured fin fish and crustaceans.

As with shallow-water mariculture, macroalgae/seaweeds will be attached to cables, and/or netting, with weights and adjustable-buoyancy floats to stay near enough to the surface to maximize sunlight on most days but sink down to avoid excessive heat or wind-driven high waves as described in 9.3, Anchors, Floats, and Infrastructure for the New Atlantis. The seaweeds will serve the following functions for the archipelago.

- Provide food for filter feeders throughout the archipelago's controlled fertile flow.
- Provide seaweed food products for sale, for feeding the archipelago crew, and for feeding caged species (see 7.9 Minilivestock: Chickens Down to Bugs, McBugs, McCrawleys, & McWigglies—Grazers of Small Prairies and Small Ponds).
- Provide food for herbivorous wild fish in the area, some of which will be wild-fish catch.
- Provide more hiding/refuge areas for young fish and crustaceans so they can grow and become part of the wild-fish catch within the archipelago and elsewhere.
- Absorb and use excessive nutrients that cause acidity and possible overgrowth of poisonous plankton (such as red tide).
- Carry nutrients in break-away pieces that will provide fertility and fish food down-current.

For direct use by people, seaweeds have been heavily used in Asia for centuries, and there are many established industries around different species.[907,908]

Many of the most well-established species for cultivation are temperate species with large-scale culturing that developed in Korea, Japan, and Western Europe. That demonstrates what can

---

[907] Dennis J. McHugh, *A Guide to the Seaweed Industry*, FAO Fisheries Technical Paper 441, Food and Agriculture Organization of the United Nations, Rome, 2003. http://www.fao.org/3/y4765e/y4765e00.htm#Contents (accessed Aug. 10, 2020)

[908] Nicholas A. Paul and Michael Borowitzka, Chapter 15, "Seaweed and Microalgae," *Aquaculture: Farming Aquatic Animals and Plants*, Third Edition, edited by John S. Lucas, Paul C. Southgate, and Craig S. Tucker, John Wiley & Sons Ltd, Hoboken, New Jersey, pp. 313–337, 2019.

also be done in more tropical waters. For instance, mariculturists in the Philippines already harvest a number of seaweeds to produce carrageenan, an additive used to thicken, emulsify, and preserve foods and drinks.[909] There are a number of other tropical species being used and many more with potential. *Codium*, *gracilaria*, *sargassum*, and *Ulva* are considered adequate both for use as food and amenable to culturing.[910]

Of course, this all goes back to eutrophic amplification[911] in a good way for reducing global warming. Fertilized production of plants, both phytoplankton and attached macroalgae, has been proposed as a means to reduce global warming in the following ways.[912]

- Draw down carbon dioxide from the air while growing plant body mass.

- Feed zooplankton, some of which will excrete waste products and/or die in deep water so carbon dioxide is carried down out of the surface layer.

- Feed larger sea animals (directly or as feed to zooplankton, which are then eaten by the larger animals), some of which will excrete waste products and/or die in deep water so carbon dioxide is carried down out of the surface layer.

- Provide thousands of tons of seafood to humans on shore who will transfer hydrocarbons to the sewage treatment facilities on shore.

- Emit the biproduct dimethyl sulfide (per the CLAW Hypothesis in Subsection 9.4.1, "Fertilizing the Bluewater Deserts and the Ammonia Economy at Sea") that breaks down into sulfate aerosols that are light-reflecting anti-greenhouse (cooling) agents.

- (Last but not least) carry some water-logged leaves and stems into the depths for long-term carbon storage and thus (again) less global-warming carbon dioxide in the atmosphere.

All these are natural oceanic processes. They will just be amplified in a few more places in the vast ocean to keep our planet's temperatures at levels to which we have grown accustomed.

The key difference from other climate-modification proposals (sometimes called geoengineering and sometimes called industrial ecology) is that, first, the archipelagoes would be working to make a profit from their activities. Profit-making ventures are self-funding; they do not require billions of dollars of government funding. Second, the archipelagoes would be widely

---

[909] Ruth Kassinger, *Slime: How Algae Created Us, Plague Us, and Just Might Save Us*, Houghton Mifflin Harcourt, 2019.

[910] Ricardo Radulovich, Schery Umanzor, Rubén Cabrera, and Rebeca Mata, "Tropical Seaweeds for Human Food, Their Cultivation and Its Effect on Biodiversity Enrichment," *Aquaculture*, vol. 436, pp. 40–46, Jan. 20, 2015.

[911] Guillem Chust, J. Icarus Allen, Laurent Bopp, et al., "Biomass Changes and Trophic Amplification of Plankton in a Warmer Ocean," *Global Change Biology*, March 7, 2014. https://doi.org/10.1111/gcb.12562 (accessed Aug. 10, 2020)

[912] Braden Allenby, "Industrial Ecology," *Inventing for the Environment*, Arthur Molella and Joyce Bedi (editors), The MIT Press, Cambridge, Massachusetts, pp. 339–372, 2003.

and diffusely spread across the oceans so that there would not be any intense changes in any given locality on land.

9.4.4.8 Cages and Rafts for Finfish, Shellfish, and Crustaceans

Production of finfish, shellfish, and crustaceans will be the greatest revenue source for each archipelago. People are gradually adopting diets with more freshwater and oceanic plants, and that is a growing market, but people outside of Asia do not pay premium prices for sea plants. People pay premium prices for sea animals, so those species make up the key market set.

As noted in Subsection 9.3, "Marine Farming and Infrastructure for the New Atlantis," there has been a rapid and ongoing increase in maricultural operations around the world since the 1970s. These operations farm a wide assortment of species:

- Marine finfish mariculture was only 7% of world finfish production in 2014 (versus 93% for freshwater aquaculture), but the value was 14% of the total because marine fish tended to be higher value, which will undoubtedly grow as populations and income levels grow. Marine finfish aquaculture includes more than 90 species from shallow-water, deep-water, and various temperature regimes. Some of the best prospects for archipelago fish caging near the surface include milkfish, European seabass, yellowtail amberjack, red sea bream, and cobia.[913]

- Shrimps (also known as/prawns) were one of the first groups of species raised on an agro-industrial scale, and by 2015 farmed shrimp provided 60% of world production.[914] Shrimp thrive in tropical saline waters, and as omnivores, they eat anything. The biggest complaint about mariculture has been destruction of environment-protecting mangrove forests along tropical coast lines, making those areas more susceptible to flooding during storms or tsunamis. From the GND perspective, some studies have concluded that destroying those mangrove forests results in a carbon footprint per unit of shrimp that is ten times that of beef.[915] Also, because coastal shrimp farms pack the animals in densely, they are susceptible to disease infestations leading to major die-offs, and the shrimp farmers have often resorted to dangerously high doses of antibiotics, parasiticides, and other questionable chemicals.[916] Those issues might allow archipelago operators to

---

[913] Wade O. Watanabe, Md Shah Alam, Patrick M. Carroll, et al., Chapter 20, "Marine Finfish Aquaculture," *Aquaculture: Farming Aquatic Animals and Plants*, Third Edition, edited by John S. Lucas, Paul C. Southgate, and Craig S. Tucker, John Wiley & Sons Ltd, Hoboken, New Jersey, pp. 437–482, 2019.

[914] Darryl Jory, Chapter 22, "Shrimps," *Aquaculture: Farming Aquatic Animals and Plants*, Third Edition, edited by John S. Lucas, Paul C. Southgate, and Craig S. Tucker, John Wiley & Sons Ltd, Hoboken, New Jersey, pp. 499–525, 2019.

[915] "Tiny Shrimp Leave Giant Carbon Footprint: Scientist," *Phys.org*, Feb. 18, 2012. https://phys.org/news/2012-02-tiny-shrimp-giant-carbon-footprint.html (accessed Sept. 9, 2020)

[916] Rupert Taylor, "The Dirty Secret of Shrimp Farming," *Owlcation*, Feb. 20, 2020. https://owlcation.com/stem/The-Dirty-Secret-of-Shrimp-Farming (accessed Sept. 8, 2020)

- Bivalve mollusks (Clams, cockles, mussels, oysters, and scallops) have a history stretching back at least to Roman oyster farms. Culturing them requires no extra feed because they are the cleanup crew of the aquatic world, feeding on bacteria and fine particulate matter of phytoplankton, zooplankton, and detritus.[917] Thus, for an archipelago, mollusks can do double duty. First, they can provide product, with scallops yielding high price and mussels being very high productivity. Second, they (like seaweeds) help prevent excess organic matter from accumulating to keep the ecology in balance.

- Abalone are large gastropod mollusks produced in temperate and tropical waters. Wild catch of abalone has drastically declined due to overfishing and habitat destruction; and abalone farming (dominated by China) has been rising to fill the market demand. Abalone are vegetarian, so feed for caged animals can be produced from readily available seaweeds. Production systems on land use single-pass, flow-through tanks, and sea-based system use cages or other enclosures.[918]

- Sea Cucumbers: are eyeless, armless, boneless animals that look much like their namesake except their bumps are sometimes spikes. These globs eat whatever marine snow or detritus of dead plants, dead animals, shed skin, and fecal matter that has fallen down from above. However, they contain a chemical that has been used to treat arthritis in East Asia. For that reason and a maybe just their exotic looks, these animals have become a delicacy in East Asia, and the spikier the better. Because they have not yet been farmed, they are being rapidly fished out, and the price of some species has approached $3,000 per kilogram ($1,400 per pound).[919] The expected large production of sea cucumbers in controlled-production areas will significantly decrease prices; however, culturing of the more exotic varieties in controlled sea cucumber cages should still command high prices. Sea cucumbers are often bottom-dwelling animals, so their culturing can be in deeper cooler waters. Furthermore, improvements in underwater robo workers will eventually make the sea bottom far below another area of controlled production of sea cucumbers and other bottom-dwelling species.

---

[917] John S. Lucas, Chapter 24, "Bivalve Molluscs," *Aquaculture: Farming Aquatic Animals and Plants*, Third Edition, edited by John S. Lucas, Paul C. Southgate, and Craig S. Tucker, John Wiley & Sons Ltd, Hoboken, New Jersey, pp. 499–525. 2019.

[918] Peter Cook, Chapter 25, "Abalone," *Aquaculture: Farming Aquatic Animals and Plants*, Third Edition, edited by John S. Lucas, Paul C. Southgate, and Craig S. Tucker, John Wiley & Sons Ltd, Hoboken, New Jersey, pp. 573–585. 2019.

[919] Nathaniel Lee and Shira Polan, "Sea Cucumbers Are So Valuable That People Are Risking Their Lives Diving for Them," *Business Insider*, Jan. 9, 2019. https://www.businessinsider.com/why-sea-cucumbers-so-expensive-seafood-2019-1 (accessed Aug. 21, 2020)

9.4.4.9 The Hatchery/Nursery, and Culturing Center

Many of the livestock species for the archipelago begin life as tiny eggs and then often evolve via one or more intermediate forms different from adulthood. In the wild there may be thousands or more eggs or seeds to get one adult individual. This is very inefficient. Hence, hatcheries and nurseries for marine plants and animal have been a key driver of both freshwater aquaculture and saltwater mariculture. (Many experts in the field are now using the term aquaculture for any agriculture in either freshwater or salt water.)

The hatchery and nursery functions tend to be labor intensive, and worse; they require a number of heavy water-filled tanks or pens. Such a facility above the waterline would be a major expensive increase in platform mass and expense. Building facilities at the waterline would increase the surface area susceptible to wave action during storms. Consequently, a case can be made for on-shoring this function.

However, facilities on shore have often been infected by various diseases of the maricultural livestock, and these diseases have led to major crashes in productivity.[920,921] It would be prudent to minimize the potential disease transfer from larval and juvenile livestock coming from various shore facilities to the archipelago.

A third option is to operate a nursery/hatchery facility below the platform waterline and below the most significant wave action. Such a facility might be about 20 meters (66 feet) below the waterline. This would be a pressure of 3 atmospheres (3 bars, 3,060 kilograms per square meter, 43 pounds per square inch), which includes a 0.6-bar (3-atmosphere, 6,100-kilograms per square meter, 8.7-pounds per square inch) safety margin compared to the 3.6-bar (3.6-atmosphere, 37,000 kilograms per square meter, 52-pounds per square inch) maximum pressure for tunnel-workers.[922]

Open-topped fish pens could be maintained with very little wave action. Also, this area could provide a docking/maintenance area for submersibles, both crewed and robotic. The submersibles would be available for transporting cultured plants and animals to the open-sea growing area, collecting deepwater produce, and performing deepwater maintenance and construction.

The nursery/hatchery workers might remain in the pressurized area for a weekly shift and decompress completely over several hours because of gas saturation in their tissues. (They might also have supplemental inert gases to reduce effects of oxygen and nitrogen under increased pressure.

---

[920] Bruce L. Nicholson, "Fish Diseases in Aquaculture," The Fish Site web page, April 4, 2007. https://thefishsite.com/articles/fish-diseases-in-aquaculture (accessed Sept. 12, 2020)

[921] Brian Austin (editor), *Infectious Disease in Aquaculture: Prevention and Control*, Woodhead Publishing Limited, Oxford, United Kingdom, 2012,

[922] E. H. Martin Herrenknecht and K. Bäppler, "Compressed Air Work with Tunnel Boring Machines," *Underground Space – the 4th Dimension of Metropolises*, Barták, Hrdina, Romancov & Zlámal (editors), Taylor & Francis Group, London, pp. 1175–1181, 2007. https://www.nordseetaucher.de/assets/PDF-Downloads/Tunneling/NST_Compressed-Air-Work-with-Tunnel-Boring-Machines.pdf (accessed Sept. 13, 2020)

Two structural expenses would result from the large air pocket of the hatchery/nursery below the water line. First, the air pocket would buoy the platform up, but a serious air leak in this structure would subtract from platform buoyancy—possibly endangering the platform. Thus, as a safeguard, the major columns of the platform could be strengthened above the water line with a lighter-than-water column (or columns) so that the level of sinking would be minimized.

This addition to the platform will increase its overall strength, but it will add to the second structural expense. The air pocket (and the additional buoyancy column or columns) will lower the center of balance, which is not a good thing. In 1628, the Swedish navy's Vasa set out from the dock on its maiden voyage, but it tipped over and sank while still in the harbor. The ship designers had added a second row of cannons above the first row to be even more deadly in combat, but it was deadly for 30 crewmembers who were trapped below decks. The lesson was that you must not let your nautical structure be top heavy.[923] The lesson was repeated by the SS Eastland in 1915 that killed 877 when it rolled over on its side while still next to the dock—not even at sea but in the Great Lakes.[924]

That leads to the second expense. There must be more weight (ballast) below the air pocket of the hatchery/nursery. Of course, that requires an even stronger structure, which also adds more weight, and those increases in weight require more flotation in the column or columns above the waterline.

The good news is that at some additional construction expense, the central platform will become tremendously tougher and more capable.

9.4.5 The Bottom Line for Green Oceans as Well as Agriculture and Energy

Again, as with advances in land agriculture and energy, oceanic advances could by themselves alone provide humanity sufficient food and halt global warming dangers. Also, again, the issue is time. Any one of the three individually might not evolve quickly enough to stave off the possible dangers predicted after 2100.

However, with the three together, governments can perform moderately priced research and encouragement of new technological avenues. Entrepreneurs can try (and often fail) as they push the envelope of what can be done—often in unexpected ways.

In the oceans, many technologies must continue to evolve over several decades before archipelagoes could become major features on our planet. However, there are a number of profitable paths that could lead to them, and profit is the key.

---

[923] Alexander Hamer, "Vasa: the Swedish Empire's doomed flagship," *Real History*, Nov. 3, 2017. https://realhistory.co/2017/11/03/vasa-ship/ (accessed Sept, 12, 2020)

[924] Debra Kelly, "The Eastland disaster: The tragedy that no one remembers," *Grunge*, Aug. 7, 2020. https://www.grunge.com/234246/the-eastland-disaster-the-tragedy-that-no-one-remembers/?utm_campaign=clip

As with the Industrial Revolution of steam engines, factories, and transportation, a myriad of distantly related ventures must succeed (or fail) before flowing together into a technological marvel. Only many competing ventures can explore and attempt the possibilities.

Together, such entities can make archipelagos inevitable unless … some competing entities find something better. Time will tell.

Major Things for Governments and Businesses to Do

- Government: Fund research to breed corals resistant to higher temperatures and higher levels of carbon dioxide.

- Government: Propose a world treaty supporting conversion of large commercial vessels (freighters and tankers) to combustion of ammonia and injection of iron salts and the ammonia exhaust products into the ocean. (Liquid natural gas tankers that run on boil-off of the stored liquid natural gas would get an exception.)

- Government: Legislate that otherwise forbidden oceanic species for commerce (corals, finfish, bivalves, and others) grown in certified non-natural environments, on land or on artificial island environments, may be considered legal produce for sale and use.

- Business: Build a dynamically positioned casino and mini tropical beach complex with oil or natural gas power and waste heat to warm the indoor tropical beach. to sail in one place in northern waters—first prospect the triangle point between New York and Boston while staying just outside the Exclusive Economic Zones of the United States and Canada.

- Business: Build and maintain a small artificial reef (away from existing reefs) to culture living fish and coral for saltwater aquariums so that these fish can be sold as certified as not wild catch.

- Business: Develop systems for culturing and harvesting sea cucumbers.

What You Can Do

- Take a cruise to enjoy a taste of a future floating city. (The 2020 corona virus has provided a painful lesson about air filtration, and probably also ultraviolet-light sterilization, to prevent illness just as the Titanic sinking supplied a lesson in evacuation safety.)

- If you like caring for exotic pets, get your own tropical saltwater aquarium—with certified sustainably produced corals and reef fish.

- Support a reef-protection and restoration organization, but only if they work with reefs rather than being more concerned with banning plastic bags or ban coal-fired electrical power plants in Utah.

9 Fears and Fixes from Sea and Sky

10 Conclusions

10.1 A Start on the Better Green New Deal

The Occasio-Cortez–Markey proposal for a Green New Deal (GND), and a frequently asked questions (FAQ) posting (both attached in Chapter 11) provided a foil, a thesis for this book to provide the counter-foil or antithesis. Ideally, the interchange of ideas will help develop policies that are the most practical in dealing with global warming and other

This book has provided detailed descriptions of many energy and climate issues that could make a GND difference. The following is a summary of the main points.

1. **Background of the GND and other national mobilizations**

    a. The key premise of the GND House resolution, and similar proposals, is that humanity is facing a climate-warming apocalypse of greater heat, rising sea levels, and increasing droughts within less than two decades unless drastic actions are taken.

    b. That putative threat would justify a vast set societal and technological changes be made by 2030 or 2040 with a set of federal programs bigger than the New Deal during the Great depression and comparable to the national mobilization of World War II. This effort would have a cost in the tens of trillions of dollars—possibly approaching one-hundred-trillion dollars.

    c. That national mobilization would include rebuilding all structures to be environmentally efficient, replacing most commercial aviation with expanded and electrified railways, largely replacing meat with vegetarian foods, and replacing all fossil-fuels use with electricity generated by solar, wind, and hydroelectric power.

    d. Similar government-managed giant mobilizations for development have often led to spectacular failures. During the two Energy Crises of the 1970s and their aftermath, American government initiatives lurched to nuclear fission, then coal, then to nothing (when oil prices collapsed in the 1980s), and then on to solar and wind. The old Soviet Union (now Russia) ran the vast farming expansion (called "The Virgin Lands") that created a dust bowl in Kazakhstan. Worst of all was "The Great Leap Forward" in late 1950s China that caused a famine killing tens of millions.

    e. Hardly mentioned is that the "national mobilization comparable to World War II would include rationing and price controls. A longer-term problem with national mobilization is that it eventually results in the diseconomies of socialism

f. Fortunately, global warming is not coming as fast or in a form with a threat as great that postulated in the GND, so there is more time to develop policies to mitigate the warming and/or adopt to a warmer planet.

g. **GND Writers Distracted by Socialist Wish List:** Furthermore, the GND contains many additional social engineering goals that are totally unrelated to climate (such as guaranteed medical care and guaranteed housing). The focus on these unrelated distractions suggests that the GND was not as serious a goal as the writers proclaimed. Rather, it was a "change-the-entire-economy thing" program," as Representative Occasio-Cortez's former chief of staff was quoted as saying. Might they have lied for that same greater good that drove the Great Leap of the Chinese communists?

h. **But There Are Real Global Warming Concerns:** That all said, global warming will probably cause painful changes in present conditions on Earth such as shifting in present climate bands and sea-level rise along the increasingly populated and built-up coastal areas.

i. Besides global-warming concerns, there have also been historically common chilling events (such as mega-volcano eruptions such as the Mount Tambora explosion of 1815) that can cause several years of crop failures with resulting famines and government collapses. The more severe eruption of Mount Toba (about 70,000 years ago) may have brought humanity to the edge of extinction. These chilling events tend to only last for a few years, but they are as threatening as global warming because such an event could happen with little or no warning.

j. As for a national mobilization, the general public has not been energized by the threat of global warming because it is still just a theoretical threat "proved" in computer models. Storms, droughts, and sea levels are not manifestly different from those two centuries ago. Thus, the people are in favor of slowing global warming—but they don't want to personally pay for expensive changes. Draconian government-enforced costs and hardships would meet active resistance at the ballot box and eventually in the streets, as the yellow-vest protests in France demonstrated.

k. **People Only Rally for a Clear Threat:** People will only support major new expenses and inconveniences if they can see a clear and present danger. Considering the threats are less immediate and less visible, it would be difficult to maintain support for national-level or world-level crash programs until major climate changes are clearly discernible. The people of the Soviet Union were not strongly anti-Nazi until German troops surged across the border. Americans were isolationist against any involvement in World War II until Japanese planes attacked the U.S. fleet in Pearl Harbor.

l. Meanwhile, there is time for government research for exploring technical possibilities and modest prizes and subsidies for demonstration projects.

m. Meanwhile, there is more time for entrepreneurs to take risks, big and small, for actually implementing a plethora of innovations needed to save the world. As always, the successful innovations must survive the test of the market, not a decision in high government or corporate management. The managers of large corporate or government managers often become stuck on failing concepts: Edison and his magnetic iron mining, the U.S. federal government's quest for nuclear fusion (and the associated Superconducting Supercollider[925]), and the California bullet train to nowhere.[926] The difference is that entrepreneurs (even the very wealthy Thomas Edison) can be reined in by a technological blind alley and mounting losses; government techno dreams can go on almost forever.

n. Entrepreneurs in relatively small companies make the revolutions. They are the Edison's of electricity, the Fords of cars, the Wright brothers of airplanes, the Jobs and Woznick's of personal computers, the Musk's of reusable space boosters, and many more. These are the people who are hungry and often desperate. As former Vice-President Al Gore said, entrepreneurs are the ones who make things happen.

o. In contrast, governments often take advice from the revolutionaries of the last revolution—the very people who will condemn the next revolution. Thomas Edison who pioneered electric current with direct current waged political war on the alternating current of Nicola Tesla. Admiral Hyman Rickover who pioneered use of light-water reactors for submarines fought against any other reactor concepts.

2. **Introduction to the Better Green New Deal**

   a. The proposed Better Green New Deal of this book focused on three major technological-advancement directions of advances in land agriculture, energy (particularly nuclear fission), and oceans development.

   b. A fourth major technological development of increasing energy efficiency was not included simply because it includes so many different complex (and often minor) improvements. Those many improvements, though often small, sum to an ongoing revolution in more productivity per unit of energy that ripples through all the other technologies to multiply their potentials. Readers wishing to explore the possibilities of increased efficiency might start with *Energy Futures* from Stobaugh and Yergin for an academic

---

[925] John G. Cramer, "The Decline and Fall of the SSC," *Analog Science Fiction & Fact Magazine*, May 1997. Available at "The Alternate View" columns of John G. Cramer, https://web.archive.org/web/19971010114852/http://www.npl.washington.edu/AV/altvw84.html (accessed Sept. 17, 2020)

[926] "Bullet Train Boondoggle Keeps Chugging," *Pasadena Star News*, updated March 9, 2020. https://www.pasadenastarnews.com/2020/03/04/bullet-train-boondoggle-keeps-chugging/ (accessed Sept. 17, 2020)

viewpoint.[927] Weizsacker, Lovins, and Lovins provide a number of case studies of how energy and materials can be used more efficiently.[928] For a fun, but insightful reading, enjoy Robert Bryce's *Smaller Faster Lighter, Denser Cheaper*[929] *and* More from Less by Andrew McAfee.[930]

3. **Food, fear and climate fixes from the land:** Land agriculture (farming) is one of the oldest and most needed occupations. If the climate were to grow much hotter, people could avoid heat in various ways, but lack food (famine) is near-term life-or-death within weeks. Thus, this book discussed increased food production together with mitigating global warming as a single topic. Fortunately, there are many ways that land agriculture can be applied for these two interwoven goals.

   a. Rebuilding soil (also known as regenerative agriculture or conservation agriculture) uses reduced plowing and other techniques to enhance natural build-up of soil. This increases land productivity, and it can help maintain productivity despite hotter and drier weather. As a byproduct, soil building draws down global-warming carbon dioxide from the air, which reduces global warming.

   b. There are thousands of plants that are potential additional crops. Adding some of those crops to the farming repertoire would increase productivity and provide greater resilience against harsher climate.

   c. Breeding of better crops helped make the Green Revolution of the 1970s and 1980s. Even better plant breeding could draw down tremendously more carbon dioxide, produce more food, and provide more resilience against a harsher climate.

   d. Changing diets to be more vegetarian and less meat can provide more usable food for people, provide a healthier diet, and cause less global warming than the world's present average diet. It would not need to be a total replacement as implied in the GND resolution, but the GND partisans have a good idea in this area.

   e. Robotic farm machinery is already making agriculture production cheaper, less environmentally intensive, and more resilient to harsher climate. In the future, it will enable revolutionary changes.

---

[927] Robert Stobaugh and Daniel Yergin (editors), *Energy Futures: The Report of the Energy Project at the Harvard Business School*, Random House, New York, 1979.

[928] Ernst von Weizsacker, Amory B. Lovins, L. Hunter Lovins, *Factor Four: Doubling Wealth, Halving Resource Use - The New Report to the Club of Rome*, Routledge, Dec. 1, 1998.

[929] , Robert Bryce, *Smaller Faster Lighter Denser Cheaper: How Innovation Keeps Proving the Catastrophists Wrong*, Public Affairs, New York, May 2013.

[930] Andrew McAfee, More from Less: The Surprising Story of How We Learned to Prosper Using Fewer Resources—and What Happens Next, Simon and Schuster, London, 2019. The author examines increasing efficiency I

f. Improved controlled-atmosphere growing areas (such as greenhouses and vertical farms) can produce more high-value crops, in smaller areas, with less expenditure of water, and less use of pesticides.

g. Sprouting seeds is an often-forgotten way to provide the better nutrition of fresh vegetable mater with less land area and regardless of season.

h. Fungi (including mushrooms) provide many ways to extend food supplies by growing on otherwise waste straw and branches and by processing foods into more nutritious forms. For instance, mushrooms can be a high-protein substitute for meat.

i. Minilivestock from chickens down to snails (escargot!), crickets, and worms can eat items that are now agricultural waste. The minilivestock can be eaten directly, or they can part of the feed for more conventional livestock and pets, thus decreasing the requirement for grain and fish.

j. Many grasses have high-protein, high-nutrient juices inside the woody plant material that is undigestible to humans and other omnivores. This juice can be squeezed out for food if crop production is ever low. The remaining woody material can be baked into charcoal-like biochar in put into the crop lands for soil building.

k. Speaking of harvesting grass, American turf grass for lawns, parks, and golf courses covers an area the size of Kentucky. Small robotic farm equipment, along with the global Positioning System and computer record keeping could harvest the clippings from turf grass for growing mushrooms, feeding minilivestock, and/or squeezing out high-protein juice with the leavings available for composting or biochar.

l. All these presently used food sources, and potential other food sources, mean that people can produce enough food despite global warming or global chilling. Moreover, changes in farming methods can help mitigate global warming.

4. **Nuclear fission (splitting atoms) is the only power source likely to support a successful GND.**

    a. Although, the so-called green energies of wind and solar will help, they are overly expensive because they are intermittent (which requires expensive back-up power) and they are dilute (which requires large land areas and expensive transmission lines to urban areas needing power).

    b. Although nuclear fusion reactors (fusing atoms together) have been proposed for nearly six decades, fusion will probably remain impractical because it is expensively complex and because it has a neutron flux so intense that it would limit reactor life.

c. Nuclear fission energy from thorium and uranium has the potential to supply all humanity's energy needs for thousands of years.

d. Nuclear fission was developed as two government crash programs to build fission bombs during World War II and then to power nuclear submarines. This developed the technology rapidly, but with the problems of being stuck in a technology with the problems of being expensive, temperamental easily capable of diversion to bomb making, and associated with nuclear bombs in the public's mind.

e. In the United States (but not in many other countries), costs and hazards were multiplied by not recycling fuel

f. Worst of all for nuclear-fission-reactor operators, fossil-fuel power plants burning coal, oil, and natural gas kept increasing in efficiency and dropping in price.

g. Then, concerns about global warming have brought the alternative energies of nuclear, solar, and wind back to the forefront.

h. The next generation of fission reactors can achieve a nuclear renaissance of lower-cost power and industrial process heat using small modular reactors, standardized designs, inherently safe designs, molten-salt reactors, and (the more abundant) thorium.

i. Lower cost nuclear fission will reduce the carbon footprint of electrical power generation.

j. Smaller and lower-cost nuclear-fission units will enable use of fission for various process-heat uses, again lowering humanity's carbon footprint.

k. One of the most important process inputs from smaller and lower-cost nuclear-fission units will be producing synthetic fuel for vehicles and small power plants, also to lower humanity's carbon footprint.

l. There are several choices to choose from in adopting a major synthetic fuel. Synthetic gasoline, methanol, and natural gas still put carbon dioxide into the air. Hydrogen is awkward to store. Ammonia might be the best option because it can store as a liquid and its combustion products include no greenhouse-warming carbon dioxide.

m. Nuclear energy and vertical-vortex wind energy can loft material into the stratosphere for mitigating global warming or chilling.

5. **There are key toolmaking technologies that apply to all three of the Better Green New Deal themes.**

a. Computer power continues growing for assorted uses including finding new catalysts, designing complex facilities, processing geological data to find oil, decoding genetic data to breed better crops and cures,

b. Besides making existing processes cheaper, are growing into capabilities that could not be done before: radiation-resistant robots for running reactors; small robots for millions of operations to culture small areas plants and tiny minilivestock (such as snails) in polyculture; microscopic assembly of those increasingly capable computers; robots for working in cold and high-pressure aquaculture operations; and distant robotic operations in outer space.

c. Advanced materials and materials processing are increasing capabilities and decreasing costs. These advances include lighter and high-temperature alloys, more capable composites, 3D printing for faster and lighter production, and prefabricated construction.

d. Replacing supercritical steam with supercritical carbon dioxide can increase the efficiency of any heat engine, from oil fired, through nuclear, and even through solar.

e. After many decades of stagnating, battery energy density radically improved to make electric cars pricey but competitive for the first time in a century. Another revolution of two- or three-times existing power density could largely replace internal-combustion-engine cars ... and make the Los-Angeles-to-Portland car practical.

f. Likewise, a revolution in thin-film photovoltaic cells just could succeed in making an GND world more practical. As a practical half-way measure, hybrid cars with advanced batteries and ultracapacitors might be able to drive from Los Angeles to Vancouver, British Columbia.

g. Those advances start to make an electric airplane practical—really. Such a change would also be a major advance against global warming because it would decrease the warming of jet-deposited soot in the stratosphere as well as decrease carbon dioxide deposited in the atmosphere.

6. **Fears and Fixes from Sea and Sky**

a. The ocean is 71% of the Earth's surface with major worries of overfishing, decreased productivity and carbon drawdown because of increased warm water in the top layer, pollution, increased sea trash, and increased human settlement in the dangerous area near sea level where warming and sea level rise is increasing the danger of flooding and tsunamis.

b. Meanwhile, humanity is increasing the crowding of expensive construction on the coastal flood plains and even into wetlands below sea level. This foolish construction is increasing risk to lives and potential flood costs even if there were no sea-level rises expected.

c. Overfishing has caused stagnating wild fish catch since the mid-1990s, but mariculture of animals and plants at the shoreline and into very shallow waters now supplies more food than wild catch, and mariculture will continue growing.

d. In addition, there are dangers of giant "megatsunamis that have not been seen in historical times but that will happen again.

e. The moderately shallow continental shelf out to about 100 meters (330 feet) is less than 8% of the ocean surface. The continental shelf is already fertile greenwater, but it could become could more fertile and support the culture of seaweed-and-animal mariculture enough to feed ten-billion people and provide them with all needed carbon-neutral fuel.

f. Moreover, seaweeds are a crucial part of the polyculture (also known as integrated multi-trophic aquaculture) because these plants suck up the phosphate and nitrate wastes from animal culturing that could causing toxic red tides. At the same time, they draw down carbon dioxide from the atmosphere subtracting from global warming.

g. Carbonation of volcanic rock into carbonate rocks can both permanently store carbon that would have gone into the atmosphere and build barrier islands along the edge of the continental shelves to mitigate tsunamis and megatsunamis. These areas would also be able to harness the increased energy difference between combustion or nuclear heat and the deep-ocean chill of nearby deep waters. For warming mitigation, rock carbonation would be a major way to permanently store carbon that would otherwise go into the sky.

h. A key development for continental shelf will be the seagoing line shack, a moored submersible work area and short-term sleeping quarters that would be a cross between a cowboy's line shack and a Winnebago camper. The seagoing will submerge during stormy weather and get air from a snorkel air tube.

i. Mariculture on the continental shelf will include many developments for gradually moving farther from shore and into deeper water for eventually moving into the deeper waters.

j. The deep ocean waters have a tremendously greater potential than the continental shelves for energy production, for food production, and for countering global warming. At 92% of the ocean surface, they are roughly twelve times the size of the shelves, and as mostly bluewater desert, their increase in fertility will be more than the shelf waters, many of which are already fertile green water.

k. Oceanic fertilization will be a key component in extending mariculture into the deep sea.

l. This fertilization will be accomplished by pumping up cold high-nutrient waters from the depths and use it for cooling thermal power plants, mostly nuclear fission plants. The nutrient-rich waters can then be released at temperatures just slightly above the ocean surface waters so that the nutrients can stay in the sunlit waters and turn bluewater desert green.

m. The fertilized deep waters will be part of a larger food and energy production system than that of the continental shelves. This will be done with a number of food and energy complexes I call archipelagos because they will each consist of a large power central platform, a few other large platforms, and many small submerged or mostly submersible platforms for maricultural production.

n. The primary focus of the archipelagos will be maricultural food production rather than energy because food earns much more money per unit of output, and an archipelago will need to ern major revenues to be affordable.

o. The central platform will be a much larger version of existing floating oil production platforms. Its functions will include power plant, synthesis of ammonia fuel for work boats and submersible platforms, riser pipe to bring up cold high-nutrient deep waters, a cannery for processing maricultural produce, a protected marina for work boats.

p. The agriculture production area will be the archipelago of smaller nearly submerged platforms with bluewater coral-atoll mariculture upstream from the artificial upwelling of the central platform and greenwater mariculture of plankton, seaweed, finfish, crustaceans, and shellfish extending tens of miles (kilometers) down-current from the upwelling nutrients until the nutrients have been largely used.

q. For GND climate mitigation, each archipelago will decrease global warming in four ways: (1) phytoplankton and seaweed will draw down carbon dioxide from the air to build plant tissue—significant portions of which will go into deeper waters; (2) phytoplankton will release greater amounts of sunlight-reflecting chemicals; (3) the nutrients from crustacean and lobster production will extend the nutrient plume farther down-current than the upwelling alone, extending the cooling effects; (4) raising water up from the depths draws down water elsewhere carrying carbon dioxide along with it and reducing the percentage of carbon dioxide in the atmosphere.

r. The evolving technologies that are making archipelagos practical include floating concrete, oil production platforms, robotics for underwater maricultural operations, remote communications so that the maximum amount of work can be on-shored to more affordable land locations, small modular nuclear fission reactors, and production of synthetic liquid fuels.

s. Some combination of the three major technological themes—agriculture, energy, and oceans—can supply our needs and protect against dangerous levels of global warming.

10.2 Beyond the Better Green New Deal?

This *Better Green New Deal* tract has been Version 2.0 of a possible Green New Deal with some suggested improvements and corrections. However, it is still only a collection of incomplete visions. Some proposals will prove to be unworkable blind alleys or at least something that will have to wait until technology evolves further. Some totally unknown concepts will be world changers.

- As late as the early 1930s, many experts thought that lighter-than-air dirigibles were the aviation wave of the future. There are a few dirigibles today, but the vast majority of aviation is heavier-than-air craft, airplanes.

- As World War II was approaching in the 1930s, military leaders were hoping that the new RAdio Detection And Ranging (radar) might burn planes out of the sky or at least cause their engines to die. No, but it was (and still is) a great way to track airplanes in both war and peace.

- During World War II in the early 1940s, the United States and Great Britain developed the first computers for cracking codes and calculating artillery firing solutions. After the war, an executive at IBM that built some of the computers said that there might be a use for perhaps a dozen computers.

- In 1962, it seemed obvious that the bottled sunlight of nuclear fusion would be everywhere within twenty years—the latest fusion predictions are that fusion is twenty years in the future—and that is the same prediction that was made in 1962.

- Conversely, more efficient appliances allow us to live comfortably with half the electric power plants that were predicted in 1970.

- In the first decade of the 2000s, a sizeable number of the energy experts expected a crisis of peak oil production followed by an economic crash.[931] Then, the Fracking Revolution happened, and the United States is now the leading oil-producing country.

- There could very well be surprising technological innovations that would be more congenial to politically correct environmentalism. Batteries with three times the power per unit weight and half the price per battery would make solar and wind power practical and make electric cars much more affordable than the internal-combustion engine. Room-temperature superconduction or practical fusion power would each be another game-changer.

---

[931] Kenneth Deffeyes, *Beyond Oil: The View from Hubbert's Peak*, Hukk and Wang, 2006.

Evolving technology always has surprises for good or ill. I am not the first to say that the greatest enemies of prediction are genius, madness, and luck.

Future Green New Deal proposals by any number of people, Versions 3.0 through n, will incorporate the goods and avoid the ills as much as possible.

10 Conclusions

11 Original Green New Deal Resolution & FAQ

11.1 Original Green New Deal Resolution

[This section is from the Library of Congress and is not copyrighted.]

116th CONGRESS
1st Session

H. RES. 109

Recognizing the duty of the Federal Government to create a Green New Deal.

IN THE HOUSE OF REPRESENTATIVES
February 7, 2019

Ms. Ocasio-Cortez (for herself, Mr. Hastings, Ms. Tlaib, Mr. Serrano, Mrs. Carolyn B. Maloney of New York, Mr. Vargas, Mr. Espaillat, Mr. Lynch, Ms. Velázquez, Mr. Blumenauer, Mr. Brendan F. Boyle of Pennsylvania, Mr. Castro of Texas, Ms. Clarke of New York, Ms. Jayapal, Mr. Khanna, Mr. Ted Lieu of California, Ms. Pressley, Mr. Welch, Mr. Engel, Mr. Neguse, Mr. Nadler, Mr. McGovern, Mr. Pocan, Mr. Takano, Ms. Norton, Mr. Raskin, Mr. Connolly, Mr. Lowenthal, Ms. Matsui, Mr. Thompson of California, Mr. Levin of California, Ms. Pingree, Mr. Quigley, Mr. Huffman, Mrs. Watson Coleman, Mr. García of Illinois, Mr. Higgins of New York, Ms. Haaland, Ms. Meng, Mr. Carbajal, Mr. Cicilline, Mr. Cohen, Ms. Clark of Massachusetts, Ms. Judy Chu of California, Ms. Mucarsel-Powell, Mr. Moulton, Mr. Grijalva, Mr. Meeks, Mr. Sablan, Ms. Lee of California, Ms. Bonamici, Mr. Sean Patrick Maloney of New York, Ms. Schakowsky, Ms. DeLauro, Mr. Levin of Michigan, Ms. McCollum, Mr. DeSaulnier, Mr. Courtney, Mr. Larson of Connecticut, Ms. Escobar, Mr. Schiff, Mr. Keating, Mr. DeFazio, Ms. Eshoo, Mrs. Trahan, Mr. Gomez, Mr. Kennedy, and Ms. Waters) submitted the following resolution; which was referred to the Committee on Energy and Commerce, and in addition to the Committees on Science, Space, and Technology, Education and Labor, Transportation and Infrastructure, Agriculture, Natural Resources, Foreign Affairs, Financial Services, the Judiciary, Ways and Means, and Oversight and Reform, for a period to be subsequently determined by the Speaker, in each case for consideration of such provisions as fall within the jurisdiction of the committee concerned

RESOLUTION

Recognizing the duty of the Federal Government to create a Green New Deal.

Whereas the October 2018 report entitled "Special Report on Global Warming of 1.5 °C" by the Intergovernmental Panel on Climate Change and the November 2018 Fourth National Climate Assessment report found that—

> (1) human activity is the dominant cause of observed climate change over the past century;
>
> (2) a changing climate is causing sea levels to rise and an increase in wildfires, severe storms, droughts, and other extreme weather events that threaten human life, healthy communities, and critical infrastructure;
>
> (3) global warming at or above 2 degrees Celsius beyond preindustrialized levels will cause—
>
>> (A) mass migration from the regions most affected by climate change;
>>
>> (B) more than $500,000,000,000 in lost annual economic output in the United States by the year 2100;
>>
>> (C) wildfires that, by 2050, will annually burn at least twice as much forest area in the western United States than was typically burned by wildfires in the years preceding 2019;
>>
>> (D) a loss of more than 99 percent of all coral reefs on Earth;
>>
>> (E) more than 350,000,000 more people to be exposed globally to deadly heat stress by 2050; and
>>
>> (F) a risk of damage to $1,000,000,000,000 of public infrastructure and coastal real estate in the United States; and
>
> (4) global temperatures must be kept below 1.5 degrees Celsius above preindustrialized levels to avoid the most severe impacts of a changing climate, which will require—
>
>> (A) global reductions in greenhouse gas emissions from human sources of 40 to 60 percent from 2010 levels by 2030; and
>>
>> (B) net-zero global emissions by 2050;

A Better Green New Deal

Whereas, because the United States has historically been responsible for a disproportionate amount of greenhouse gas emissions, having emitted 20 percent of global greenhouse gas emissions through 2014, and has a high technological capacity, the United States must take a leading role in reducing emissions through economic transformation;

Whereas the United States is currently experiencing several related crises, with—

> (1) life expectancy declining while basic needs, such as clean air, clean water, healthy food, and adequate health care, housing, transportation, and education, are inaccessible to a significant portion of the United States population;
>
> (2) a 4-decade trend of wage stagnation, deindustrialization, and antilabor policies that has led to—
>
>> (A) hourly wages overall stagnating since the 1970s despite increased worker productivity;
>>
>> (B) the third-worst level of socioeconomic mobility in the developed world before the Great Recession;
>>
>> (C) the erosion of the earning and bargaining power of workers in the United States; and
>>
>> (D) inadequate resources for public sector workers to confront the challenges of climate change at local, State, and Federal levels; and
>
> (3) the greatest income inequality since the 1920s, with—
>
>> (A) the top 1 percent of earners accruing 91 percent of gains in the first few years of economic recovery after the Great Recession;
>>
>> (B) a large racial wealth divide amounting to a difference of 20 times more wealth between the average white family and the average black family; and
>>
>> (C) a gender earnings gap that results in women earning approximately 80 percent as much as men, at the median;

Whereas climate change, pollution, and environmental destruction have exacerbated systemic racial, regional, social, environmental, and economic injustices (referred to in this preamble as "systemic injustices") by disproportionately affecting indigenous peoples, communities of color, migrant communities, deindustrialized communities, depopulated rural communities, the poor, low-income workers, women, the elderly, the unhoused, people with disabilities, and youth (referred to in this preamble as "frontline and vulnerable communities");

Whereas, climate change constitutes a direct threat to the national security of the United States—

   (1) by impacting the economic, environmental, and social stability of countries and communities around the world; and

   (2) by acting as a threat multiplier;

Whereas the Federal Government-led mobilizations during World War II and the New Deal created the greatest middle class that the United States has ever seen, but many members of frontline and vulnerable communities were excluded from many of the economic and societal benefits of those mobilizations; and

Whereas the House of Representatives recognizes that a new national, social, industrial, and economic mobilization on a scale not seen since World War II and the New Deal era is a historic opportunity—

   (1) to create millions of good, high-wage jobs in the United States;

   (2) to provide unprecedented levels of prosperity and economic security for all people of the United States; and

   (3) to counteract systemic injustices: Now, therefore, be it

Resolved, That it is the sense of the House of Representatives that—

   (1) it is the duty of the Federal Government to create a Green New Deal—

      (A) to achieve net-zero greenhouse gas emissions through a fair and just transition for all communities and workers;

      (B) to create millions of good, high-wage jobs and ensure prosperity and economic security for all people of the United States;

      (C) to invest in the infrastructure and industry of the United States to sustainably meet the challenges of the 21st century;

      (D) to secure for all people of the United States for generations to come—

         (i) clean air and water;

         (ii) climate and community resiliency;

         (iii) healthy food;

         (iv) access to nature; and

(v) a sustainable environment; and

(E) to promote justice and equity by stopping current, preventing future, and repairing historic oppression of indigenous peoples, communities of color, migrant communities, deindustrialized communities, depopulated rural communities, the poor, low-income workers, women, the elderly, the unhoused, people with disabilities, and youth (referred to in this resolution as "frontline and vulnerable communities");

(2) the goals described in subparagraphs (A) through (E) of paragraph (1) (referred to in this resolution as the "Green New Deal goals") should be accomplished through a 10-year national mobilization (referred to in this resolution as the "Green New Deal mobilization") that will require the following goals and projects—

(A) building resiliency against climate change-related disasters, such as extreme weather, including by leveraging funding and providing investments for community-defined projects and strategies;

(B) repairing and upgrading the infrastructure in the United States, including—

(i) by eliminating pollution and greenhouse gas emissions as much as technologically feasible;

(ii) by guaranteeing universal access to clean water;

(iii) by reducing the risks posed by climate impacts; and

(iv) by ensuring that any infrastructure bill considered by Congress addresses climate change;

(C) meeting 100 percent of the power demand in the United States through clean, renewable, and zero-emission energy sources, including—

(i) by dramatically expanding and upgrading renewable power sources; and

(ii) by deploying new capacity;

(D) building or upgrading to energy-efficient, distributed, and "smart" power grids, and ensuring affordable access to electricity;

(E) upgrading all existing buildings in the United States and building new buildings to achieve maximum energy efficiency, water efficiency, safety, affordability, comfort, and durability, including through electrification;

(F) spurring massive growth in clean manufacturing in the United States and removing pollution and greenhouse gas emissions from manufacturing and industry as much as is technologically feasible, including by expanding renewable energy manufacturing and investing in existing manufacturing and industry;

(G) working collaboratively with farmers and ranchers in the United States to remove pollution and greenhouse gas emissions from the agricultural sector as much as is technologically feasible, including—

  (i) by supporting family farming;

  (ii) by investing in sustainable farming and land use practices that increase soil health; and

  (iii) by building a more sustainable food system that ensures universal access to healthy food;

(H) overhauling transportation systems in the United States to remove pollution and greenhouse gas emissions from the transportation sector as much as is technologically feasible, including through investment in—

  (i) zero-emission vehicle infrastructure and manufacturing;

  (ii) clean, affordable, and accessible public transit; and

  (iii) high-speed rail;

(I) mitigating and managing the long-term adverse health, economic, and other effects of pollution and climate change, including by providing funding for community-defined projects and strategies;

(J) removing greenhouse gases from the atmosphere and reducing pollution by restoring natural ecosystems through proven low-tech solutions that increase soil carbon storage, such as land preservation and afforestation;

(K) restoring and protecting threatened, endangered, and fragile ecosystems through locally appropriate and science-based projects that enhance biodiversity and support climate resiliency;

(L) cleaning up existing hazardous waste and abandoned sites, ensuring economic development and sustainability on those sites;

(M) identifying other emission and pollution sources and creating solutions to remove them; and

(N) promoting the international exchange of technology, expertise, products, funding, and services, with the aim of making the United States the international leader on climate action, and to help other countries achieve a Green New Deal;

(3) a Green New Deal must be developed through transparent and inclusive consultation, collaboration, and partnership with frontline and vulnerable communities, labor unions, worker cooperatives, civil society groups, academia, and businesses; and

(4) to achieve the Green New Deal goals and mobilization, a Green New Deal will require the following goals and projects—

(A) providing and leveraging, in a way that ensures that the public receives appropriate ownership stakes and returns on investment, adequate capital (including through community grants, public banks, and other public financing), technical expertise, supporting policies, and other forms of assistance to communities, organizations, Federal, State, and local government agencies, and businesses working on the Green New Deal mobilization;

(B) ensuring that the Federal Government takes into account the complete environmental and social costs and impacts of emissions through—

(i) existing laws;

(ii) new policies and programs; and

(iii) ensuring that frontline and vulnerable communities shall not be adversely affected;

(C) providing resources, training, and high-quality education, including higher education, to all people of the United States, with a focus on frontline and vulnerable communities, so that all people of the United States may be full and equal participants in the Green New Deal mobilization;

(D) making public investments in the research and development of new clean and renewable energy technologies and industries;

(E) directing investments to spur economic development, deepen and diversify industry and business in local and regional economies, and build wealth and community ownership, while prioritizing high-quality job creation and economic, social, and environmental benefits in frontline and vulnerable communities, and deindustrialized communities, that may otherwise struggle with the transition away from greenhouse gas intensive industries;

(F) ensuring the use of democratic and participatory processes that are inclusive of and led by frontline and vulnerable communities and workers to plan, implement, and administer the Green New Deal mobilization at the local level;

(G) ensuring that the Green New Deal mobilization creates high-quality union jobs that pay prevailing wages, hires local workers, offers training and advancement opportunities, and guarantees wage and benefit parity for workers affected by the transition;

(H) guaranteeing a job with a family-sustaining wage, adequate family and medical leave, paid vacations, and retirement security to all people of the United States;

(I) strengthening and protecting the right of all workers to organize, unionize, and collectively bargain free of coercion, intimidation, and harassment;

(J) strengthening and enforcing labor, workplace health and safety, antidiscrimination, and wage and hour standards across all employers, industries, and sectors;

(K) enacting and enforcing trade rules, procurement standards, and border adjustments with strong labor and environmental protections—

    (i) to stop the transfer of jobs and pollution overseas; and

    (ii) to grow domestic manufacturing in the United States;

(L) ensuring that public lands, waters, and oceans are protected and that eminent domain is not abused;

(M) obtaining the free, prior, and informed consent of indigenous peoples for all decisions that affect indigenous peoples and their traditional territories, honoring all treaties and agreements with indigenous peoples, and protecting and enforcing the sovereignty and land rights of indigenous peoples;

(N) ensuring a commercial environment where every businessperson is free from unfair competition and domination by domestic or international monopolies; and

(O) providing all people of the United States with—

    (i) high-quality health care;

    (ii) affordable, safe, and adequate housing;

    (iii) economic security; and

    (iv) clean water, clean air, healthy and affordable food, and access to nature.

11.2 Original Green New Deal Fact Frequently Asked Questions (FAQs)

This is the fact sheet and FAQ posted on the morning of February 7, 2019 (and then quickly taken down) from the official Congressional website of Representative Alexandria Ocasio-Cortez and her Green New Deal.

LAUNCH: Thursday, February 7, [2019] at 8:30 AM.

Overview

- **We will begin work immediately on Green New Deal bills to put the nuts and bolts on the plan described in this resolution (important to say so someone else can't claim this mantle).**
- **This is a massive transformation of our society with clear goals and a timeline.**
  - The Green New Deal resolution a 10-year plan to mobilize every aspect of American society at a scale not seen since World War 2 to achieve net-zero greenhouse gas emissions and create economic prosperity for all. It will:
    - Move America to 100% clean and renewable energy
    - Create millions of family supporting-wage, union jobs
    - Ensure a just transition for all communities and workers to ensure economic security for people and communities that have historically relied on fossil fuel industries
    - Ensure justice and equity for frontline communities by prioritizing investment, training, climate and community resiliency, economic and environmental benefits in these communities.
    - Build on FDR's second bill of rights by guaranteeing:
      - A job with a family-sustaining wage, family and medical leave, vacations, and retirement security
      - High-quality education, including higher education and trade schools
      - Clean air and water and access to nature
      - Healthy food
      - High-quality health care
      - Safe, affordable, adequate housing
      - Economic environment free of monopolies

- Economic security for all who are unable or unwilling to work
- **There is no time to waste.**
  - IPCC Report said global emissions must be cut by Merkley 40-60% by 2030. US is 20% of total emissions. We must get to 0 by 2030 and lead the world in a global Green New Deal.
- **Americans love a challenge. This is our moonshot.**
  - When JFK said we'd go to the by the end of the decade, people said impossible.
  - If Eisenhower wanted to build the interstate highway system today, people would ask how we'd pay for it.
  - When FDR called on America to build 185,000 planes to fight World War 2, every business leader, CEO, and general laughed at him. At the time, the U.S. had produced 3,000 planes in the last year. By the end of the war, we produced 300,000 planes. That's what we are capable of if we have real leadership
- **This is massive investment in our economy and society, not expenditure.**
  - We invested 40-50% of GDP into our economy during World War 2 and created the greatest middle class the US has seen.
  - The interstate highway system has returned more than $6 in economic productivity for every $1 it cost
  - This is massively expanding existing and building new industries at a rapid pace – growing our economy
- **The Green New Deal has momentum.**
  - 92 percent of Democrats and 64 percent of Republicans support the Green New Deal
  - Nearly every major Democratic Presidential contender say they back the Green New deal including: Elizabeth Warren, Cory Booker, Kamala Harris, Jeff Merkeley, Julian Castro, Kirsten Gillibrand, Bernie Sanders, Tulsi Gabbard, and Jay Inslee.
  - 45 House Reps and 330+ groups backed the original resolution for a select committee
  - Over 300 local and state politicians have called for a federal Green New Deal
  - New Resolution has 20 co-sponsors, about 30 groups (numbers will change by Thursday).

Why 100% clean and renewable and not just 100% renewable? Are you saying we won't transition off fossil fuels?

Yes, we are calling for a full transition off fossil fuels and zero greenhouse gases. Anyone who has read the resolution sees that we spell this out through a plan that calls for eliminating greenhouse gas emissions from every sector of the economy. Simply banning fossil fuels immediately won't build the new economy to replace it – this is the plan to build that new economy and spells out how to do it technically. We do this through a huge mobilization to create the renewable energy economy as fast as possible. We set a goal to get to net-zero, rather than zero emissions, in 10 years because we aren't sure that we'll be able to fully get rid of farting cows and airplanes that fast, but we think we can ramp up renewable manufacturing and power production, retrofit every building in America, build the smart grid, overhaul transportation and agriculture, plant lots of trees and restore our ecosystem to get to net-zero.

Is nuclear a part of this?

A Green New Deal is a massive investment in renewable energy production and would not include creating new nuclear plants. It's unclear if we will be able to decommission every nuclear plant within 10 years, but the plan is to transition off of nuclear and all fossil fuels as soon as possible. No one has put the full 10-year plan together yet, and if it is possible to get to fully 100% renewable in 10 years, we will do that.

Does this include a carbon tax?

The Green New Deal is a massive investment in the production of renewable energy industries and infrastructure. We cannot simply tax gas and expect workers to figure out another way to get to work unless we've first created a better, more affordable option. So we're not ruling a carbon tax out, but a carbon tax would be a tiny part of a Green New Deal in the face of the gigantic expansion of our productive economy and would have to be preceded by first creating the solutions necessary so that workers and working class communities are not affected. While a carbon tax may be a part of the Green New Deal, it misses the point and would be off the table unless we create the clean, affordable options first.

Does this include cap and trade?

The Green New Deal is about creating the renewable energy economy through a massive investment in our society and economy. Cap and trade assumes the existing market will solve this problem for us, and that's simply not true. While cap and trade may be a tiny part of the larger Green New Deal plan to mobilize our economy, any cap and trade legislation will pale in comparison to the size of the mobilization and must recognize that existing legislation can incentivize companies to create toxic hotspots in frontline communities, so anything here must ensure that frontline communities are prioritized.

Original Green New Deal Resolution and FAQs

Does a GND ban all new fossil fuel infrastructure or nuclear power plants?

The Green New Deal makes new fossil fuel infrastructure or nuclear plants unnecessary. This is a massive mobilization of all our resources into renewable energies. It would simply not make sense to build new fossil fuel infrastructure because we will be creating a plan to reorient our entire economy to work off renewable energy. Simply banning fossil fuels and nuclear plants immediately won't build the new economy to replace it – this is the plan to build that new economy and spells out how to do it technically.

Are you for CCUS [Carbon Capture, Utilization, and Storage]?

We believe the right way to capture carbon is to plant trees and restore our natural ecosystems. CCUS technology to date has not proven effective.

How will you pay for it?

The same way we paid for the New Deal, the 2008 bank bailout and extended quantitative easing programs. The same way we paid for World War II and all our current wars. The Federal Reserve can extend credit to power these projects and investments and new public banks can be created to extend credit. There is also space for the government to take an equity stake in projects to get a return on investment. At the end of the day, this is an investment in our economy that should grow our wealth as a nation, so the question isn't how will we pay for it, but what will we do with our new shared prosperity.

Why do we need a sweeping Green New Deal investment program? Why can't we just rely on regulations and taxes and the private sector to invest alone such as a carbon tax or a ban on fossil fuels?

- The level of investment required is massive. Even if every billionaire and company came together and were willing to pour all the resources at their disposal into this investment, the aggregate value of the investments they could make would not be sufficient.

- The speed of investment required will be massive. Even if all the billionaires and companies could make the investments required, they would not be able to pull together a coordinated response in the narrow window of time required to jump-start major new projects and major new economic sectors. Also, private companies are wary of making massive investments in unproven research and technologies; the government, however, has the time horizon to be able to patiently make investments in new tech and R&D, without necessarily having a commercial outcome or application in mind at the time the investment is made. Major examples of government investments in "new" tech that subsequently spurred a boom in the private section include DARPA projects, the creation of the internet - and, perhaps most recently, the government's investment in Tesla.

- Simply put, we don't need to just stop doing some things we are doing (like using fossil fuels for energy needs); we also need to start doing new things (like overhauling whole industries or retrofitting all buildings to be energy efficient). Starting to do new things requires some upfront investment. In the same way that a company that is trying to change how it does business may need to make big upfront capital investments today in order to reap future benefits (for e.g., building a new factory to increase production or buying new hardware and software to totally modernize its IT system), a country that is trying to change how its economy works will need to make big investments today to jump-start and develop new projects and sectors to power the new economy.

- Merely incentivizing the private sector doesn't work - e.g. the tax incentives and subsidies given to wind and solar projects have been a valuable spur to growth in the US renewables industry but, even with such investment promotion subsidies, the present level of such projects is simply inadequate to transition to a fully greenhouse gas neutral economy as quickly as needed.

- Once again, we're not saying that there isn't a role for private sector investments; we're just saying that the level of investment required will need every actor to pitch in and that the government is best placed to be the prime driver.

Resolution Summary

- **Created in consultation with multiple groups from environmental community, environmental justice community, and labor community.**
- **5 goals in 10 years:**
    - Net-zero greenhouse gas emissions through a fair and just transition for all communities and workers
    - o Create millions of high-wage jobs and ensure prosperity and economic security for all
    - Invest in infrastructure and industry to sustainably meet the challenges of the 21st century o Clean air and water, climate and community resiliency, healthy food, access to nature, and a sustainable environment for all
    - Promote justice and equity by stopping current, preventing future, and repairing historic oppression of frontline and vulnerable communities
- **National mobilization our economy through 14 infrastructure and industrial projects. Every project strives to remove greenhouse gas emissions and pollution from every sector of our economy:**
    - Build infrastructure to create resiliency against climate change-related disasters
    - Repair and upgrade U.S. infrastructure. ASCE estimates this is $4.6 trillion at minimum.

- Meet 100% of power demand through clean and renewable energy sources
- Build energy-efficient, distributed smart grids and ensure affordable access to electricity
- Upgrade or replace every building in US for state-of-the-art energy efficiency
- Massively expand clean manufacturing (like solar panel factories, wind turbine factories, battery and storage manufacturing, energy efficient manufacturing components) and remove pollution and greenhouse gas emissions from manufacturing
- Work with farmers and ranchers to create a sustainable, pollution and greenhouse gas free, food system that ensures universal access to healthy food and expands independent family farming
- Totally overhaul transportation by massively expanding electric vehicle manufacturing, build charging stations everywhere, build out highspeed rail at a scale where air travel stops becoming necessary, create affordable public transit available to all, with goal to replace every combustion-engine vehicle
- Mitigate long-term health effects of climate change and pollution
- Remove greenhouse gases from our atmosphere and pollution through afforestation, preservation, and other methods of restoring our natural ecosystems
- Restore all our damaged and threatened ecosystems
- Clean up all the existing hazardous waste sites and abandoned sites
- Identify new emission sources and create solutions to eliminate those emissions
- Make the US the leader in addressing climate change and share our technology, expertise and products with the rest of the world to bring about a global Green New Deal

- **Social and economic justice and security through 15 requirements:**
  - Massive federal investments and assistance to organizations and businesses participating in the green new deal and ensuring the public gets a return on that investment
  - Ensure the environmental and social costs of emissions are taken into account
  - Provide job training and education to all
  - Invest in R&D of new clean and renewable energy technologies
  - Doing direct investments in frontline and deindustrialized communities that would otherwise be hurt by the transition to prioritize economic benefits there
  - Use democratic and participatory processes led by frontline and vulnerable communities to implement GND projects locally

A Better Green New Deal

- Ensure that all GND jobs are union jobs that pay prevailing wages and hire local
- Guarantee a job with family-sustaining wages
- Protect right of all workers to unionize and organize
- Strengthen and enforce labor, workplace health and safety, antidiscrimination, and wage and hour standards
- Enact and enforce trade rules to stop the transfer of jobs and pollution overseas and grow domestic manufacturing
- Ensure public lands, waters, and oceans are protected and eminent domain is not abused
- Obtain free, prior, and informed consent of Indigenous peoples
- Ensure an economic environment free of monopolies and unfair competition
- Provide high-quality health care, housing, economic security, and clean air, clean water, healthy food, and nature to all

Original Green New Deal Resolution and FAQs

12 Annotated Bibliography

Many people have studied fears of potential doom and dreams for possible wonders. Often, they are focused in one narrow discipline. A Green New Deal provides a connecting theme to bring them together.

If you want to learn more, the list of annotated references may be of use to you now or in the future. Some are heavy going. Many are fun to read and inspiring as well as being fact filled.

Enjoy!

A Climate of Crisis: America in the Age of Environmentalism, Patrick Allitt, Penguin Books; March 2015 (original release 2014). The author broadly surveys the history of many environmental issues with sympathy for both sides. Although he criticizes the relentless crisis mentality in the environmental area, he also notes environmental improvements and calls for more improvements.

Advances in Small Modular Reactor Technology Developments, A Supplement to: IAEA Advanced Reactors Information System (ARIS), International Atomic Energy Agency 2018 Edition, Sept. 2018. https://aris.iaea.org/Publications/SMR-Book_2018.pdf (accessed July 31, 2019) This document shows the range of small modular reactors that have been seriously proposed. Hopefully, some of them will get funded, built, and successfully operated.

After Geoengineering: Climate Tragedy, Repair, and Restoration, Holly Jean Buck, Verso, 2019. The author has a central theme that geoengineering (especially solar radiation management) should be explored along with other proposed remedies for dealing with global warming. She makes the case for possible uses of geoengineering along with regenerative agriculture, marine agriculture, carbon capture, and other technologies. She also provides an interesting perspective about megaprojects being social constructs with potential for failure and provides related references. However, for good but also ill, she directs much of her rhetorical case building to the Marxist wing of the environmental movement. Those discussions are sometimes afflicted with turgid obscure prose and tangents about how society should be changed as well as a bias against any business initiatives for mitigating global warming. Nuclear fission apparently does not exist in her universe.

The Agriculture Manifesto, Ten Key Drivers That Will Shape Agriculture in the Next Decade, Robert Saik, ISBN 9781499382709, 2014. This obviously self-published manifesto is hardly larger than a pamphlet. Yet, it presents a wonderful summary of the most important trends in current agriculture. The author is a Canadian agricultural consultant with apparently many years of experience.

Anasazi Architecture and American Design, editors: Baker H. Morrow, V. B. Price, and Robert C. Heyder, University of New Mexico Press, Albuquerque, New Mexico, Nov. 1997. The authors and editors detail the passive architecture and terratecture that were developed independently by the Anasazi of the North American Southwest.

Annotated Bibliography

The Answer: Why Only Inherently Safe, Mini Nuclear Power Plants Can Save Our World, Reese Palley, The Quantuck Lane Press, New York, 2011. Do not be intimidated by Appendixes A through G at the back. The main text is a very readable 193 pages that flows nicely from global-warming fears related to global warming, why alternative energies will not prevent warming, why fission reactors developed into the gigawatt-sized behemoths that he considers unsafe, and his proposal for safer small reactors and mini-reactors.

Apocalypse: A History of the End of Time, John Michael Greer, Quertas, London, England, 2012. Greer follows the self-replicating idea (meme) that the end is (often) near from its known origin in Mithraism, through the Hebrews' end of the world, through John's book of *Revelation*, through centuries of Christian millennialists, through unidentified flying object (UFO) warnings to wayward Earthlings, to the misquoted Mayan inscription that supposedly foretold a major disaster in December 2012. Unfortunately, Greer drank some of the apocalyptic Kool Aid ™ and then wrote about the new *Dark Age America: Climate Change, Cultural Collapse, and the Hard Future Ahead* that he claims will be precipitated by uncontrolled climate change and resource depletion.

Aquaculture: Farming Aquatic Animals and Plants, 3rd Edition. John S. Lucas, Paul C. Southgate, Craig S. Tucker (editors), Wiley-Blackwell, Jan. 22, 2019. This is the definitive academic book on water agriculture. (It uses aquaculture as the term for both freshwater and marine agriculture.) This is book for you if you are an academic or a professional in the field. Otherwise, it will probably be too big and too expensive.

The Art and Science of Grazing: How Grass Farmers Can Create Sustainable Systems for Healthy Animals and Farm Ecosystems, Sarah Flack and Hubert Karreman (foreword), Chelsea Green Publishing, June 7, 2016. The authors provide a detailed manual for grazing cattle and sheep—probably too detailed unless you are a working farmer or deeply into the subject.

Apocalypse, A Natural History of Global Disasters, Bill McGuire, Blandford; First Edition, May 2000. The author connects a number of potential dooms to events in geologic and human history. However, as a geologist, McGuire's doom focus is limited to global warming, asteroid impacts, and major volcanoes, earthquakes, and tsunamis.

Aquaculture: Farming Aquatic Animals and Plants, Third Edition, John S. Lucas, Paul C. Southgate, and Craig S. Tucker, editors, Wiley Blackwell, 2019. This is the (expensive) reference for professional reference and college textbooks. The latest edition is

Aquatic Photosynthesis, Third Edition, Paul G. Falkowski, and John A. Raven, University Press, Princeton, New Jersey, 2007. The authors provided the definite guide to micro0algae and macro-algae, but it is highly technical.

Beyond Oil and Gas: The Methanol Economy, George Olah, Alain Goeppert, and Surya Prakesh, Wiley, March 2006. Nobel prize-winning chemist George Olah and his coauthors examine the limits of the hydrogen economy (especially low energy density, storage issues, and a required new distribution infrastructure). From there, they propose the use of methanol (CH_3OH) as an alternate carrier of energy (including from renewable sources) because it is denser, stores more easily, and would require much less adjustment of vehicles and the distribution infrastructure.

Beyond Oil: The View from Hubbert's Peak [also see *Hubbert's Peak*], Kenneth Deffeyes, Hill and Wang, 2006. Deffeyes continued Hubbert's resource peaking analysis to alternative energy sources and the possible issues with their development. Interestingly, he avoided trying to make any estimates about natural gas because gas use on-site often does not go into production data so calculating trends is very difficult. Even more deliciously, he advised against fission breeder reactors because according to his peaking analysis, "… there's so much uranium out there." [My view is that the breeders and reprocessing are worth the extra cost to avoid leaving massive stores of lethal material behind for millennia.]

The Biochar Revolution: Transforming Agriculture & Environment, Paul Taylor, Hugh McLaughlin, and Tim Flannery, Global Publishing Group, Dec. 2010. The authors have compiled a number of articles on the details of biochar production and why they feel it is important for soil building and reducing carbon dioxide in the atmosphere.

The Biochar Solution: Carbon Farming and Climate Change, Albert Bates, New Society Publishers; 2nd edition, 1994. The author provides an excellent and entertaining background of the Amazonian society before the conquistadores that used charcoal (biochar) and organic materials to generate the black soils that that are still fertile to this day. He also argues for much greater use of biochar.

Bold: How to Go Big, Create Wealth and Impact the World (Exponential Technology, Peter H. Diamondis and Steven Kotler, Simon & Schuster; paperback reprint edition, Feb. 23, 2016. The authors supply a fun read with excellent vignettes on innovation perils of failing to innovate, profiles of recent super entrepreneurs, and crowdfunding.

The Box: How the Shipping Container Made the World Smaller and the World Economy Bigger, Marc Levinson, Princeton University Press, Jan. 27, 2008. The author argues persuasively that the humble standardized shipping container revolutionized world trade making transportation of goods so much cheaper that cities such as Chicago and Shanghai are now in direct competition. As such, the container might be the most powerful invention of the Twentieth Century.

Burn: Using Fire to Cool the Earth, Albert Bates and Kathleen Draper, Chelsea Green Publishing, White River Junction, Vermont, 2018. The authors display a wide range of technologies and business practices to convert present waste streams from organic materials into useful materials that will not become carbon dioxide in the sky. Examples include cooking food or chemicals from plant wastes; cooking down the remainder into biochar charcoal that can improve soil quality; and using charcoal in cement, road asphalt, and cardboard.

Burning the Sky: Operation Argus and the Untold Story of the Cold War Nuclear Tests in Outer Space, Mark Wolverton, The Overlook Press, New York, 2018. The author provides history that often reads like a thriller novel. In 1958, the United States began a series of nuclear bomb tests in the high atmosphere and in space. The tests were to see if nuclear explosions might disable intercontinental ballistic missiles or have other major effects or not. The results were reported to be minor, which is true … or not.

Cahokia: Ancient America's Great City on the Mississippi, Timothy R. Pauketat, Penguin Group, July 1, 2009. Cahokia appears to be the source area and cultural center for the mound-building civilization that flourished from in the present area of the U.S. Midwest and Southeast. This book is an entertaining and fact-filled account that updates Robert Silverberg's *The Mound Builders*. It still does not have a definitive answer on why the civilization collapsed.

Clean Disruption of Energy and Transportation: How Silicon Valley Will Make Oil, Nuclear, Natural Gas, Coal, Electric Utilities and Conventional Cars Obsolete by 2030, publisher: Tony Seba, May 20, 2014. The author is biased toward solar and wind, and he may be several years optimistic on the adoption of electric autonomous ridesharing; but he makes a strong case that it will happen and that it will strongly decrease the use of fossil hydrocarbons both as fuels and as energy for vehicles for which the world fleet will be much smaller. Additionally, electric vehicles should last several times the distance of combustion engines, so that will further reduce hydrocarbons used for materials preparation and fabrication of the vehicles.

Climate of Fear: Why We Shouldn't Worry about Global Warming, Thomas Gale Moore, The Cato Institute, 1998. This relaxed and largely evenhanded book has stood up well to time. In 1998, the author looked from an economist's viewpoint at the key issues of the warming controversy. Would the costs be greater than the benefits of a massive switch away from fossil fuels? Would the developing countries forego progress to avoid the supposed dangers of coal power? Might agricultural production increase in a global-warming world. Last but not least, might the warming effects in models be exaggerated?

Climate, History and the Modern World, 2nd Edition, Hubert H. Lamb, Routledge; 2 edition, Aug. 17, 1995. The author, who was one of the first climatologists to propose that climate might be variable, provides a wide range of scientific and historical data for both times of warming and cooling over thousands of years. The book is exhaustive. Readers starting on climate change throughout history might warm up first with some of Brian Fagan's works such as *The Great Warming, The Long Summer, The Little Ice* Age, and *The Elixir of Life*.

Climate: A New Story, Chapter 8, "Regeneration," Charles Eisenstein, North Atlantic Books, Berkeley, California, 2018. The author argues for a better ecological focus of preserving or regaining what was good rather than focusing on narrow quantitative goals such as carbon dioxide levels. He argues that this is a less fight oriented and more solution-oriented view that is more likely to solve the problems. His new age-style views might help get a more coordinated positive approach for maintaining prosperity and the present climate.

Climatopolis: How Our Cities Will Thrive in the Hotter Future," Matthew E. Kahn, Basic Books, New York, NY, 2010. The author subscribes to the theory of major global-greenhouse warming, and he considers it inevitable because of the technological and social inertia preventing a major noncarbon energy economy or any method to sweep carbon dioxide out of the atmosphere. However, he sees future cities as not just surviving, but thriving, in the warmer climate because they will be able to site on higher ground and adapt more energy efficient ways to remain cool.

The Cold and the Dark: The World After Nuclear War, Paul R. Ehrlich, Carl Sagan, Donald Kennedy, et al., W. W. Norton & Company, 1985. The authors exaggerated the supposed effects of nuclear winter to the point of falsehood, ... but they apparently had a major part in ending the Cold War. The loyal communist comrades of the Soviet won the horrendous battles of World War II against the Germans, so they were ready for a possible World War III. But being from the north and east of the Eurasian continent, they knew winters and feared a super winter. Still, a major nuclear exchange anywhere on the planet would produce worldwide radiation and probably a milder version of the winter threatened in this book.

Collapse: How Societies Choose to Fail or Succeed, Jared Diamond, Penguin Books; Revised edition, January 2011. The author explores geographic and environmental reasons why some human populations have flourished and others have collapsed. Prime examples cited include the Anasazi of the American Southwest the Viking colonies of Greenland, as well as more contemporary examples such as Rwanda and Montana. The author frets about eco-meltdowns historic and potential, but he also describes ways to mitigate the dangers.

The Cooling: Has the Next Ice Age Already Begun? Lowell Ponte, Prentice-Hall; 1st Printing edition, 1976. The journalist-author hyped a story that was being told by a significant number of climate researchers at the time—more than two decades of cooling suggested a major cooling climate change. One of the ironies is that the cooling trend had already begun to reverse when the book was released in 1976. Another irony is that the whiplash of global cooling to global warming gave Ponte a very cynical view of future climate and environmental predictions.

Climate Crash: Abrupt Climate Change and What It Means for Our Future, John D. Cox, Joseph Henry Press; May 2, 2005. Cox argues that alterations in our climate can happen quickly and dramatically. Slow, creeping climate variations between ice ages and hothouse ages have been punctuated by far more rapid changes.

Concrete Planet: The Strange and Fascinating Story of the World's Most Common Man-Made Material, Prometheus Books; Robert Courtland, Nov. 22, 2011. Who knew that concrete could be exciting … and scary? Innovations in concrete contributed to some of humanity's great advances. Conversely, the rusting of rebar reinforcement for many of the concrete structures built before the 1980s means there will be a need for major reconstructions.

Cows Save the Planet: And Other Improbable Ways of Restoring Soil to Heal the Earth, Judith D. Schwartz, Laura Jorstad, and Gretel Ehrlich, Chelsea Green Publishing, April 2013. The authors discuss soil-regenerating agricultural with emphasis on large grazing animals particularly on land too steep for planted crops.

From Cupcakes to Chemicals: How the Culture of Alarmism makes Us Afraid of Everything and How to Fight Back, Julie Gunlock, IWF Press, Oct. 31, 2013. The author explains and provides examples of how scaring people with FUD (fear, uncertainty, and doubt) makes money and power for some people and causes their victims to spend money on things they don't need.

Comet/Asteroid Impacts and Human Society: An Interdisciplinary Approach, Peter Bobrowski and Hans Rickman, editors, Springer, 2007 edition, Jan. 8, 2007. The editors provide a compendium of reports from leading specialists from various disciplines who participated in a multidisciplinary workshop on comet/asteroid impacts. This book is an excellent report on the subject, but it is highly technical and expensive.

The Deep Range, Arthur C. Clarke, Harcourt, Brace & World, Inc., 1957. Although Clarke was best known for things related to outer space (such as *2001: A Space Odyssey* and invention of the geosynchronous communication satellite concept), he had an abiding interest in the ocean, and he spent much of his life on the island nation of Sri Lanka where he could scuba dive amid beautiful coral reefs. This science fiction adventure novel has a background large-scale mariculture with plankton harvesting and herding of baleen whales that overawes the story. Unfortunately, nobody has yet proposed a system for filtering out plankton (as the baleen filters of whale sharks and baleen whales do) without getting clogged to make a practical plankton harvester.

Dirt: The Erosion of Civilizations, David R. Montgomery, University of California Press, Berkeley, California, 2007. The author summarizes the crucial role of soil use and abuse in the history of Mesopotamia, Ancient Greece, the Roman Empire, China, Central America, and the American frontiers. This synthesis of geology, archeology, and agricultural history set the stage for his later proposals in *Growing a Revolution: Bringing Our Soil Back to Life*.

Dirt to Soil: One Family's Journey into Regenerative Agriculture, Gabe Brown, Chelsea Green Publishing; Oct. 1, 2018. The author is a North Dakota rancher and farmer who has spent more than two decades developing regenerative agriculture in his operation and then being a proponent of such methods. He provides a very readable book that starts with the evolution of his operation and then goes into the concepts behind such operations.

The Doubly Green Revolution: Food for All in the Twenty-First Century, Gordon Conway, Cornell University Press; Feb. 17, 1999. The author's central theme is that it is possible to raise yields three-fold on most smallholder farms worldwide by practicing sustainable agriculture. As an architect of the original Green Revolution he can acknowledge its failings (and its successes) better than most.

Drawdown: The Most Comprehensive Plan Ever Proposed to Reverse Global Warming, Paul Hawken, Penguin Books, April 2017. Hawken and his contributors are probably much too optimistic in many areas, but they get credit for saying that global warming can be stopped. It this book, they catalog the most complete list of the possible drawdown methods so far and provide descriptions of each. [Concepts from this book's three areas of food production, energy, and ocean are included.]

The Earth's Biosphere: Evolution, Dynamics, and Change, Vaclav Smil, MIT Press, ISBN 978-0-262-69298-4, 2003. This clear and succinct book ties life and its interaction with Earth and its atmosphere.

East of Eden, John Steinbeck, The Viking Press, Sept. 19, 1952. This is the book that Steinbeck considered his magnum opus. It is a sweeping story of the agricultural area around Salinas, California. For the purposes of this book, the key theme in *East of Eden* is entrepreneur Adam Trask's maniacal quest to ship refrigerator carloads of lettuce east to New York. Through the lens of fiction, the reader experiences Adam's concept, his dream, and the danger of risking a fortune on that dream. The final lesson in this thread of the novel is that entrepreneurs should be appreciated. Even a good idea (and Adam had a good idea) can fail for any number of reasons. Any number of failures may happen before success is achieved.

An Edible History of Humanity, Tom Standage, (originally published by Walker Publishing Co., New York) Bloomsbury USA, May 3, 2010. The author views history through the lens of food from the transition of hunter-gather societies to farming, to the breeding of crops, to the drive for spices that built empires, … to guesses about what the future may bring.

Edible Insects: Future prospects for food and feed security, Arnold van Huis, Joost Van Itterbeeck, Harmke Klunder, et al., Food and Agriculture Organization, United Nations, Rome, 2013. http://www.fao.org/docrep/018/i3253e/i3253e00.htm (accessed Dec. 13, 2015) The preface begins with the words. "Insects are often considered a nuisance to human beings and mere pests for crops and animals. Yet this is far from the truth. Insects provide food at low environmental cost, contribute positively to livelihoods, and play a fundamental role in nature." This 200-page work is a compendium of research to do just that. However, it's just a start.

Elixir: A History of Water and Humankind, Brian Fagan, (first published Bloomsbury Publishing PLC, June 7, 2011. Fagan entertainingly weaves the history of aqueducts, irrigation, climate variation, and the resulting shifts in food production that built and destroyed great empires in history, particularly emphasizing those of antiquity (e.g., ancient Egypt, the Fertile Crescent, and Peru).

The Emperor's New Hydrogen Economy, Darryl McMahon, iUniverse, Inc., 2006. The author details why the much-heralded Hydrogen Economy would not work as advertised, and he suggest a number of more practical approaches than the hydrogen economy.

Empires of Food: Feast, Famine and the Rise and Fall of Civilizations, Evan D. D. Fraser and Andrew Rimas, Counterpoint, March 20, 2012. The authors work maybe a little too hard in tying food in as a key part of human history, and arguing the eco-gloom angle. It has some entertaining spots, and sometimes hops around incoherently.

A Better Green New Deal

The End of Doom: Environmental Renewal in the Twenty-first Century, Ronald Bailey, St. Martin's Press, New York, 2015. The author considers a number of supposed environmental dooms and notes ways society is working toward (or already has) found ways to transcend those dooms. As Matt Ridley, author of the *Rational Optimist* said, "Ronald Bailey sets out factually and simply the unassailable, if inconvenient, truth: that if you care for this planet, technological progress and economic enterprise are the best means of saving it."

Energy Futures: The Report of the Energy Project at the Harvard Business School, Robert Stobaugh and Daniel Yergin (editors), Random House, New York, 1979. The editors were backed by dozens of researching students in producing an encyclopedic analysis of possible ways to increase available energy. Their massively referenced conclusion was that energy efficiency is the most cost-effective way to provide the increased energy needs of the future. Even though old, the book is an excellent reference.

Energy Myths and Realities: Bringing Science to the Energy Policy Debate, Vaclav Smil, AEI Press, Washington, D.C., Aug. 16, 2010. This is one of Smil's best books. He covers a wide range of material but manages to do it clearly and entertainingly.

Energy Victory: Winning the War on Terror by Breaking Free of Oil, Robert Zubrin, Prometheus, April 10, 2009. The author released the book in 2009, shortly after oil had reached its peak in 2008, so the price was down because of the Great Recession, but prices were expected to go back up, perhaps to even more expensive highs, when the economy recovered. Based on that, he proposed much greater use of both ethanol and methanol with flexfuel cars that could reset for burning either of the two alcohols or gasoline depending on market conditions. There are also a number of vignettes about the importance of energy in recent history.

Eruptions That Shook the World, Clive Oppenheimer, Cambridge University Press. The author provides an overview of volcanoes and then follows the tangled history of how major eruptions lead to several years of cold, crop failures, and frequent disease and government turmoil. This is a frequently recurring theme throughout history.

Factor Four: Doubling Wealth, Halving Resource Use - The New Report to the Club of Rome, Ernst von Weizsacker, Amory B. Lovins, L. Hunter Lovins, Routledge, Dec. 1, 1998. The authors provide a number of detailed case studies on making various technologies more efficient, along with the market and tax incentives around them. The examples stretch from power poles (concrete versus steel), to highly efficient hypercars," to water-efficient drip irrigation.

Famine 1975: America's Decision: Who Will Survive? Paul, Jr., and William Paddock, 1967. The Paddock brothers were two of the best-known Malthusians in the second half of the Twentieth Century. They wrote several books predicting worldwide famines and political collapse—predictions then echoed by entomologist-turned population-control expert Paul Ehrlich in *The Population Bomb*. The Paddocks were two of the premiere experts of the day, and their predictions were highly likely—except that other people worked to produce the Green Revolution. Once again, the worst enemies of prediction are genius, madness, and chance.

Farmers of Forty Centuries: Organic Farming in China, Korea, and Japan, Franklin Hiram King, (originally published 1911) Dover Publications, March 19, 2004. The author was forced out of the U.S. Department of Agriculture in the early 1900s because he thought production resulted from the nutrients that were available to plants rather than total nutrients in the soil and/or provided by fertilizers. In searching for a better way, he toured East Asia. His book detailed methods (such as return of organic wastes to the soil, cover crops, and intercropping) that could be adopted in

2000s regenerative agriculture. http://library.umac.mo/ebooks/b30796635.pdf (accessed Feb. 9, 2019)

A feasibility study of implementing an Ammonia Economy, Final Report, Jeffrey R. Bartels and Michael B. Pate, Iowa State University, Ames Iowa, Dec. 2008. https://lib.dr.iastate.edu/cgi/viewcontent.cgi?article=2119&context=etd (accessed Oct. 2, 2019) This report summarizes the feasibility of a potential ammonia economy in terms of synthesis, storage, and use. It also makes comparisons with a potential hydrogen economy.

Feeding Everyone No Matter What: Managing Food Security After Global Catastrophe, David Denkenberger and Joshua M. Pearce, Elsevier, 2015. The authors look at alternate food supplies through the lens of dealing with some major catastrophe, they provide a summary that may be too top level but is wide ranging from increased sea harvests to fungi, bugs, processed petroleum, and many others. Their references lead on to a number of good areas.

Feeding the World, Vlacov Smil, The MIT Press; Cambridge, MA, 2000. As a reviewer said in *Scientific American*, "Smil's message is that it can be done. He sees "no insurmountable biophysical reasons why we could not feed humanity in decades to come while at the same time easing the burden that modern agriculture puts on the biosphere." That says it all. Smil has good analysis backed by solid referencing.

Fixing Climate: What Past Climate Changes Reveal About the Current Threat-and How to Counter It, Wallace S. Broecker and Robert Kunzig, Hill and Wang; Reprint edition March 31, 2009. The authors provide essentially two books. The first a summary of the theories and personalities that led to our present understanding of climate-change issues, and the second summarizes their proposals for mitigating global warming. The first is fabulous, and often funny; the second is still a work in progress.

Fixing the Sky: The Checkered History of Weather and Climate Control, James Rodger Fleming, Columbia University Press, New York, 2010. The author provides an academic survey of weather modification ideas from Greek mythology, through rainmakers, to the present proposals and political arguments.

Food, Climate, and Carbon Dioxide: The Global Environment and World Food Production, Sylvan H. Wittwer, CRC Press, Inc., Boca Raton, FL. The author expressed a less worried view about global warming and food than many environmentalists. He expected warming would be associated with increased levels of carbon dioxide in the atmosphere, so that he expected productivity of to increase significantly.

The Food Police: A Well-Fed Manifesto About the Politics of Your Plate, Jayson Lusk, Crown Forum; April 2013. The author blows the lid off the hypocrisy of the self-proclaimed experts who attack business in the food industry and propose regulatory solutions that could cause disasters. His major theme is that the present production of food is more affordable and safer than any time in history, and that moving to all organic would be much more expensive. He speaks from a farm background and as an academic in agricultural science.

Food Politics: What Everyone Needs to Know, Robert Paarlberg, Oxford University Press, 2nd edition, Sept. 25, 2013. The author has been a renowned food-policy expert for decades. The book uses a readable question-and-answer format for a wide assortment of food issues from Malthus, to the Green Revolution, to genetically engineered crops, and many more.

A Furious Sky: The Five-Hundred-Year History of America's Hurricanes, Eric Jay Dolin, Liveright, New York, Aug. 4, 2020. The author manages to combine meteorology and history along with human triumphs and tragedies. For the general reader. This is a good comprehensive discussion of hurricanes; other books are usually focused on particular storms or particular regions.

Future Food: Alternate Protein for the Year 2000, Barbara Ford, William Morrow & Co., New York, 1978. This wonderful book is desperately in need of updating because after all these years it still describes the widest assortment of human food sources at the time and potential human food sources from alfalfa, to earthworms, to mushrooms, to single-celled cultures, to processed feathers, to dozens more—the book describes immense sources of protein. And of course, every weight unit of new protein can free up three to eight times the grain that would have been livestock feed.

The Future of Nuclear Energy in a Carbon-Constrained World, Massachusetts Institute of Technology, Jacopo Buongiorno, John Parsons, Jacopo Buongiorno, et al., Cambridge, Massachusetts, 2018. http://energy.mit.edu/wp-content/uploads/2018/09/The-Future-of-Nuclear-Energy-in-a-Carbon-Constrained-World.pdf (accessed Feb. 3, 2019)

Geopolitical Constructs: The Mulberry Harbours, World War Two, and the Making of a Militarized Transatlantic, Colin Flint, Rowman & Littlefield Publishers, Lanham, Maryland, Sept. 19, 2016. The author describes not a wartime battle but the design and coordination of probably the largest oceanic project of all times, and it has some lessons for future offshore projects. Copies of this book may be hard to get.

The Genesis Strategy: Climate and Global Survival, Stephen Schneider, Springer, 1976. This title and primary theme of this book is that humanity should have a significant reserve of food to prepare for variations in weather and climate. The title alludes to a story in the book of Genesis in The Bible, in which the ancient Egyptian ruler laid in a 7-year surplus of grain to prepare for 7 "lean years" of drought. Schneider was one of the first major proponents of global warming as an oncoming concern. However, his book also provides a good summary of many minor weather and climate trends and backtracks such as the Roman Warm Period, the Medieval Warm Period, the Little Ice Age (that only ended in the mid-1800s), and shorter variations such as warming in the 1920s and 1930s and cooling from the 1940s into the 1970s.

Global Crisis: War, Climate Change & Catastrophe in the Seventeenth Century, Geoffrey Parker, Yale University Press, October 21, 2014. Parker argues in massive detail about the terrible effects of global cooling in the 1600s. The combination of the already ongoing Little Ice Age and further cooling from a number of volcanic eruptions contributed to one of the worst periods of wars, famines and plagues in human history.

A Global Green New Deal: Rethinking the Economic Recovery, Edward B. Barbier, Cambridge University Press, 2010. The author led a team that developed this theme for a United Nations report, and this book is that report put into a more accessible and updated form. It contains many of the ideas in the Green New Deal, with a great deal more detail, but without the eminent-doom implications. Its theme is that the next century would be much better if the proposed changes were implemented. Interestingly, the book had a sense of foreboding that the Great Recession might evolve into another Great Depression. Also, as in many expert reports of the time, there was no inkling of the Fracking Revolution that was to come several years later.

Golden Rice: The Imperiled Birth of a GMO Superfood, Ed Regis, Johns Hopkins University Press; Oct, 2019. The author provides a somewhat dispassionate view of the angry arguments on both sides about golden rice that could decrease deficiency diseases from lack of vitamin A in poor countries. The author describes how efforts to develop and distribute this rice strain have been delayed by environmentalists and even more by government regulations. Worse, those for and against the rice strains have developed political attacks and counter attacks against each other to a fever pitch.

The Great Warming: Climate Change and the Rise and Fall of Civilizations, Brian Fagan, Bloomsbury Press; Reprint edition (March 10, 2009. This is another complementary view to Fagan's *The Long Hot Summer* and *The Little Ice Age*.

The Green New Deal and Beyond: Ending the Climate Emergency While We Still Can, Stan Cox, City Light Books, San Francisco, 2020. The author contends that the Green New Deal resolution is not radical enough. After a couple pages to some details the resolution did not mention—rationing, price controls, and additional costs for providing a large share of a world green new deal. Also suggested is that material production must be decreased and America must share the decline in living standards must be shared with poorer countries because (of course) without them we are all doomed.

The Green New Deal: Economics and Policy Analytics, Benjamin Zycher, American Enterprise Institute, April 2019. If you don't like the Green New Deal, this is the book for you. By the author's economic analyses, the Green New Deal would impose massive costs, and even if fully adopted, it would only reduce global warming by a fraction of degree.

Growing a Revolution: Bringing Our Soil Back to Life, David A. Montgomery, W. W. Norton & Company, 2017. The author provides a very readable introduction to the possibilities of soil building using cover crops, perennials rather than annuals, and minimal plowing. The exposition examines the concepts by verbal sketches of operations by farmers and agricultural experts teaching their proposed methods, which allows it to give more details in a more learnable manner—"Facts tell; stories sell." The book provides a more optimistic complement to the more pessimistic and more academic *Dirt: The Erosion of Civilizations*. Best of all, it is non-judgmental; there are comments about reducing the costs for fertilizer when possible and "the great reset button of plowing" when all else fails in controlling weeds.

Hack the Planet: Science's Best Hope—or Worst Nightmare—for Averting Climate Catastrophe, Eli Kiintisch, John Wiley & Sons, Inc., Hoboken, New Jersey, 2010. The author provides a readable account of the major proposals for geoengineering climate to prevent global warming. Those include underground injection of carbon dioxide, shading the world with various materials injected in the atmosphere, ocean fertilization, and direct extraction of carbon dioxide from the atmosphere. The text includes enough technical detail to explain the concepts as well as the personalities and politics involved. The material is still current because politics and great cost have delayed most work on these concepts.

Hacking Planet Earth: How Geoengineering Can Help Us Reimagine the Future, Thomas M. Kostigen, Tarcher Perigee, March 24, 2020.

Hothouse Earth, Howard A. Wilcox, Praeger Publishers, New York, 1975. The author actually ran an ambitious U.S. Navy program in the early 1970s aimed at developing kelp farms to generate synthetic fuel, produce food, and remove carbon dioxide from the atmosphere. This is an earlier version of Tim Flannery's *Sunlight and Seaweed* of 2017 but with more practical details. The biggest complaint is that Wilcox doubtless had much more, but he spent a good part of the book arguing his concern about global greenhouse warming and his arguments against the other major way to reduce greenhouse warming, nuclear fission. Back at the kelp farm, funding collapsed along with oil prices in the 1980s.

How Innovation Works: And Why It Flourishes in Freedom, Matt Ridley, Harper, May 20, 2020. The author argues that innovation is crucially different from invention because it comprises the more complex subprocesses of transforming inventions into practical and affordable items. It happens mainly in just a few active areas of the world at any one time. It is difficult to predict, and government policies often stifle rather encourage it. In a very entertaining manner, the author connects innovation concepts and events in a superbly entertaining fashion.

How to Cool the Planet: Geoengineering and the Audacious Quest to Fix Earth's Climate, Jeff Goodell, (originally published by Houghton Mifflin Harcourt Publishing Company, 2010) Mariner Books; 2011. The *Rolling Stone* contributor author provides an entertaining romp through geoengineering with the theme that there might be an unexpected surge in global warming that would require use of geoengineering. The chapters tend to focus on key individuals involved in geoengineering proposals such as James Lovelock and David Keith.

Hubbert's Peak: The Impending World Oil Shortage, Kenneth S. Deffeyes, Princeton University Press, New Jersey, 2001. Deffeyes extended M. King Hubbert' predictions of U.S. peak oil production with better mathematical analysis, which he used to predict a world peak in oil production in about 2005. Stop laughing right now! The analysis was good, but did not include hydrofracturing (or fracking) that was an entirely new invention. (The three great enemies of trend extrapolation are genius, madness, and chance.) Nevertheless, this analysis, and one example of its failure, are useful for other resource issues that may seem insurmountable … unless the playing field changes radically … and it often does.

The Hype About Hydrogen: Fact and Fiction in the Race to Save the Climate, Joseph J. Romm, Island Press, (originally published in 2004) New Edition, July 15, 2005. The author is an ardent environmentalist, yet he provides a litany of reasons why the hydrogen economy will not work and may even be counterproductive in the short and even medium term. Much of the excellent analysis is still current because the progress has been slow on the two biggest issues of getting usable hydrogen at an affordable price and transporting it also at an affordable price. [This book suggests next generation fission for the first and combining hydrogen into a more transportable fuel such as ammonia for the second.] If you get it, be sure you get the second edition, referred to as the "New Edition."

An Inconvenient Truth: The Crisis of Global Warming, Rodale Books, Al Gore, 2006. Al Gore' documentary film and associated book arguing that global warming could have major calamitous result, particularly if there were major slide-offs of ice from Greenland or the West Antarctic Ice Shelf. The most severe affects are not considered likely near term, and they have not been seen in historical times, but they are not impossible.

The Infinite Resource: The Power of Ideas on a Finite Planet, Ramez Naam, UPNE; First Edition, April, 2013. The author, in his innocence, accepts many environmental scare stories of peak resources, looming climate apocalypse, etc. Then he says that innovation is the way to deal with those limitations. For instance, a hunter gatherer needed 3000 acres to survive, but a person today is fed by one third of an acre—an improvement of a factor of 10,000, and the author proposes that we do it again in many areas.

Insects as Sustainable Food Ingredients: Production, Processing and Food Applications, Aaron T. Dossey, Juan A. Morales-Ramos, and M. Guadalupe Rojas, editors, Elsevier, Inc., 2016. The authors and editors describe how insects can be mass produced and incorporated into our food supply at an industrial and cost-effective scale, providing valuable guidance on how to build the insect-based agriculture and the food and biomaterial industry.

Annotated Bibliography

Justinian's Flea: The First Great Plague and the End of the Roman Empire, William Rosen, Penguin Books, 2007. The author provides an entertaining history, and it does detail the famine and plague. The complaint is that he spends much time on the history of Justinian, his diplomacy, and his military campaigns.

"Keynote at Farming for the Future 2020," Gabe Brown, *YouTube*, March 13, 2020. https://www.youtube.com/watch?v=ExXwGkJ1oGI (accessed Dec. 5, 2020) The speaker, Gabe Brown, is the apostle of regenerative farming. As a North Dakota rancher and farmer who has spent more than two decades developing a regenerative operation, he clearly describes the major issues involved along with examples from his operation and from others.

Krakatoa: The Day the World Exploded: August 27, 1883, Simon Winchester, HarperCollins, April 4, 2003. This is a popularized geological description of the famous eruption and an historical account of the background and social results from it. It is quite entertaining, and it provides a clear inkling of what a much larger eruption would be like.

Large Floating Structures: Technological Advances, C. M. Wang and B. T. Wang (editors), Springer, Singapore, 2015. This collection of articles describes recent large floating structures starting with completed or under-construction docks, bridges, wind turbines, storage terminals and deep-sea drilling platforms. It also extends to design proposals for ocean-thermal energy conversion, mobile offshore military bases, and even floating cities. This is a high-tech document produced by marine engineers and other experts. It contains excellent description and color images of structures being described.

Late Victorian Holocausts: El Niño Famines and the Making of the Third World, Mike Davis, Verso; First Edition July 2002. Davis delivers a powerful polemic claiming that colonial responses to weather variations of *El Niño and La Niña* contributed to famines and the pauperization of the Third World.

The Limits to Growth, Donella H. Meadows, Dennis L. Meadows, Jorgen Randers, and William H. Behrens, III, Signet; Oct. 1972. The authors, working with Jay Forrester at the Massachusetts Institute of Technology, used a computer simulation to show an interplay among mineral resources, energy, population, and environment. Most of their scenarios ended in some form of world disaster. There has been increasing skepticism (deniers) since a number of the proposed mineral exhaustions did not happen. See *Models of Doom: A Critique of the Limits to Growth*.

Limits to Growth: The 30-Year Update, Donella H. Meadows, Jorgen Randers, and Dennis L. Meadows, Chelsea Green Publishing; 3rd edition, June 2004. The authors continue their simulations although they had to modify for a few resources that did not run out. If their simulations were to be correct, the population of Earth had already overshot by 20 percent and a world collapse is impending. However, as before, models reflect the assumptions going in to multiple decimal points. Also, one economist who managed to access the detailed code, noted that introducing market forces overcame most of simulation dooms. For example, high oil prices led to the development of hydraulic fracturing and a 21st Century boom in American oil and gas production.

The Little Ice Age: How Climate Made History 1300-1850, Brian Fagan, Basic Books; 1 edition, Dec. 27, 2001. Fagan makes history entertaining even with the grim story of cold, storms, and famines in the Little Ice Age. This is the cold counterpoint to Fagan's *The Great Warming: Climate Change and the Rise and Fall of Civilizations.*

The Long Summer: How Climate Changed Civilization, Brian Fagan, Basic Books; Reprint edition, Dec. 29, 2004. Basic Books; Reprint edition (December 29, 2004. Humanity evolved when glaciers covered much of the world. But starting 15,000 years ago, temperatures began climbing. Civilization and all of recorded history occurred in this warm period, the era known as the Holocene-the long summer of the human species.

Lukewarming: The New Climate Science that Changes Everything, Patrick J. Michaels and Paul C. Knappenberger, Cato Institute, 2016. The authors' detail two major themes. First, rather than the false dichotomy of "alarmist" catastrophic human-caused global warming versus no warming of "deniers," there is a middle ground of generally beneficial moderate warming—lukewarming. Second, they explore the sociology of politicized science and the pursuit of continued funding fueling the stories of maximum danger.

Making Aquatic Weeds Useful: Some Perspectives for Developing Countries, (originally National Academy of Sciences [U.S.], 1984), Books for Business, June 1, 2002. This report examines methods for controlling aquatic weeds and using them to best advantage. It emphasizes techniques for converting weeds to feed, food, fertilizer, and energy production.

Meals to Come: A History of the Future of Food, Warren Belasco, University of California Press, October 2006. As Homer said, "A hungry stomach will not allow its owner to forget it, whatever his cares and sorrows." Thus, he focuses quite entertainingly on the story of food from famines of Malthusian doom to "cornucopian" optimism to the present and on to future possibilities.

Models of Doom: A Critique of the Limits to Growth, edited by H.S.D. Cole, Christopher Freeman, Marie Jahoda, and K.L.R. Pavitt, Universe Books, New York, 1973. The Sussex Research Unit examined the *Limits to Growth* computer simulations point by point and found many of them overwrought and/or not confirmed because of simulation weaknesses. Nearly a half century later, it would appear that the skeptics of simulation were correct.

More from Less: The Surprising Story of How We Learned to Prosper Using Fewer Resources—and What Happens Next, Andrew McAfee, Simon and Schuster, London, 2019. The author argues from a sociological perspective that the inherent capitalist drive for efficiency is driving technology toward a much more energy-efficient society.

The Mound Builders: The Archaeology of a Myth, Robert Silverberg, 2nd Edition, Ohio University Press; May 1, 1986. Science fiction writer Silverberg produced a good concise reference on the archaeological data and theories about the related set of cultures in eastern North America before the European settlement. The book is somewhat dated because it expands on Silverberg's 1969 article in American Heritage article on origins of the Mound Builders:
Also, Robert Silverberg, "…and The Mound-builders Vanished from the Earth" *American Heritage,* vol. 20, issue 4, June 1969. https://www.americanheritage.com/and-mound-builders-vanished-earth#1

Mycelium Running: How Mushrooms Can Help Save the World, Paul Stamets, Ten Speed Press; 1st Edition, Oct. 1, 2005. The author has wonderful descriptions of mushroom capabilities for food, pharmaceuticals and bioremediation of land although the great points are sometimes buried in the details.

The Nature of Crops: How We Came to Eat the Plants We Do, John Warren, CABI, June 2015. The author describes the adoption and adaption of fifty crops with interesting (and sometimes eccentric) history of the processes. Yet, the most eccentric stories may be the most important for developing new crops when needed.

Annotated Bibliography

Nuclear 2.0: Why a Green Future Needs Nuclear Power, Mark Lynas, UIT Cambridge, Ltd., Cambridge, England, 2013. This is a very clear and succinct little book arguing for a major resurgence of nuclear fission power. The author is a committed environmentalist, firmly believing in and concerned about global warming. Thus, he proposes replacing fossil fuels with noncarbon energy sources. However, as a practical matter, he has concluded that the alternate energies of solar and wind are entirely insufficient to make that replacement—leaving nuclear fission as the only practical alternative. As an environmentalist himself, Lynas makes the most telling arguments that anti-nuclear activism is often based on misconceptions, poor science, and even blatant disinformation, spread by well-meaning but ideologically narrow-minded activists.

The Nuclear Energy Option: An Alternative for the 90s, Bernard Leonard Cohen Springer, softcover reprint of the original 1990 edition, 1990. Despite its age, this is still one of the best books written about nuclear power. The highly technical information is very accessible for ordinary readers. It fairly presents the arguments both for and against.

Our Livable World: Creating the Clean Earth of Tomorrow, Marc Schaus, Diversion Books, October 13, 2020. Although the author starts with the usual climate-doom worries, he soon switches to a broad survey of ongoing and potential future innovations that could mitigate or even stop the threat of global warming and associated problems. The treatment is generally supportive with introductions to many technologies. He is maybe a little over-optimistic about solar, wind, fusion, and the hydrogen economy. He is a bit too doubtful about nuclear fission, but he describes next-generation fission and hopes for the best. Meanwhile, he provides an entertaining read with wide-ranging list of technologies from regenerative agriculture, vertical farms, more-efficient homes, and many others.

The Path Between the Seas: The Creation of the Panama Canal, 1870-1914, David McCullough, Simon & Schuster, New York, 1977. This prize-winning account details the engineering and the medical advances that made the canal building in Panama a success. It is also a great epic read.

Pickled, Potted, and Canned: How the Art and Science of Food Preserving Changed the World, Sue Shepard, Simon & Schuster, 2001. Again, here is a history through the lens of food. Only this one is focused on how food preservation methods changed history and will doubles change it again.

Plant Factory: An Indoor Vertical Farming System for Efficient Quality Food Production, by Toyoki Kozai, Genhua Niu, and Michiko Takagaki (editors), Academic Press; 1st edition, November 9, 2015. In four short years, Despommier's wild-eyed maybe-it-could work proposal has morphed into detailed research tomes like this analyzing details of a "plant factory." Don't try this book unless you really want to run a plant factory.

Plastiki Across the Pacific on Plastic: An Adventure to Save Our Oceans, David de Rothschild, Chronicle Books, March 16, 2011. The author and his compatriots are typical eco-activists with their catalog of environmental woes. What is impressive is their construction of a boat largely from recycled plastic and sailing it across the Pacific Ocean. Ah hah! There is another use for discarded plastic.

Plentiful Energy: The Story of the Integral Fast Reactor, Charles E. Till and Yoon Il Chang, CreateSpace, 2011. The authors follow the work, triumphs, and then 1994 cancellation of a next generation nuclear reactor program for integral fast reactors. The goals of the IFR project were to increase the efficiency of uranium usage by breeding plutonium and reprocessing into fuel so that no transuranic isotopes would ever to leave the site there would be much less nuclear waste for long-term storage. The Clinton administration and the U.S. Congress closed the program with the rationale that it might contribute to proliferation.

Plows, Plagues, & Petroleum: How Humans Took Control of Climate, William F. Ruddiman, Princeton University, 2007. Ruddiman argues that human agriculture has been generally warming the climate for some 8,000 years and that population die-offs may have caused cooling as forests regenerated and drew down carbon dioxide from the atmosphere. If the climate were found to be that sensitive to human activities, the potential for climate-mitigating innovation would be much greater.

Plowman's Folly, Edward H. Faulkner, University of Oklahoma Press, Norman, Oklahoma, July 1943. The author's heretical comments disparaging the sacrosanct plow were not well received by agricultural experts at the time, but his ideas are now being adopted in the trend to minimize plowing to reduce costs, reduce erosion, and regenerate more soil in the ground.

The Population Bomb, Paul Ehrlich, Ballantine 1970, (first published by Sierra Club/Ballantine Books, 1968). Ehrlich was a zoologist who studied mites and butterflies until he began writing and lecturing on population control in 1967. He dressed Malthusian doomsaying in modern cloths and added bombast that the good Rev. Malthus would probably not have used. He predicted that hundreds of millions would die in famines in the 1970s, that aid to counties such as India should be ended because such countries were doomed anyway. Years later when asked about the great famines of the 1970s, Ehrlich said that he only gave "scenarios." Ehrlich's unfortunate screed is a classic example of failing to consider the power of innovation.

The Potential of U.S. Cropland to Sequester Carbon and Mitigate the Greenhouse Effect, 1st Edition, Rattan Lal, John M. Kimble, Ronald F. Follett, and C. Vernon Cole, 1st Edition, CRC Press; 1 edition, Aug. 1, 1998. The authors make a case that soil building can remove carbon dioxide from the air and store it as carbon in sufficient amounts of new soil to reduce the net amount of carbon dioxide entering the atmosphere each year from farming operations and perhaps a good deal more. This book is a bit dated, but it is a good introduction to soil regeneration issues.

Powering the Future: How We Will (Eventually) Solve the Energy Crisis and Fuel the Civilization of Tomorrow, Robert H. Laughlin, Basic Books; Sept. 27, 2011. The physics Nobel laureate describes options for a noncarbon energy future because of climate concerns and/or fossil-fuels exhaustion. He does so in a very clear and understandable way. Interestingly and ironically, the 2011 release date did not have any comments about fracking for hydrocarbons (which has caused a revolution in hydrocarbon supplies) or small modular reactors (which may be a revolution in fission nuclear reactors). Both of these revolutions were largely unexpected in 2011. What other energy revolutions may come in the next ten years?

Power to Save the World: The Truth About Nuclear Energy, Gwyneth Cravens and Richard Rhodes (introduction), Vintage; Reprint edition (October 14, 2008. The authors provide a very understandable book for nontechnical readers about nuclear energy with a special emphasis on radiation safety. They consider it a totally viable and practical solution to global warming.

The Privatization of the Oceans, Rögnvaldur Hannesson, The MIT Press, Cambridge, Massachusetts, 2006. The author compares the development of property rights in the world's fisheries with the enclosures and clearances of common lands in England and Scotland. He details the benefits of controlled access to ocean production and some of the rules that might be imposed.

The Prize: The Epic Quest for Oil, Money & Power, Daniel Yergin, (originally published by Simon and Schuster, 1992), Free Press; Reissue Edition. Dec. 23, 2008. Yergin, a renowned energy expert takes the reader on oil's wild ride from before the first American oil well in 1859 through Saddam Hussein's invasion of Kuwait in 1990. He chronicles a series of panics about impending oil exhaustion followed by the inevitable price collapses.

Profiles of the Future: An Inquiry into the Limits of the Possible, Arthur C. Clarke, Harper & Rowe, February 1963. This wide-ranging collection of essays looked at both potential innovations and why predictions often fall short. Clarke was one who would know. He essentially invented the concept of the communications satellite, but he never patented the concept because he never imagined it happening in his lifetime. The majority of his comments are still useful and still entertaining after all these years.

The Rational Optimist: How Prosperity Evolves, Matt Ridley, (originally released 2010, Harper Perennial; Reprint edition, June 7, 2011. Ridley makes the case for an economics of hope, arguing that the benefits of commerce, technology, innovation, and change—cultural evolution—will [almost] inevitably increase human prosperity. Better yet, he does it in a charming and entertaining manner.

Redesigning the American Lawn: A Search for Environmental Harmony, F. Herbert Bormann, Diana Balmori, and Gordon T. Geballe, Yale University Press, 1993. The authors describe the American lawn as a major cost to the environment in terms of water, fertilizer, equipment used, and energy by what they call the industrial lawn. They propose to replace it with the freedom lawn of much less watering, less fertilizing, and more flowers and shrubs that require fewer inputs. In dryer areas, they propose shifting to a less carefully preserved lawn and more dry-land plants.

Rethinking Aquaculture to Boost Resource and Production Efficiency: Sea and Land-Based Aquaculture Solutions for Farming High Quality Seafood, Version 1.1, Pia Klee, editor in chief, The Rethink Water network and Danish Water Forum White Papers, Copenhagen. Available at www.rethinkwater.dk (accessed Nov. 2, 2019) Yes, this white paper is one big advertisement for Danish aquacultural companies. Still, it provides an excellent introduction to both land-based and oceanic aquaculture with great sensitivity to ecological concerns.

Rising Seas: Past, Present, Future, Vivien Gornitz, Columbia University Press, New York, 2013. The author provides a wonderfully readable, while still encyclopedic, of sea levels sinking with glaciers, rising in warming times, effects on shoreline human settlements, and suggested policies.

Roosters of the Apocalypse: How the Junk Science of Global Warming Nearly Bankrupted the Western World, Rael Jean Isaac, Revised and Expanded Edition, CreateSpace Independent Publishing Platform; 2 edition, Nov. 25, 2013. The author starts with an account of how a millennialist cult of one tribe in South Africa slaughtered their cattle and burned their crops in hopes this would cause their ancestors to rise up, drive European settlers away, and usher in a golden age. Of course, the result was a genocidal famine inflicted by themselves. Isaac uses this as a metaphor for the roosters of environmentalist apocalypse who attack the denier owls who are skeptical of killing our energy cattle for their greater cause.

Road to Survival, William Vogt, William Sloane Associates, 1948. This is a carefully reasoned and researched environmental thesis arguing that the technological advances of the Twentieth Century have many negative consequences. This book is the basis of many environmental concepts held today. It is the polar opposite of techno optimism.

Roosters of the Apocalypse: How the Junk Science of Global Warming Nearly Bankrupted the Western World, Rael Jean Isaac, Revised and Expanded Edition, CreateSpace Independent Publishing Platform; 2

edition, Nov. 25, 2013. The author starts with an account of how a millennialist cult in South Africa slaughtered their cattle and burned their crops in hopes this would drive European settlers away and usher in a golden age. Of course, the result was disaster. Isaac uses this as a metaphor for the roosters of environmentalist apocalypse who attack the denier owls who are skeptical of killing our energy cattle for their greater cause.

Road to Survival, William Vogt, William Sloane Associates, 1948. This is a carefully reasoned and researched environmental thesis arguing that the technological advances of the Twentieth Century have many negative consequences. This book is the basis of many environmental concepts held today. It is the polar opposite of techno optimism.

Sea Change: How Markets and Property Rights Could Transform the Fishing Industry, Richard Wellings, editor, The Institute of Economic Affairs, London, England, 2017. The authors of this short book summarize of how "the tragedy of the commons" often leads to depletion of fish stocks and eventual collapses of fisheries. They describe further how government regulation often results in subsidized fishing fleets that destroy fisheries faster, time limits that are hazardous to the fishing personnel, and throwing dead bycatch back into the sea when they cannot be legally taken by fishers.

Seasteading: How Floating Nations Will Restore the Environment, Enrich the Poor, Cure the Sick, and Liberate Humanity from Politicians, Joe Quirk and Patri Friedman, Free Press, March 21, 2017. The authors approach oceanic settlement from a libertarian perspective. As with Kropotkin, their political vision may or may not succeed, but it gives them a drive to settle a new frontier just as many of the European colonists in the new lands of Australia and the Americas were radicals and refugees getting farther away from central authorities at one time or another. As such, they are part of a small but fervent intellectual movement that is exploring new concepts for ocean settlement and is actually working to build things that have been only engineers dreams for decades.

The Seavegetable Book, Judith Cooper Madlener, Outlet, July 1, 1977. The author guide to 52 species of edible sea plants including sketches and complete descriptions of each, instructions for foraging and preparation, and recipes drawn from many traditions around the world.

Seeing the Light: The Case for Nuclear Power in the 21st Century, Scott L. Montgomery and Thomas Graham, Jr.., Cambridge University Press, Sept. 14, 2017. The authors provide an optimistic case for a major increase in the use of nuclear fission to get more usable energy with less greenhouse warming. They discuss of next-generation reactors such as molten salt reactors.

Shadow Cities: A Billion Squatters, a New Urban World, Robert Neuwirth, Routledge; Dec. 1, 2004. The author actually spent time living in several squatter/slum neighborhoods. His account puts human faces on a surprisingly vibrant engine of development. Incidentally, the trend to greater urbanization increases land available for crop production.

Silent Spring, Rachel Carson, First published Houghton Mifflin, 1962. The author provided a warning about overuse of pesticides, which led to more careful use of broad-spectrum and long-lasting pesticides that attack many insects to use of more narrow spectrum pesticides formulated to break down in shorter periods of time. Unfortunately, she also pioneered the use of scare stories and personal attacks with statements that DDT had been invented as a poison gas to kill people and that it is a carcinogen. The scare stories caused total bans on DDT use leading to resurgent malaria in many areas and millions of deaths.

The Skeptical Environmentalist: Measuring the Real State of the World, Bjørn Lomberg, Cambridge University Press, 2001. The author examines a number of environmental concerns (including global warming, food shortages, pollution, and mineral shortages) and concludes that many are greatly exaggerated. Moreover, he concludes that the skewed concerns have diverted funds from more important issues. For instance, money spent to reduce greenhouse warming could save lives by providing more potable water in poor countries.

Slide Rule: The Autobiography of an Engineer, Nevil Shute, House of Stratus, Jan. 12, 2008 (first published by William Heinemann Ltd., London, 1954). Later in his life, Shute became a famous writer, but in the 1920s he was the lead stress engineer for designing and building a radically new airship (the R100) for a private firm in competition with an entirely-government competing project the (R101). Shute's book details many of the challenges faced by both development programs, including the ultimately disastrous crash of the R101. For future reference, the story details many of the problems in government technology development that have repeated many times.

Slime: How Algae Created Us, Plague Us, and Just Might Save Us, Ruth Kassinger, Houghton Mifflin Harcourt, 2019. The author assumes no previous knowledge as she introduces the collection of organisms that make up algae, their geological history, their present and future uses, toxicity issues, brief descriptions of people working with algae, and even references and recipes in the back. Amazingly, the book is both entertaining and thorough.

Smaller Faster Lighter Denser Cheaper: How Innovation Keeps Proving the Catastrophists Wrong, Robert Bryce, Public Affairs, New York, May 2013. Robert Bryce shows himself to be the 21st Century's Arthur C. Clarke on the sweep of technological development, particularly in energy.

Soft Energy Paths: Towards a Durable Peace, Amory B. Lovins, (originally published by) Ballinger Publishing Co, Cambridge, Mass, 1977. In simple language, Lovins argued that there are a number of simple "soft" energy-saving technologies that would be much cheaper than the massive increase in power generating capacity that was predicted in the 1970s. With this book, along with a number of speeches and articles, Lovins was able to encourage a movement for researching energy efficiency.

Sprouts: The Miracle Food: The Complete Guide to Sprouting, 6th Edition, Steve Meyerowitz, July 1998. The author provides an introduction to sprouting, the wide variety of seeds that can be sprouted, the ways sprouts can be prepared for eating, and the nutritional benefits.

Starved for Science: How Biotechnology Is Being Kept Out of Africa, Robert Paarlberg and foreword by Norman Borlaug, Harvard University Press; Aug. 5, 2009. The author argues that rich environmentalist pressure groups from well-fed advanced economies have used economic pressure to keep a hungry Africa from adopting genetically engineered crops that the Africans need.

State of Innovation: The U.S. Government's Role in Technology Development, Fred L. Block and Mathew Keller, Routledge, Nov. 17, 205. The authors provide essays and case studies supporting their theme that innovation requires both government and private entrepreneurship playing central roles with the critical need being positive synergies between the two.

The State of the World's Land and Water Resources for Food and Agriculture: Managing Systems at Risk, Food and Agricultural Organization of the United Nations with Earthscan, New York, 2011. Especially, see Chapter 4, "Technical Options for Sustainable Land and Water Management." http://www.fao.org/docrep/017/i1688e/i1688e.pdf (accessed Feb. 14, 2019)

Storing Energy: with Special Reference to Renewable Energy Sources, Trevor H. Letcher, editor, Elsevier; Amsterdam, May 11, 2016. A Green New Deal Future will get a substantial portion of its power from intermittent energy sources, such as solar and wind. This highly technical reference details why that is and details many of the most likely storage methods from industrial-scale batteries, to heat storage, to underground or underwater storage of compressed air, to flywheels, and more.

Super Fuel: Thorium, the Green Energy Source for the Future, Richard Martin, St. Martin's Press LLC, New York, NY, 2012. The author provides a very reader-friendly argument that thorium molten salt reactors are the technology revolution that the designers of uranium/plutonium reactors hoped to achieve in the 1950s and 1960s.

Sudden Sea: The Great Hurricane of 1938, R. A. Scott, Little, Brown and Company, 2003. The author supplies a detailed and entertaining account of the storm with many personal stories. There is a fair amount of storm science and a good bibliography tucked within the historical and personal accounts.

Sunlight and Seaweed: An Argument for How to Feed, Power and Clean up the World, Tim Flannery, The Text Publishing Company, Melbourne, Victoria, Australia, 2017. Flannery popularizes two proposed ways to reduce carbon dioxide in the atmosphere, regional-sized ocean agriculture and concentrating solar collectors. The seaweed mariculture theme echoes Wilcox's *Hothouse Earth* from forty years before. Much of the concentrating solar text dwells on the mass-production of relatively small units that can be dispersed around the electrical grids for more grid resilience and les power transmission cost.

Surviving Armageddon: Solutions for a Threatened Planet, Bill McGuire, Oxford University Press, 2005. After McGuire's earlier book, *Apocalypse, A Natural History of Global Disasters,* he was chastised for being too grim, Furthermore, he had a happy face that caused cognitive dissonance with his recurring doom. Surrendering to the genetics that caused a jovial exterior, he assembled a number of mitigations for many such disasters. However, the potential doom mitigations are limited to global warming and major volcanoes, earthquakes, and tsunamis.

Terrestrial Energy: How Nuclear Power Will Lead the Green Revolution and End America's Energy Odyssey, William Tucker, Bartleby Press, 2008. The author supplies a highly readable account for the general reader on why nuclear fission is much less dangerous than generally believed, the limitations of solar and wind, and the potential of fusion. It is a little dated with no mention of the Fracking revolution or small modular reactors and only limited mention of thorium.

The Third Horseman: A Story of Weather, War, and the Famine History Forgot, William Rosen, Viking, May 15, 2014. The author provides an interesting history that connects a colder climate to famines, plagues and more wars. However, it spends a great deal of space detailing the maneuvers of various European rulers.

This Is the Way the World Ends: How Droughts and Die-offs, Heat Waves and Hurricanes Are Converging on America, Jeff Nesbitt, Thomas Dunne Books, Sept. 25, 2018. If you want a shrill repetitive polemic, this is for you.

Thorium: Energy Cheaper Than Coal, Robert Hargraves, CreateSpace Independent Publishing Platform, July 25, 2012. The author provides the most complete and most readable description of progressing and potential next-generation nuclear fission power reactors, especially thorium molten salt reactors.

Tombstone: The Great Chinese Famine, 1958-1962, Yang Jisheng (author), Edward Friedman (editor and introduction), Stacy Mosher (editor and translator), et al., Farrar, Straus and Giroux; Nov. 19, 2013. The lead author lived through what is euphemistically called the "Three Years of Natural

Disaster," but as his book details, the deaths were not natural disasters. The gruesome calamities were all man made by leaders in a totalitarian society that is prone to "historical amnesia."

Tomorrow's Table: Organic Farming, Genetics, and the Future of Food, Pamela C. Ronald and Raoul W. Adamchak, Oxford University Press; 2nd edition, April 10, 2018. The authors argue that a judicious blend of genetic engineering and organic farming-are key to helping feed the world's growing population in an ecologically balanced manner.

Topsoil and Civilization, 2nd edition, Vernon Gill and Tom Dale, University of Oklahoma Press, (first edition 1955) Jan. 1974. The author links topsoil and its loss to the rise and fall of civilizations. See also David R. Montgomery's *Dirt: The Erosion of Civilizations*.

Triumph of the City: How Our Greatest Invention Makes Us Richer, Smarter, Greener, Healthier, and Happier, Edward Glaeser, Penguin Press, New York, 2011. Glaeser argues the case that cities have been the focus points, the nexi, for the advancement of civilization. He argues that tactile and olfactory connections trump the wider connections of the Internet wide virtual city—time will tell.

The Turning Point: Creating Resilience in a Time of Extremes, Gregg Braden. Hay House, 2014. The author keeps looking for ways ancient wisdom can be used today. Along the way, he describes evolution of areas out of the main stream, such as the green glows from cell phones in the cells of Buddhist monks in Tibet.

Twilight in the Desert: The Coming Saudi Oil Shock and the World Economy, Mathew Simmons, John Wiley & Sons, 2005. This book is a case study of how secrecy, trend extrapolation, and a fevered imagination can run amok. Many details of Saudi Arabian oil production are kept as state secrets. Simmons, who was an oil trade analyst, saw some items that appeared suspicious. He concluded that the Saudis were covering up stagnating oil production, conflated and that their production was about to fall, which would cause an even greater spike in oil prices. Since Simmons' book was released. Saudi Arabia has reduced production a number of times, but it has usually been in efforts to avoid price declines.

Uncertain Harvest: The Future of Food on a Warming Planet, Ian Mosby, Sarah Rotz, and Evan D. G. Fraser, University of Regina Press, Saskatchewan Canada, 2020. This often lighthearted, but quite insightful, book suggests that there are many ways to feed humanity in the future while focusing on seven case studies of algae, caribou, kale, millet, tuna, crickets, milk, and rice.

Underexploited Tropical Plants with Promising Economic Value, (originally from The National Research Council [U.S.], 1971), Books for Business, June 1, 2002. Scientists from around the world picked and described 36 tropical plants that they believed had the most potential for increasing tropical food production—something that may also become useful for semi-tropical areas (such as the American Southeast) if climate warms significantly.

Underexposed: What If Radiation Is Actually Good for You? Ed Hiserodt, Laissez Faire Books, Oct. 31, 2005. The author provides a surprisingly readable, but well documented, case for saying that low levels of radiation actually improve health because of the phenomenon of hormesis whereby a large amount of something can be poisonous whereas a trace amount can be a vital nutrient.

Unnaturally Delicious: How Science and Technology Are Serving Up Super Foods to Save the World, Jayson Lusk, St. Martin's Press, March 22, 2016. This book focuses on innovations being made in food production as written by someone who definitely knows farming. This is a definite complement to Lusk's *Food Police*.

The Unnatural History of the Sea, Callum Roberts, (originally Shearwater Books, 2007) Island Press, 2008. The author provides a grim centuries-long narrative of fisheries being opened up, exploited, and eventually crashing. He describes government effort to preserve fisheries as generally too little and too late.

Use of Yeast Biomass in Food Production, Anna Halász and Radamir Lászity, CRC Press, 1991. This is a major text on yeast for food production.

The Vertical Farm: Feeding the World in the 21st Century, Dickson Despommier, Picador, 2011. The author provides an introduction to the advantages and concepts of vertical farms (indoor artificially lit growing that can use multiple levels). This was the first wide-eyed wild-guess major description of the concept. There are serious gaps and misconceptions, as might be expected. Obviously, we need more books as the field advances.

Waste: Uncovering the Global Food Scandal, Tristram Stuart, W. W. Norton & Company, Oct, 12, 2009. Stuart is a partisan of New Agey waste-not on food, but he has also been a food marketer and trader all his life, so he appreciates and understands the markets. He provides a good summary of food production and marketing in both the developed and the developing worlds. Along the way, he emphasizes his signature issue, reducing food waste, which presently consumes roughly a third of world food production.

We Can't Eat Grass: For Food Security Trade Your Lawn for an Ecologically Sustainable Victory Garden, Kevin Thomas Morgan, independently published, May 4, 2020. The author had hungry times in Britain during world War II and after. Thus, he would rather replace all the lawns with victory gardens. He supplies a simple gardening manual geared to his yard in the cooler part of North Carolina. He touches on diplomacy, but convincing many of the city zoning boards and neighborhood associations will be difficult … unless food were to grow scarce.

The West without Water: What Past Floods, Droughts, and Other Climatic Clues Tell Us about Tomorrow, B. Lynn Ingram and Frances Malamud-Roam, University of California Press, Aug. 1, 2013. The authors detail the correlation between climate and crucial changes in moisture for North America. The Little Ice Age that harmed Europe brought more water for North America. Conversely, the Medieval Warm Period that was great for Europe brought drought to much of North America.

When China Ruled the Seas: The Treasure Fleet of the Dragon Throne, 1405-1433, by Louise Levathes, Oxford University Press; Revised edition January 9, 1997. The author provides a fabulous introduction to the brief but amazing period of time when powerful Chinese fleets dominated the Indian Ocean and much of the western Pacific Ocean ordered for imperial grandeur and then abandoned when the imperial benefactor died.

When Humans Nearly Vanished: The Catastrophic Explosion of the Toba Volcano, Donald R. Prothero, Smithsonian Books, Oct. 16, 2018. The author provides a general introduction and historical summary of major volcanic events. This all leads to his central thesis that the giant eruption of Mount Toba about 74,000 years ago brought humans to the brink of extinction and that future super eruptions could do the same.

Why We Need Nuclear Power: The Environmental Case, Michael H. Fox, Oxford University Press, New York, 2014. The author makes the case that solar and wind are much more costly and less reliable than fossil fuel and that only nuclear fission can provide a practical low-carbon future.

Whole Earth Discipline: An Ecopragmatist Manifesto, Stewart Brand, Viking Penguin Group, New York, New York, 2009. Stewart Brand has been a pioneer in the environmental movement since it began, and his *Whole Earth Catalog* was a major part of that movement. In this book, he describes global warming as a major threat. He argues that two major transformations are the best defense: the greater environmental efficiency of growing urbanization and biotechnology. Brand argues that environmentalist must embrace genetic engineering rather than fighting it.

Why Not Eat Insects, Vincent M. Holt, Intl Specialized Book Services; Revised edition, June 1, 1988 (original 1885). Holt's quirky slightly quaint book makes his points well and even includes sample insect menu items, in French as though at a five-star restaurant. I would be wary about some of the raw items in the recipes due to possible parasites.

Why Nuclear Power has been a Flop at Solving the Gordian Knot of Electricity Poverty and Global Warming, Jack Devanney, Book Baby, Nov. 11, 2020. The author helped design large crude-oil tankers, and is the principal engineer of ThorCon US, Inc. with their proposal for a barge-mounted molten-salt nuclear fission reactor.

Windfall: The Booming Business of Global Warming, McKenzie Funk, Penguin Press, New York, 2014. The author is concerned about global warming, but he notes that many people (particularly in colder climates) may benefit from such changes. He has a great deal of first-hand knowledge about colder areas, and he provides it with sympathy for all sides.

The Winged Bean a High Protein Crop for the Tropics, Second Edition, Richard Evans Schultes, National Academy Press, 1981. The author details the merits of the little-known crop that is high production per unit land and adopted for a damp tropical climate—something that might be very important in a significantly warmer world.

The Wizard and the Prophet: Two Remarkable Scientists and Their Dueling Visions to Shape Tomorrow's World, Charles C. Mann, Knopf; 1st edition, January 23, 2018. The author attempts to contrast, without taking a side, the techno optimism of Norman Borlaug (developer of improved strains of grain that made "the Green Revolution") with the environmental pessimism of William Vogt. He argues that these two individuals embody two world views that are both needed by the world.

The Worst Hard Time: The Untold Story of Those Who Survived the Great American Dust Bowl, Timothy Egan, Houghton Mifflin Harcourt, 2006. The author supplies searing portraits of land-sale promoters selling land prone to periodic drought, use of bad farming practices, and desperate conditions when the drought struck and continued. He does less well in describing how better farming practices, such as tree strips, might have helped then and afterwards.

The Year Without Summer: 1816 and the Volcano That Darkened the World and Changed History, William K. Klingaman and Nicholas P. Klingaman, St. Martin's Press, New York, 2013. The authors focus on the severe world-wide effects of the volcano-induced cooling of the earth and the resulting crop losses, famines, uprisings, and disease outbreaks.

Zheng He: China and the Oceans in the Early Ming Dynasty, 1405-1433, Edward K, Dreyer, Pearson; 1 edition, Library of World Biography Series) 1st Edition, May 13, 2006. Dreyer goes into the great what if—what if imperial China had continued with the great nautical endeavor? That, of course, leads to the other great question, why did they abandon it?

Acknowledgements

I give my heartfelt thanks to my friend and former coworker, Bob Wells, who found many of the most important areas of confusion, repetitions, and multiple spelling and grammatical errors. All remaining errors are items that crept in during the multiple revisions are strictly my responsibility.

Thanks also to Craig Duswalt, Maurice DiMino, and Michael Stevenson who are all writers, speakers, and experts in their various fields. They provided vital instruction on organizing, generating, and releasing a finished copy of a book. More importantly, they described the power of such a written document for spreading ideas and doing a myriad of other things. Most importantly, they shared the stories of how they created books or their own, and that showed that I could too.